BAYESIAN ASTROPHYSICS

Bayesian methods are increasingly being employed in many different areas of physical sciences research. In astrophysics, models are used to make predictions to compare to observations that are incomplete and uncertain, so the comparison must be pursued by following a probabilistic approach. With contributions from leading experts, this volume covers the foundations of Bayesian inference, a description of the applicable computational methods, and recent results from their application to areas such as exoplanet detection and characterisation, image reconstruction, and cosmology. With content that appeals both to young researchers seeking to learn about Bayesian methods and to astronomers wishing to incorporate these approaches into their research, it provides the next generation of researchers with tools of modern data analysis that are becoming standard in astrophysical research.

ANDRÉS ASENSIO RAMOS is an astrophysicist working as a researcher at the Instituto de Astrofísica de Canarias, Tenerife, Spain, where he also completed his Ph.D. His current research focuses on the interpretation of spectropolarimetric signals in the Sun and other stars to infer properties about the magnetic field. His interests in astrophysics are spectropolarimetry, solar and stellar physics, and Bayesian inference.

IÑIGO ARREGUI is an astrophysicist working at the Instituto de Astrofísica de Canarias. He completed a Ph.D. in Physics at Universitat de les Illes Balears with a thesis on MHD waves in the solar atmosphere. His research focuses on the interpretation and modeling of wave activity in the solar atmosphere, designing tools for remote sensing of solar atmospheric plasmas, and the study of wave based plasma heating mechanisms.

Canary Islands Winter School of Astrophysics

Volume XXVI
Bayesian Astrophysics

Series Editor
Rafael Rebolo, Instituto de Astrofísica de Canarias

Previous Volumes in This Series

I. Solar Physics
II. Physical and Observational Cosmology
III. Star Formation in Stellar Systems
IV. Infrared Astronomy
V. The Formation of Galaxies
VI. The Structure of the Sun
VII. Instrumentation for Large Telescopes: A Course for Astronomers
VIII. Stellar Astrophysics for the Local Group: A First Step to the Universe
IX. Astrophysics with Large Databases in the Internet Age
X. Globular Clusters
XI. Galaxies at High Redshift
XII. Astrophysical Spectropolarimetry
XIII. Cosmochemistry: The Melting Pot of Elements
XIV. Dark Matter and Dark Energy in the Universe
XV. Payload and Mission Definition in Space Sciences
XVI. Extrasolar Planets
XVII. 3D Spectroscopy in Astronomy
XVIII. The Emission-Line Universe
XIX. The Cosmic Microwave Background: From Quantum Fluctuations to the Present Universe
XX. Local Group Cosmology
XXI. Accretion Processes in Astrophysics
XXII. Asteroseismology
XXIII. Secular Evolution of Galaxies
XXIV. Astrophysical Applications of Gravitational Lensing
XXV. Cosmic Magnetic Fields

BAYESIAN ASTROPHYSICS

Edited by

ANDRÉS ASENSIO RAMOS

Instituto de Astrofísica de Canarias, Tenerife

and

IÑIGO ARREGUI

Instituto de Astrofísica de Canarias, Tenerife

CAMBRIDGE
UNIVERSITY PRESS

CAMBRIDGE
UNIVERSITY PRESS

University Printing House, Cambridge CB2 8BS, United Kingdom

One Liberty Plaza, 20th Floor, New York, NY 10006, USA

477 Williamstown Road, Port Melbourne, VIC 3207, Australia

314–321, 3rd Floor, Plot 3, Splendor Forum, Jasola District Centre, New Delhi – 110025, India

79 Anson Road, #06–04/06, Singapore 079906

Cambridge University Press is part of the University of Cambridge.

It furthers the University's mission by disseminating knowledge in the pursuit of education, learning, and research at the highest international levels of excellence.

www.cambridge.org
Information on this title: www.cambridge.org/9781107102132
DOI: 10.1017/9781316182406

First published 2018

Printed in the United Kingdom by TJ International Ltd. Padstow Cornwall

A catalogue record for this publication is available from the British Library.

Library of Congress Cataloging-in-Publication Data
Names: Canary Islands Winter School of Astrophysics (26th : 2014 : La Laguna, Canary Islands) | Asensio Ramos, Andres, editor. | Arregui, Inigo, editor.
Title: Bayesian astrophysics / edited by Andres Asensio Ramos (Instituto de Astrofisica de Canarias, Tenerife), Inigo Arregui (Instituto de Astrofisica de Canarias, Tenerife).
Other titles: Canary Islands Winter School of Astrophysics (Series) ; v. XXVI.
Description: Cambridge, United Kingdom ; New York, NY : Cambridge University Press, 2018. | Series: Canary Islands Winter School of Astrophysics ; volume XXVI | Lectures presented at the XXVI Canary Islands Winter School of Astrophysics, held in La Laguna, Tenerife, Spain, Nov. 3-14, 2014. | Includes bibliographical references and index.
Identifiers: LCCN 2017060539 | ISBN 9781107102132 (hardback ; alk. paper) | ISBN 1107102138 (hardback ; alk. paper) | ISBN 9781107499584 (pbk. ; alk. paper) | ISBN 1107499585 (pbk. ; alk. paper)
Subjects: LCSH: Astrophysics–Mathematical models–Congresses. | Bayesian statistical decision theory–Congresses. | Astronomy–Mathematical models–Congresses. | Cosmology–Mathematical models–Congresses.
Classification: LCC QB462.3 .C26 2018 | DDC 523.0101/519542–dc23
LC record available at https://lccn.loc.gov/2017060539

ISBN 978-1-107-10213-2 Hardback

Contents

Contributors

José M. Bernardo, Universidad de Valencia, Spain

Brendon J. Brewer, Department of Statistics, University of Auckland, New Zealand

Phil Gregory, University of British Columbia, Canada

J.-L. Starck, Service d'Astrophysique, CEA/Saclay, France

Roberto Trotta, Imperial College London, United Kingdom

Participants

1. Andrés Asensio Ramos	Instituto de Astrofísica de Canarias, Spain
2. Andrea Cunial	Università degli Studi di Padova, Italy
3. Ando Ratsimbazafy	University of the Western Cape, South Africa
4. Alessandro Ridolfi	Max-Planck-Institut für Radioastronomie, Germany
5. Federica Ricci	Università degli Studi Roma Tre, Italy
6. Janez Kos	University of Ljubljana, Slovenia
7. Elizabeth Martínez Gómez	Instituto Tecnológico Autónomo de México, Mexico
8. Johannes Rothe	Princeton University / TU München, Germany
9. Francisco Nogueras Lara	Instituto de Astrofísica de Andalucía, Spain
10. David Fernández	INAOE, Mexico
11.	
12. Phil Gregory	University of British Columbia, Canada
13. J.-L. Starck	Service d'Astrophysique (SAp)CEA-Saclay, France
14. Thomas J. Loredo	Cornell University, United States of America
15. Alfredo Mejía-Narváez	Centro de Investigaciones de Astronomía, Venezuela
16. Maciej Zapiór	Universitat de les Illes Balears, Spain
17. Sergio Velasco	Instituto de Astrofísica de Canarias, Spain
18. Gregor Traven	University of Ljubljana, Slovenia
19. Fabiola Hernández	Centro de Investigaciones de Astronomía, Venezuela
20. Dustin Lang	Carnegie Mellon University, United States of America
21. Jorge Zavala	INAOE, Mexico
22. Antonio Ferragamo	Instituto de Astrofísica de Canarias, Spain
23. Davide Amato	Technical University of Madrid, Spain
24. David López Fernández Nespral	Instituto de Astrofísica de Canarias, Spain
25. Inés Camacho Iñesta	Instituto de Astrofísica de Canarias, Spain
26. Pablo Martín-Fernández	Universidad de Granada, Spain
27. Magdalena Gryciuk	Space Research Centre Polish Academy of Sciences, Poland
28. Elaheh Hamraz	IPM, Iran
29. Brendon J. Brewer	University of Auckland, New Zealand
30. Héctor Vázquez Ramió	CEFCA, Spain
31. Francisco Jimenez-Forteza	Universitat de les Illes Balears, Spain
32. Marusa Zerjal	University of Ljubljana, Sloveniaa
33. Sara Rodríguez Berlanas	Instituto de Astrofísica de Canarias, Spain
34. Denis Tramonte	Instituto de Astrofísica de Canarias, Spain
35. Gonzalo Holgado Alijo	Instituto de Astrofísica de Canarias, Spain
36. Neus Águeda	University of Barcelona, Spain
37. ZengHua Zhang	Instituto de Astrofísica de Canarias, Spain
38. Mariela Martínez Paredes	INAOE, Mexico
39. Iñigo Arregui	Instituto de Astrofísica de Canarias, Spain
Not pictured. José M. Bernardo	Universidad de Valencia, Spain
Not pictured. Mateusz Janiak	N. Copernicus Astronomical Center, Poland
Not pictured. María Jesús Martínez González	Instituto de Astrofísica de Canarias, Spain

Not pictured. Paulo Miles	Instituto de Astrofísica de Canarias, Spain
Not pictured. Grzegorz Nowak	Instituto de Astrofísica de Canarias, Spain
Not pictured. Venkatessh Ramakrishnan	Aalto University Metsähovi Radio Observatory, Finland
Not pictured. Pedro Rubén Rivera	Instituto de Ciencias Nucleares, UNAM, Mexico
Not pictured. Alejandro Suarez Mascareño	Instituto de Astrofísica de Canarias, Spain
Not pictured. Roberto Trotta	Imperial College London, United Kingdom

Preface

Our view of the universe is imperfect because information from observations is always incomplete and uncertain. We are, however, not completely blind, since some a priori knowledge about the models we use to explain astrophysical processes can be used in our reasoning. These aspects of scientific practice are natural ingredients of Bayesian inference, a tool to update the probability of hypotheses conditional on observations that has started to show its extreme power in recent years with an exponential growth in the number of applications to astrophysical problems.

This book provides an overview of the fundamentals of Bayesian inference with applications to astrophysics, such as extrasolar planet detection, cosmology, or image reconstruction. The aim is to provide the next generation of researchers with the tools of modern data analysis that are already becoming standard in modern astrophysical research. Each chapter is written by a world-leading expert. They are based on lectures given at the XXVI Canary Islands Winter School of Astrophysics, organised by the Canary Islands Institute of Astrophysics (IAC) on a yearly basis since 1989 and devoted in the 2014 edition to Bayesian astrophysics.

The school took place at the IAC headquarters in La Laguna, 3–14 November 2014, and introduced young researchers to the use of Bayesian inference as a tool in their work. The lectures were given by Thomas Loredo (Introduction to Bayesian Inference), Brendon J. Brewer (Bayesian Inference and Computation), Phil Gregory (Bayesian Inference for Exoplanets), José M. Bernardo (Objective Bayesian Statistics), Roberto Trotta (Bayesian Cosmology), J.-L. Starck (Inverse Problems), and Dustin Lang (Large-Scale Surveys). The students also presented their work in the form of poster presentations, which were discussed in special sessions.

We are grateful to Antonio Aparicio, head of the Graduate Studies Division of the IAC, and to Lourdes González for her dedication to the organisation of the activities of the school. The personnel at the IAC headquarters (Monique Gómez, Carlos Westendorp, Luis Fernando Rodríguez, and Carlos Martín) and at the Teide Observatory (Cristina Protasio, Pere L. Pallé, Ricardo Génova, and Carsten Denker) provided valuable assistance during the guided tours organised for participants.

1 Bayesian Inference and Computation: A Beginner's Guide

BRENDON J. BREWER

1.1 Introduction

Most scientific observations are not sufficient to give us definite answers to all our questions. It is rare that we get a dataset which *totally* answers every question with certainty. Even if that did happen, we would quickly move on to other questions. What a dataset usually *can* do is make hypotheses more or less plausible, even if we do not achieve total certainty. Bayesian inference is a model of this reasoning process, and also a tool we can use to make quantitative statements about how much uncertainty we should have about our conclusions. This takes the mystery out of data analysis, because we no longer have to come up with a new method every time we face a new problem. Instead, we simply specify exactly what information we are going to use, and then compute the results. In the last 2 decades, Bayesian inference has become immensely popular in many fields of science, and astrophysics is no exception. Therefore it is becoming increasingly important for researchers to have at least a basic understanding of these methods.

Accessible textbooks for those with a physics background include those by Gregory (2005) and Sivia and Skilling (2006), and parts of the textbook by MacKay (2003).[1] The online tutorial by Vanderplas is also useful.[2] For those with a strong statistics background, I recommend the books by O'Hagan and Forster (2004) and Gelman et al. (2013). I also maintain a set of lecture notes for an undergraduate Bayesian statistics course.[3] The aim of this chapter is to present a fairly minimal yet widely applicable set of techniques to allow you to start using Bayesian inference in your own research.

Any particular application of Bayesian inference involves making choices about what data you are analysing, what questions you are trying to answer, and what assumptions you are willing to make. Data analysis problems in astronomy vary widely, so in this chapter we cannot cover a huge variety of examples. Instead, we only study a single example, but spend a lot of time looking at the methods and thinking that go into such an analysis, which will be applicable in other examples. The specific assumptions we make in the example will not always be appropriate, but they should be sufficient to show you the points at which assumptions are needed, and what you need to consider when you work on a particular problem.

In principle, it is usually best to work with your data in the most raw form possible, although this is often too difficult in practice. Therefore, most scientists work with data that has been processed (by a 'pipeline') and reduced to a manageable size. While many Bayesian practitioners often have strong ideals about data analysis, a large dose of pragmatism is still very necessary in the real world.

To do Bayesian inference, you need to specify what *prior information* you have (or are willing to assume) about the problem, in addition to the data. Prior information is necessary; what you can learn from a dataset depends on what you know about how it was produced. Once you have your data, and have specified your prior information, you

[1] The MacKay text is freely available online at `www.inference.phy.cam.ac.uk/itila`.
[2] Available online at `jakevdp.github.io/blog/2014/03/11/frequentism-and-bayesianism-a-practical-intro`.
[3] Available at `www.github.com/eggplantbren/STATS331`.

are faced with the question of how to calculate the results. Usually you want to calculate the *posterior distribution* for some unknown quantities (also known as 'parameters') given your data. This posterior distribution describes your uncertainty about the parameters, but takes the data into account.

Since we are often dealing with (potentially) complicated probability distributions in high-dimensional spaces, we need to *summarise* the posterior distribution in an understandable way. In certain problems, the summaries can be calculated analytically, but numerical methods are more general, so are the focus of this chapter. The most popular and useful numerical techniques are the Markov Chain Monte Carlo methods, often abbreviated as MCMC.[4] The rediscovery of MCMC in the 1990s is one of the main reasons why Bayesian inference is so popular now. While there were many strong philosophical arguments in favour of a Bayesian approach before then, many people were uncomfortable with the subjective elements involved. However, once MCMC made it easy to compute the consequences of Bayesian models more easily, people simply became more relaxed about these subjective elements.

A large number of MCMC methods exist, and it would be unwise to try to cover them all here. Therefore I focus on a small number of methods that are relatively simple to implement, yet quite powerful and widely applicable. I try to emphasise methods that are *general*, i.e. methods that work on most problems you might encounter. One disadvantage of this approach is that the methods we cover are not necessarily the most efficient methods possible. If you are mostly interested in one specific application, you will probably be able to achieve better performance by using a more sophisticated algorithm, or by taking advantage of the particular properties of your problem. There are many popular software packages (and many more unpopular ones) available for Bayesian inference, such as JAGS (Just Another Gibbs Sampler), Stan, emcee, MultiNest, my own DNest4, and many more. Please see the appendix for a brief discussion of the advantages and disadvantages of some of these packages.

1.2 Python

Due to its popularity and relatively shallow learning curve, I have implemented the algorithms in this chapter in the Python language. The code is written so that it works in either Python 2 or 3. The programs make use of the common numerical library `numpy`, and the plotting package `matplotlib`. Any Python code snippets in this chapter assume that the following packages have been imported:

```
import numpy as np
import numpy.random as rng
import matplotlib.pyplot as plt
import copy
import scipy.special
import numba
```

Full programs implementing the methods (and the particular problems) used in this chapter are provided online.[5]

[4] Some people claim that MCMC stands for Monte Carlo Markov Chains, but they are wrong.
[5] `https://github.com/eggplantbren/NSwMCMC`.

1.3 Parameter Estimation

Almost all data-analysis problems can be interpreted as *parameter estimation* problems. The term *parameter* has a few different meanings, but you can usually think of it as a synonym for *unknown quantity*. When you learned how to solve equations in high school algebra, you found the value of an unknown quantity (often called x) when you had enough information to determine its value with certainty. In science, we almost never have enough information to determine a quantity without any uncertainty, which is why we need probability theory and Bayesian inference.

Here, we denote our unknown parameters by θ, which could be a single parameter or perhaps a vector of parameters (e.g. the distance to a star and the angular diameter of the star). To start, we need to have some idea of the set of possible values we are considering. For example, are the parameters integers? Real numbers? Positive real numbers? In some examples, the definition of the parameters already restricts the set of possible values. For example, *the proportion of extrasolar planets in the Milky Way that contain life* cannot be less than 0 or greater than 1. Strictly speaking, it has to be a rational number, but it probably will not make much difference if we just say it is a real number between 0 and 1 (inclusive). The distance to a star (measured in whatever units you like) is presumably a positive real number, as is its angular diameter. The set of possible values you are willing to consider is called the *hypothesis space*.

To start using Bayesian inference, you need to assign a *probability distribution* on the hypothesis space, which models your initial uncertainty about the parameters. This probability distribution is called the *prior*. We then use Bayes' rule, a consequence of the product rule of probability, to calculate the *posterior* distribution, which describes our updated state of knowledge about the values of the parameters, after taking the data D into account.

For a prior distribution $p(\theta|I)$ (read as 'the probability distribution for θ given I'), and a sampling distribution $p(D|\theta, I)$, Bayes' rule allows us to calculate the *posterior distribution* for θ:

$$p(\theta|D, I) = \frac{p(\theta|I)p(D|\theta, I)}{p(D|I)}. \tag{1.1}$$

The I in Equation 1.1 refers to background information and assumptions; basically, it stands for everything you know about the problem apart from the data. The I appears in the background of all of the terms in Equation 1.1, and is often omitted. Note also that the notation used in Equation 1.1 is highly simplified but conventional among Bayesians; see the appendix for a discussion of notation. For brevity we can suppress the I (remove it from the right-hand side of all equations) and just write:

$$p(\theta|D) = \frac{p(\theta)p(D|\theta)}{p(D)}. \tag{1.2}$$

The result is a probability distribution for θ which describes our state of knowledge about θ after taking into account the data. The denominator, since it does not depend on θ, is a normalising constant, usually called the *marginal likelihood* or alternatively the *evidence*. Since the posterior is a probability distribution, its total integral (or sum, if the hypothesis space is discrete) must equal 1. Therefore we can write the marginal likelihood as:

$$p(D|I) = \int p(\theta|I)p(D|\theta, I)\, d\theta, \tag{1.3}$$

where the integral is over the entire N-dimensional parameter space.

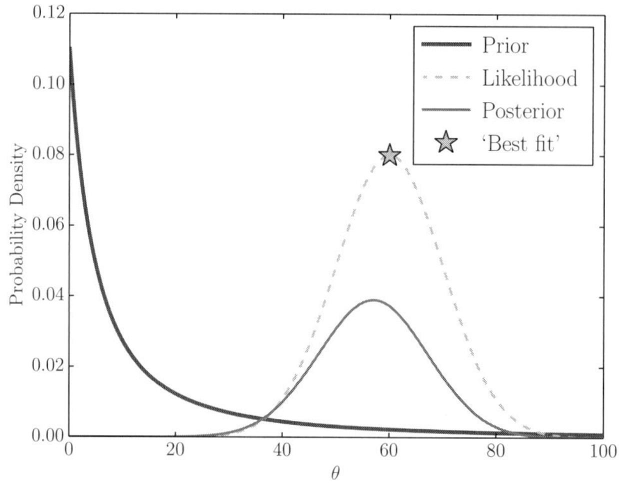

FIGURE 1.1 An example prior distribution for a single parameter (blue) gets updated to the posterior distribution (red) by the data. The data enters through the likelihood function (cyan dotted line). Many traditional 'best fit' methods are based on finding the maximum likelihood estimate, which is the peak of the likelihood function, here denoted by a star.

The posterior distribution is usually narrower than the prior distribution, indicating that we have learnt something from the data, and our uncertainty about the value of the parameters has decreased. See Figure 1.1 for an example of the qualitative behaviour we usually see when updating from a prior distribution to a posterior distribution.

1.4 Transit Example

To become more familiar with Bayesian calculations, we will work through a simple curve-fitting example. Many astronomical data-analysis problems can be viewed as examples of curve fitting. Consider a transiting exoplanet, like one observed by the Kepler space observatory. The light curve of the star shows an approximately constant brightness as a function of time, with a small dip as the exoplanet moves directly in front of the star. Clearly, real Kepler data is much more complex than this example, as stars vary in brightness in complicated ways, and the shape of the transit signal itself is more complex than the model we use here. Nevertheless, this example contains many of the features and complications that arise in a more realistic analysis.

The dataset, along with the true curve, is shown in Figure 1.2. The equation for the true curve is:

$$\mu(t) = \begin{cases} 10, & 2.5 \leq t \leq 4.5 \\ 5, & \text{otherwise.} \end{cases}$$

Assume we do not know the equation for the true curve (as we would not in reality), but we at least know that it is a function of the following form:

$$\mu(t) = \begin{cases} A - b, & (t_c - w/2) \leq t \leq (t_c + w/2) \\ A, & \text{otherwise,} \end{cases}$$

where A is the brightness away from the transit, b is the depth of the transit, t_c is the time of the centre of the transit, and w is the width of the transit.

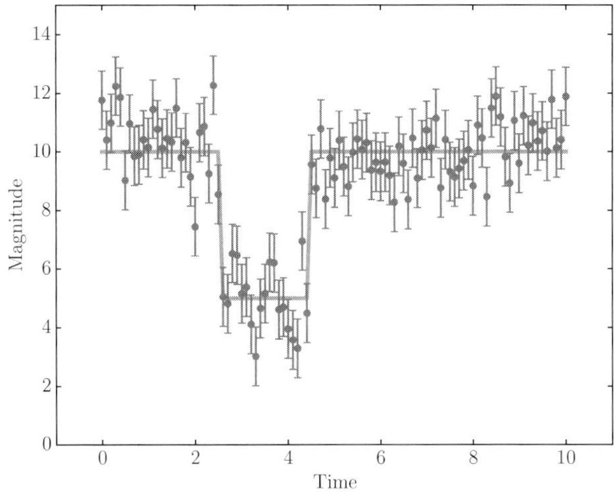

FIGURE 1.2 The 'transit' dataset. The red curve shows the model prediction based on the true values of the parameters, and the blue points are the noisy measurements.

Thus, the problem has been reduced from not knowing the true curve $\mu(t)$ to not knowing the values of 4 quantities (parameters) A, b, t_c, and w. Applying Bayesian inference to this problem involves calculating the posterior distribution for A, b, t_c, and w, given the data D. For this specific setup, Bayes' rule states:

$$p(A, b, t_c, w|D) = \frac{p(A, b, t_c, w)p(D|A, b, t_c, w)}{p(D)}. \tag{1.4}$$

So, in order for the posterior distribution to be well defined, we need to choose a prior distribution $p(A, b, t_c, w)$ for the parameters, and a sampling distribution $p(D|A, b, t_c, w)$ for the data. Since the denominator $p(D)$ is not a function of the parameters, it plays the role of a normalising constant that ensures the posterior distribution integrates to 1, as any probability distribution must.

1.4.1 Sampling Distribution

The *sampling distribution* is the probability distribution we would assign for the data if we knew the true values of the parameters. A useful way to think about the sampling distribution is to write some code whose input is the true parameter values, and whose output is a simulated dataset. Whatever probability distribution you use to simulate your dataset is your sampling distribution.

In many situations, it is conventional to assign a normal distribution (also known as a Gaussian distribution) to each data point, where the mean of the normal distribution is the noise-free model prediction, and the standard deviation of the normal distribution is given by the size of the error bar. Later, we will see how to relax these assumptions in a useful way. The probability density for the data given the parameters (i.e. the sampling distribution) is:

$$p(D|A, b, t_c, w) = \prod_{i=1}^{N} \frac{1}{\sigma_i\sqrt{2\pi}} \exp\left[-\frac{1}{2\sigma_i^2}\left(D_i - \mu(t_i)\right)^2\right]. \tag{1.5}$$

This is a product of N terms, one for each data point, and is really a probability distribution over the N-dimensional space of possible datasets. We have assumed that each data point is *independent* (given the parameters). That is, if we knew the parameters and a subset of the data points, we would use *only* the parameters (not the data points) to predict the remaining data points.

When the dataset is known, Equation 1.5 becomes a function of the parameters only, known as the *likelihood function.* The curve predicted by the model, here written as $\mu(t_i)$ (where I have suppressed the implicit dependence on the parameters), provides the mean of the normal distribution. Remember that the independence assumption is not an assumption about the actual dataset, but an assumption about our prior information about the dataset. It does not make sense to say that a particular dataset is or is not independent. Independence is a property of probability distributions.

Equation 1.5 is the sampling distribution (and the likelihood function) for our problem, but it is fairly cumbersome to write. Statisticians have developed a shorthand notation for writing probability distributions. This is extremely useful for communicating your assumptions without having to write the entire probability-density equation. To communicate Equation 1.5, we can simply write:

$$D_i \sim \mathcal{N}\left(\mu(t_i), \sigma_i^2\right), \tag{1.6}$$

i.e. each data point has a normal distribution (denoted by $\mathcal{N}$) with mean $\mu(t_i)$ (which depends on the parameters) and standard deviation σ. For the normal distribution, it is traditional to write the *variance* (standard deviation squared) as the second argument, but since the standard deviation is a more intuitive quantity (being in the same units as the mean), we often literally write the standard deviation, squared (e.g. 3^2). For other probability distributions the arguments in the parentheses are whatever parameters make sense for that family of distributions.

1.4.2 Priors

Now we need a prior for the unknown parameters A, b, t_c, and w. This is a probability distribution over a 4-dimensional parameter space. To simplify things, we can assign independent priors for each parameter, and multiply these together to produce the joint prior:

$$p(A, b, t_c, w) = p(A)p(b)p(t_c)p(w). \tag{1.7}$$

This prior distribution models our uncertainty about the parameters before taking into account the data. The independence assumption implies that if we were to learn the value of one of the parameters, this would not tell us anything about the others. This may or may not be realistic in a given application, but it is a useful starting point.

Another useful starting point for priors is the uniform distribution, which has a constant probability density between some lower and upper limit. Use 4 uniform distributions for our priors:

$$A \sim U(-100, 100) \tag{1.8}$$

$$b \sim U(0, 10) \tag{1.9}$$

$$t_c \sim U(t_{\text{min}}, t_{\text{max}}) \tag{1.10}$$

$$w \sim U(0, t_{\text{max}} - t_{\text{min}}). \tag{1.11}$$

The full expression for the joint prior probability density is:

$$p(A, b, t_c, w) = \begin{cases} \frac{1}{2000(t_{\text{max}} - t_{\text{min}})^2}, & (A, b, t_c, w) \in S \\ 0, & \text{otherwise,} \end{cases} \tag{1.12}$$

where S is the set of allowed values. Even more simply, we can ignore the normalising constant and the prior boundaries and just write:

$$p(A, b, t_c, w) \propto 1, \tag{1.13}$$

although if we use this shortcut, we must remember that the boundaries are implicit.

Now that we have specified our assumed prior information in the form of a sampling distribution and a prior, we are ready to go. By Bayes' rule, we have an expression for the posterior distribution immediately:

$$p(A, b, t_c, w|D) \propto p(A, b, t_c, w)p(D|A, b, t_c, w), \tag{1.14}$$

which is proportional to the prior times the likelihood. In the likelihood expression we substitute the actual observed dataset into the equation, so that it is a function of the parameters only. The main problem with using Bayes's rule this way is that a mathematical expression for a probability distribution in a 4-dimensional space is not very easy to understand intuitively. For this reason, we usually calculate summaries of the posterior distribution. The main computational tool for doing this is MCMC.

1.5 Markov Chain Monte Carlo

Monte Carlo methods allow us to calculate any property of a probability distribution that is an expectation value. For example, if we have a single variable x with a probability density $p(x)$, the expected value is

$$\mathbb{E}(x) = \int_{-\infty}^{\infty} xp(x)\,dx \tag{1.15}$$

which is a measure of the 'centre of mass' of the probability distribution.

If we had a set of points $\{x_1, x_2, \ldots, x_N\}$ 'sampled from' $f(x)$, we could replace the integral with a simple average:

$$\mathbb{E}(x) \approx \frac{1}{N}\sum_{i-1}^{N} x_i. \tag{1.16}$$

In 1 dimension, this may not seem very useful. Evaluating a 1-dimensional integral analytically is often possible, and doing it numerically using the trapezoidal rule (or a similar approximation) is quite straightforward. However, Monte Carlo really becomes useful in higher-dimensional problems. For example, consider a problem with five unknown quantities with probability distribution $p(a, b, c, d, e)$, and suppose we want to know the probability that a is greater than $b + c$. We could do the integral

$$P(a > b + c) = \int p(a, b, c, d, e)\mathbb{1}\,(a > b + c)\; da\, db\, dc\, dd\, de \tag{1.17}$$

where $\mathbb{1}\,(a > b + c)$ is a function that is equal to 1 where the condition is satisfied and 0 where it is not. However, if we could obtain a sample of points in the five-dimensional space, the Monte Carlo estimate of the probability is simply

$$P(a > b + c) \approx \frac{1}{N}\sum_{i=1}^{N} \mathbb{1}\,(a_i > b_i + c_i), \tag{1.18}$$

which is just the fraction of the samples that satisfy the condition.

Another important use of Monte Carlo is marginalisation. Suppose again that we have a probability distribution for 5 variables, but we only care about 1 of them. For example, the marginal distribution of a is given by

$$p(a) = \int p(a, b, c, d, e)\, db\, dc\, dd\, de \tag{1.19}$$

which describes your uncertainty about a, rather than your uncertainty about all of the variables. This integral might be analytically intractable. With Monte Carlo, if you have samples in the five-dimensional space but you only look at the first coordinate, then you have samples from $p(a)$. This is demonstrated graphically in Figure 1.3.

In Bayesian inference, the most important probability distribution is the posterior distribution for the parameters. We would like to be able to generate samples from the posterior, so we can compute probabilities, expectations, and other summaries easily. MCMC allows us to generate these samples.

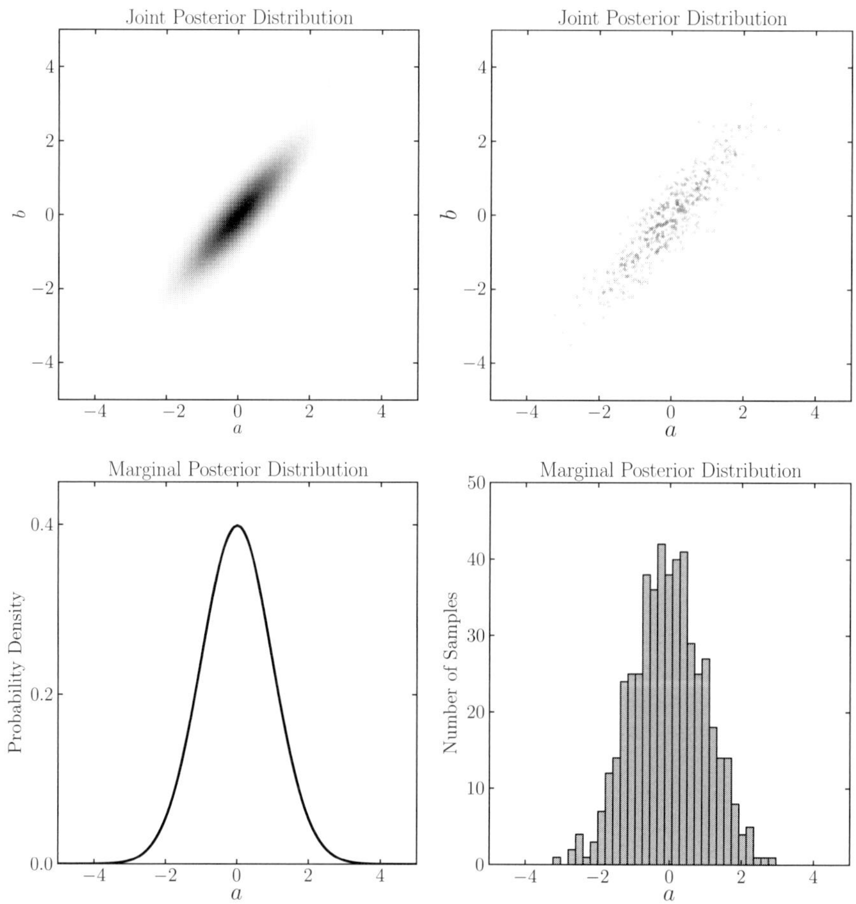

FIGURE 1.3 An example posterior distribution for 2 parameters a and b, taken from my STATS 331 undergraduate lecture notes. The full joint distribution is shown in the top left, and the marginal distribution for a (bottom left) is calculated by integrating over all possible b values, a potentially non-trivial calculation. The top right panel has points drawn from the joint posterior. Points drawn from the marginal posterior (bottom right) are obtained by ignoring the b values of the points, a trivial operation.

1.5.1 The Metropolis Algorithm

The Metropolis-Hastings algorithm, also known as the Metropolis algorithm, is the oldest and most fundamental MCMC algorithm. It is quite straightforward to implement, and works well on many problems. The version of the Metropolis algorithm presented here is sometimes called *random walk* Metropolis. More sophisticated choices are possible, but usually require problem-specific knowledge.

Consider a problem with unknown parameters θ. If the prior is some density $\pi(\theta)$ and the likelihood function is $L(\theta)$, then the posterior distribution will be proportional to $\pi(\theta)L(\theta)$. The marginal likelihood $Z = \int \pi(\theta)L(\theta)\,d\theta$ is unknown, but the Metropolis algorithm does not need to know it: all we need is the ability to evaluate π and L at a given position in the parameter space. The Metropolis algorithm tells us how to move a 'particle' around the parameter space so that we eventually sample the posterior distribution. That is, the amount of time spent in any particular region of parameter space will be approximately proportional to the posterior probability in that region. Note the change of notation from $p(\theta)$, $p(D|\theta)$, and $p(D)$ to π, L, and Z, respectively. This is a convention when discussing computational methods (as opposed to discussing priors, datasets, etc.).

The Metropolis algorithm can be summarised as follows:

(i) Choose a starting position θ, somewhere in the parameter space.
(ii) Generate a proposed position θ' from a proposal distribution $q(\theta'|\theta)$. A common choice is a 'random walk' proposal, where a small perturbation is added to the current position.
(iii) With probability $\alpha = \min\left(1, \frac{q(\theta|\theta')}{q(\theta'|\theta)}\frac{\pi(\theta')}{\pi(\theta)}\frac{L(\theta')}{L(\theta)}\right)$, accept the proposal (i.e. replace θ with θ'). Otherwise, do nothing (i.e. remain at θ).
(iv) Repeat steps (i)–(iii) until you have enough samples.

When a proposed move is rejected (and the particle remains in the same place), it is important to count the particle's position again in the output. This is how the algorithm ends up spending more time in regions of high probability: moves *into* those regions tend to be accepted, whereas moves *out* of those regions are often rejected. The Metropolis algorithm is quite straightforward to implement and I encourage you to attempt this yourself if you haven't done so before.

Python code implementing the Metropolis algorithm is given below (this code has been stripped of features for keeping track of the output, and shows just the algorithm itself). Note several features. First, the functions used to measure the prior density and likelihood of any point, and the function to generate a proposal in the first place, are problem specific and assumed to have been implemented elsewhere. Second, for numerical reasons we deal with the (natural) log of the prior density, the likelihood, and the acceptance probability. Third, note how the `log_prior` and `log_likelihood` functions only need to be called once per iteration, not twice as one might naively think. Finally, no q ratio is required in the acceptance probability: we assume that we are working with a *symmetric* proposal distribution, where the probability of proposing a move to position a given the current position is b is the same as the probability of the reverse (proposing b when at position a).

```
# Generate a starting point (if you have a good guess, use it)
# In the full version of the code, the initial point is drawn
# from the prior.
params = np.array([1., 1., 1., 1.])
logp, logl = log_prior(params), log_likelihood(params)
```

```
# Total number of iterations
steps = 100000

# Main loop
for i in range(0, steps):
  # Generate proposal
  new = proposal(params)

  # Evaluate prior and likelihood for the proposal
  logp_new = log_prior(new)
  logl_new = -np.Inf
  # Only evaluate likelihood if prior prob isn't zero
  if logp_new != -np.Inf:
    logl_new = log_likelihood(new)

  # Acceptance probability
  log_alpha = (logl_new - logl) + (logp_new - logp)
  if log_alpha > 0.:
    log_alpha = 0.

  # Accept?
  if rng.rand() <= np.exp(log_alpha):
    params = new
    logp = logp_new
    logl = logl_new
```

The 'random walk' proposal generates a proposed value θ' from a normal (Gaussian) distribution centred around the current position θ. The user is free to choose the width of the normal distribution. Here is a Python code snippet showing a proposal with width L:

```
# Generate a proposal
proposal = theta + L*rng.randn()
```

The performance of the Metropolis algorithm depends quite strongly on the width of the proposal distribution. If the width is too small, most moves will be accepted, but will not move very far. If the width is too large, most moves will be rejected, so you end up stuck in 1 place. Some authors recommend using preliminary runs to find an optimal width. Instead, I recommend that you use a mixture of widths. Basically, every time we make a proposal, the width is drawn from some range, rather than being constant. The biggest possible width we would ever want should be roughly the order of magnitude of the width of the prior (since the posterior is usually narrower than the prior). It is rare that we would need proposals many orders of magnitude smaller than that. My default suggestion is to randomise the logarithm of the step size, as in this code snippet:

```
# A heavy-tailed proposal distribution
# Generate a standard deviation
L = 10.**(1.5 - 6.*rng.rand())

# Use the standard deviation for the proposal
proposal = theta + L*rng.randn()
```

With this proposal, the minimum width is $10^{-4.5} \approx 3.16 \times 10^{-5}$, and the maximum width is $10^{1.5} \approx 31.6$. The effective proposal distribution is now very heavy tailed. Simply ensure that a good width is somewhere within your range. Remember, this is not an optimal suggestion, but a fail-safe conservative one. It is possible to spend time (doing preliminary runs) to eventually save time (by having a more efficient final run), but my personal preference is usually not to do this.

When there are multiple parameters (almost always, since MCMC is not necessary on single-parameter problems!), we need to decide how to construct the proposal. There are 2 main ways to do this. The first is that the proposal is to change all of the parameters simultaneously. This tends to be inefficient in high dimensions, because the only proposals that are likely to be accepted are those that change the parameter values only slightly: in a high-dimensional space, there are many bad directions to travel, and not very many good ones. Usually, it is better to propose to change a subset of the parameters, or even just a single parameter. A Python function that takes a `numpy` array of parameters and input and returns a proposed value for the parameters is specified below.

```
def proposal(params):
  """
  Generate new values for the parameters.
  The proposal for the Metropolis algorithm.
  """
  # Copy the parameters
  new = copy.deepcopy(params)

  # Which one should we change?
  which = rng.randint(num_params)
  new[which] += jump_sizes[which] \
                *10.**(1.5 - 6.*rng.rand())*rng.randn()
  return new
```

This function relies on `num_params` being the number of parameters, and an array `jump_sizes`, of length `num_params`, which specifies the prior width for each parameter.

For numerical reasons, instead of writing a function to evaluate the likelihood (and the prior density) at a particular point in parameter space, we usually deal with the logarithms of these quantities. For the transit example, the log prior function can be implemented like so:

```
def log_prior(params):
  """
  Evaluate the (log of the) prior distribution
  """
  A, b, tc, width = params[0], params[1], params[2], params[3]

  # Minus infinity, if out of bounds
  if A < -100. or A > 100.:
    return -np.Inf
  if b < 0. or b > 10.:
    return -np.Inf
  if tc < t_min or tc > t_max:
```

```
        return -np.Inf
    if width < 0. or width > t_range:
        return -np.Inf

    return 0.
```

Since we chose uniform priors, the prior density is some constant if the parameters (here passed to the function as a `numpy array` of 4 floating point values) are within the bounds of the uniform priors. Otherwise, the prior density is 0. We do not need to know the normalising constant of the prior density: we returned 0 for the log density, but the progress of the Metropolis algorithm would be the same if we had returned any other finite value, since the Metropolis algorithm only ever uses ratios of densities (i.e. differences in log density).

The log likelihood function is given below. This implements the logarithm of Equation 1.5. As with the log prior function, we could ignore the normalisation constants – in this case any term that is not a function of the parameters. However, I have included them for completeness.

```
def log_likelihood(params):
    """
    Evaluate the (log of the) likelihood function
    """
    # Rename the parameters
    A, b, tc, width = params[0], params[1], params[2], params[3]

    # First calculate the expected signal
    mu = A*np.ones(N)
    mu[np.abs(data[:,0] - tc) < 0.5*width] = A - b

    # Normal/gaussian distribution
    return -0.5*N*np.log(2.*np.pi) - np.sum(np.log(data[:,2])) \
        -0.5*np.sum((data[:,1] - mu)**2/data[:,2]**2)
```

1.5.2 Useful Plots

After running the Metropolis algorithm (or any other MCMC method), there are several useful plots that you should make. The first is known as a 'trace plot' (Figure 1.4), and is just a plot of one parameter over time as the algorithm ran. Trace plots are the single most useful diagnostic of whether your algorithm is working well. In the attached code, the MCMC results are stored in a 2-dimensional `numpy` array called `keep`, each row of which is the parameter vector at a particular iteration. It is trivial to make a trace plot from this output:

```
# Trace plot of parameter zero
plt.plot(keep[:,0])
```

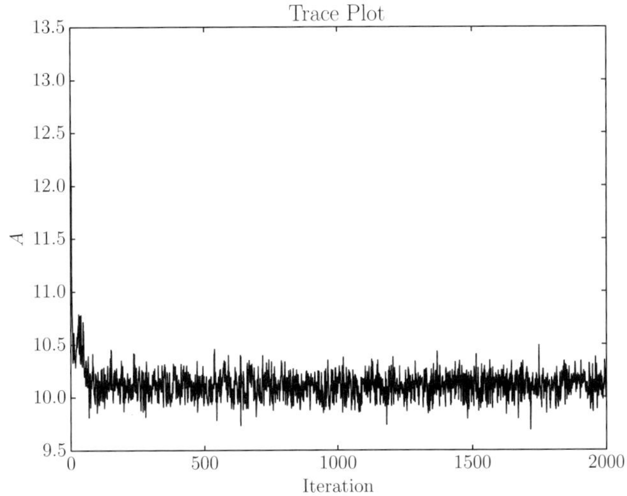

FIGURE 1.4 A 'trace plot'. At the beginning, because of the initial conditions of the algorithm, the results start in a region of parameter space that has very low probability. This initial phase is often called the 'burn-in', and should be excluded from any subsequent calculations.

The result is shown in Figure 1.4. When everything is working well, a trace plot should look like white noise when zoomed out. If this is the case, then things are *probably* working well (although you can never be 100% certain of this – like optimisation methods, MCMC methods can get stuck in local maxima). In addition, if your MCMC run was initialised at a point in parameter space that is an 'outlier' with respect to the posterior distribution, the first part of the run will be a transient feature where the MCMC chain (hopefully) moves towards the important regions of the space. This transient period is called the burn-in, and should usually be excluded from the sample, otherwise your Monte Carlo summaries will give too much importance to the part of the space the MCMC happened to go through during burn-in. When using Metropolis, it is worthwhile to monitor the fraction of the proposed moves that are accepted. This fraction should not be too close to 0 or 1, and generally somewhere between 15% and 50% is usually advised.

As an exercise, try running the Metropolis algorithm on the transit problem with a proposal that is far too small, a proposal that is far too large, and an initial condition that is very far from the bulk of the posterior distribution. Look at the resulting trace plots and compare them to the healthy one in Figure 1.4.

Other useful plots are histograms of single parameters (showing the marginal distribution for each parameter), and scatter plots showing the joint posterior distribution for a pair of parameters (with all of the other parameters marginalised out). See Figures 1.5 and 1.6 for examples. These allow us to visualise the uncertainty about the parameters, including some of the dependences that may exist in the uncertainties. For example, there is a slight correlation between A and b in the posterior, so if we obtain further information about A from elsewhere, this would affect our knowledge of b. In models with 2–10 interesting parameters, it is common to plot a grid of such plots, showing each parameter versus each other parameter. These are sometimes called 'corner' or 'triangle' plots, and a convenient Python package for using them is `corner.py` by Dan Foreman-Mackey.[6]

[6] Foreman-Mackey, D. (2016), corner.py: Scatterplot matrices in Python, *J. Open Source Soft.*, 1(2), 24, doi:10.21105/joss.00024?.

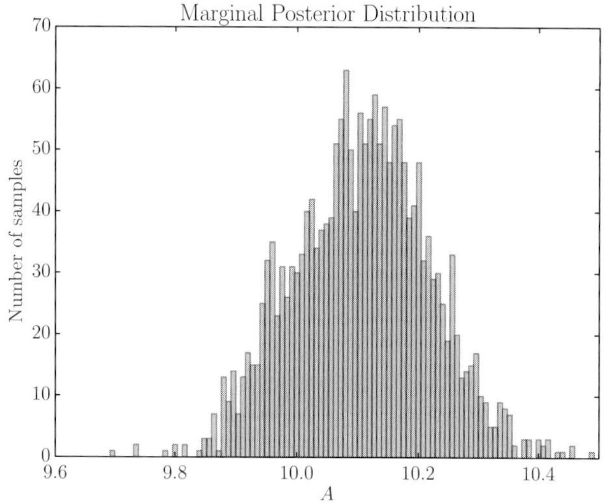

FIGURE 1.5 The marginal posterior distribution for the parameter A constructed from MCMC output.

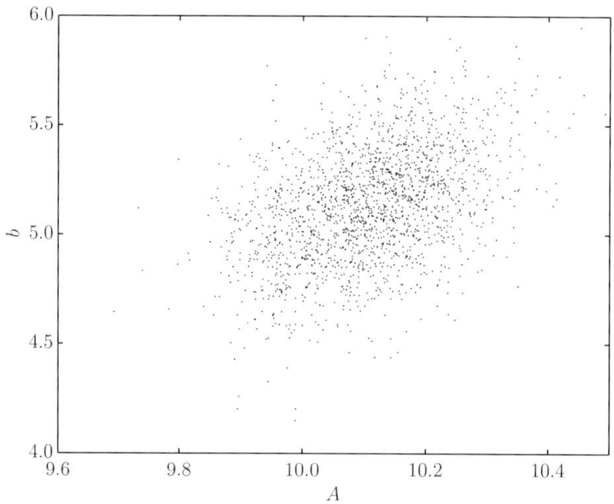

FIGURE 1.6 The joint posterior distribution for A and b constructed from the MCMC output. This is simply a scatterplot. Some authors prefer to apply some smoothing and approximate density contours.

1.5.3 Posterior Summaries

A probability distribution, such as the posterior, can potentially be arbitrarily complicated. For communication purposes, it is usually easier to summarise the distribution by a few numbers. The most common summaries are point estimates (i.e. a single estimate for the value of the parameter, such as 'we estimate $\theta = 0.43$') and intervals (e.g. the probability that $\theta \in [0.3, 0.5]$ is 68%).

Thankfully, most of these summaries are trivial to calculate from Monte Carlo samples, such as those obtained from MCMC. The most popular point estimate is the posterior mean, which is approximated by the arithmetic mean of the samples. The posterior median can also be easily approximated by the arithmetic median of the samples.

```
post_mean = np.mean(keep[:,0])
post_median = np.median(keep[:,0])
```

There are some theoretical arguments, based on decision theory, that provide guidance about which point estimate is better under which circumstances. To compute a credible interval (the Bayesian version of a confidence interval), you find quantiles of the distribution. For example, if we want to find an interval that contains 68% of the posterior probability, the lower end of the interval is the parameter value for which 16% of the samples are lower. Similarly, the upper end of the interval is the value for which 16% of the samples are higher (i.e. 84% are lower). The simplest way to implement this calculation is by sorting the samples, as shown below.

```
sorted_samples = np.sort(samples)
# Left and right end of interval
left = sorted_samples[int(0.16*len(sorted_samples))]
right = sorted_samples[int(0.84*len(sorted_samples))]
```

This kind of interval is sometimes called a *centred* credible interval, because the same amount of probability lies outside the interval on each side. In astronomy, the 68% credible interval is very popular because it is equivalent to 'plus or minus 1 posterior standard deviation' if the posterior is Gaussian. In other statistical fields such as opinion polling, psychology, and medical science, 95% credible intervals are more conventional.

1.6 Assigning Prior Distributions

There are only really two open problems in Bayesian inference. The first is how to assign sensible prior distributions (and sampling distributions) in different circumstances, and the second is how to calculate the results efficiently. With regard to the first problem, it is sometimes said that there are two different kinds of Bayesian inference, *subjective* and *objective*, with different methods for choosing priors. My view is that Bayesian describes a *hypothetical* state of prior knowledge, held by an idealised reasoner. When we apply Bayesian inference, we are studying how the idealised reasoner would update their state of knowledge based on the information we explicitly put into the calculation.

In many cases, we can just use simple 'default' choices for the prior and the sampling distribution (e.g. a uniform prior for a parameter, and a 'Gaussian noise' assumption for the data). A lot of analyses work just fine with these assumptions, and taking more care would not change the results in any important way. However, occasionally it makes sense to spend a lot of time and effort thinking about the prior distributions. So-called subjective Bayesians, who are not necessarily experts in the fields of their clients, conduct elaborate interviews with experts to try and create a prior that models the experts' beliefs well. This process is called *elicitation*. For example, consider the Intergovernmental Panel on Climate Change, which periodically writes immense reports summarising humanity's state of knowledge about global warming. Consider the question, 'How much will the global average temperature increase in the next 100 years?' This is exactly the kind of situation where elicitation of an expert's probabilities is very important: a lot is known, and a lot is at stake. In this situation, it would not be very wise to rely on convenient 'vague' priors!

On the other hand, so-called objective Bayesians search for principles based on symmetry or other arguments, which can help choose a prior that is appropriate to describe a large amount of ignorance. Some examples include the principle of indifference, the Jeffreys prior, reference priors, default priors, transformation groups, maximum entropy, and entropic priors.

1.6.1 Probability Distributions Have Consequences

When you assign prior distributions, it is easier than you might expect to build in an assumption that you do not really agree with, which can end up affecting your results in ways you may not have predicted (but which are ultimately understandable). This is especially true in high-dimensional problems, which is why 'hierarchical models' (beyond the scope of this chapter) are so useful. Here, we look at some common issues that can arise in non-hierarchical models.

Imagine that you want to infer the mass M of a galaxy. It might seem reasonable to assume 'prior ignorance' about M, using a uniform distribution between 10^5 and 10^{15} solar masses:

$$M \sim U(10^5, 10^{15}). \tag{1.20}$$

However, this has an unfortunate feature: it implies that the prior probability of M being greater than 10^{14} is 0.9, and the probability of M being greater than 10^{12} is 0.999, which seems overly confident when we are trying to describe ignorance! In astronomy, we are often in the situation of having to 'put our error bars in the exponent'. A prior that has this property is the 'log-uniform' distribution (named by analogy with the lognormal distribution, and sometimes incorrectly called a Jeffreys prior), which assigns a uniform distribution to the logarithm of the parameter. If we replace the uniform prior by

$$\log_{10}(M) \sim U(\log(10^5), \log(10^{15})) \tag{1.21}$$

then the prior probabilities are more moderate: $P(M > 10^{14}) = 0.1$, $P(M > 10^{12}) = 0.3$, and so on. The log-uniform prior is appropriate for positive parameters whose uncertainty spans multiple orders of magnitude. The probability density, in terms of M, is proportional to $1/M$.

There are two main ways to implement the log-uniform prior in the Metropolis algorithm. One is to keep M as a parameter, and take the non-uniform prior into account in the acceptance probability (i.e. implement the $1/M$ prior in your `log_prior` function). The other is to treat $\ell = \log_{10}(M)$ as the parameter, in which case the prior is still uniform. You just need to compute M from ℓ before you can use it in the likelihood function. The second approach (parametrising by ℓ instead) is generally a better idea.

Now we discuss some consequences of the sampling distribution, for which we used a normal distribution with known standard deviation for each data point, and asserted that the measurements were independent. In many applications (such as discrete-valued photon count data) other distributions such as the Poisson may be more appropriate. However, even for real-valued 'model plus noise' situations the normal distribution has some consequences which may be undesirable. For example, there is a high probability that the noise vector (i.e. all of the actual differences between the true curve and the data points) looks macroscopically like white noise, with little correlation between the data points. To see this, try generating simulated datasets from the sampling distribution for a particular setting of the parameters, and you will see that almost all datasets that you generate have this property. If this is not realistic, correlated noise models are possible (e.g. using Gaussian processes), but we will not discuss these here.

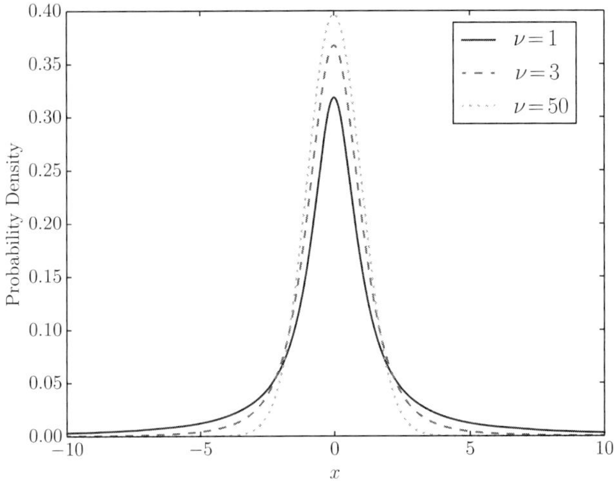

FIGURE 1.7 Three t-distributions, with $\nu = 1$, $\nu = 3$, and $\nu = 50$. Lower ν implies heavier tails and a greater probability for outliers. All 3 of these distributions have $\mu = 0$ and $\sigma = 1$.

Another implication of our sampling distribution is that we would not expect very many measurements to depart from the model by more than a few standard deviations. Normal distributions have very 'light tails', implying a very low probability for outliers. If the dataset does in fact contain outliers, the model will try to fit them very closely because the normal distribution assumption is telling it to believe the data points. This problem also occurs in least-squares fitting, and switching to Bayesian inference only resolves the problem if we use something other than a normal distribution for the sampling distribution. (See Hogg et al. (2010) for a detailed discussion of this issue.) A simple alternative to the normal distribution that is more appropriate if outliers are possible is the 'Student-t' distribution. Like the normal distribution, this has a 'location' parameter (the centre of the distribution) and a 'scale' parameter (the width), but there is a third parameter ν (often called the 'degrees of freedom') which controls the shape of the distribution. When ν is high ($\gtrsim$30) the shape is very close to Gaussian, but when ν is low (0–10) it has much heavier tails. Three t distributions with different values of ν are shown in Figure 1.7.

For a single variable x, the probability density function is:

$$p(x|\nu, \mu, \sigma) = \frac{\Gamma\left(\frac{\nu+1}{2}\right)}{\Gamma\left(\frac{\nu}{2}\right)\sigma\sqrt{\pi\nu}}\left[1 + \frac{1}{\nu}\frac{(x-\mu)^2}{\sigma^2}\right]^{-\frac{\nu+1}{2}} \tag{1.22}$$

where μ is the location parameter, σ is the scale parameter, and ν is the degrees of freedom. Γ is the gamma function. This can be used as a drop-in replacement for the normal distribution in Equation 1.5.

Student-t distributions are most well known because they arise naturally as the solution to some analytical problems in statistics, but they are also useful for allowing the possibility of outliers (as we are doing here), and for specifying informative priors. Traditionally, normal priors are frequently used when we have a lot of prior knowledge about the value of a parameter. The heavier tails of the Student-t distribution can be a more fail-safe option in this situation.

1.6.2 Should We Trust the Error Bars?

In astronomy many datasets are accompanied by error bars, which give some idea of the accuracy of a measurement. Without knowing all the details of how the error bars are produced, we do not really know to what extent to trust them, and hence whether it is a good idea to incorporate them literally into a sampling distribution like we did in Equation 1.5.

Instead of trusting the error bars, we can add an extra parameter which describes the degree to which we should trust the error bars. Let K be a constant by which we should multiply all the error bars before using them in the likelihood function. If $K = 1$, this corresponds to complete trust, and if $K > 1$ this means the error bars should have been bigger by a factor K. We will not allow $K < 1$. Let us use the following prior density for K:

$$p(K) = \frac{1}{2}\delta(K-1) + \frac{1}{2}\begin{cases} \exp\left[-(K-1)\right], & K > 1 \\ 0, & K \leq 1. \end{cases} \tag{1.23}$$

This is a 50/50 *mixture* of a Dirac delta function at $K = 1$, and an exponential distribution (with scale length 1) for $K > 1$. This expresses the idea that there is a 50% chance that $K = 1$ precisely, and if not, then it is likely to be fairly close to 1, and very unlikely to be greater than 5. This prior is plotted in Figure 1.8.

To implement this prior in MCMC, we can use a similar trick to the method of implementing the log-uniform prior. That is, we can introduce another parameter with a uniform prior (call it u_K), and then transform it to produce K. If we let $u_K \sim U(-1, 1)$, and let $K = 1$ if $u_K < 0$, then we have the desired 50% probability for $K = 1$. To obtain the exponential distribution part of the prior, we can set $K = 1 - \log(1 - u_K)$ if $u_K > 0$. This is the *inverse transform sampling* method, where the inverse of the cumulative

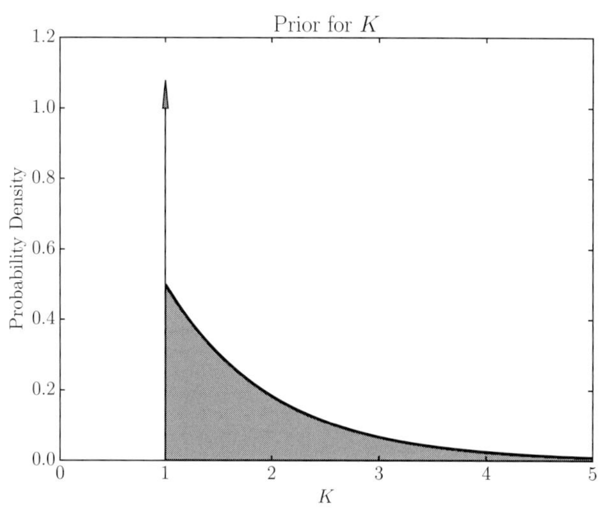

FIGURE 1.8 The prior for K, the amount by which we should scale the error bars given with the dataset. There is a 50% probability that $K = 1$, and a 50% probability that $K > 1$. Given that $K > 1$ the prior distribution is exponential (with unit scale length), so we do not expect K to be greater than 1 by an order of magnitude.

distribution function[7] is used to transform a variable from a $U(0, 1)$ distribution to some other distribution.

We can now implement the new `log_likelihood` function for the transit model with the Student-t distribution replacing the normal distribution. The code is given below, and demonstrates the technique used to implement both the log-uniform prior for ν (assuming `params[4]` has a uniform prior) and the mixture prior for K.

```
def log_likelihood(params):
   """
   Evaluate the (log of the) likelihood function
   """
   # Rename the parameters
   A, b, tc, width, log_nu, u_K = params[0], params[1], params[2]\
                                 ,params[3], params[4], params[5]

   # Parameter is really log_nu
   nu = np.exp(log_nu)

   # Compute K and 'inflated' error bars
   if u_K < 0.:
     K = 1.
   else:
       K = 1. - np.log(1. - u_K)
   sig = K*data[:,2]

   # First calculate the expected signal
   mu = A*np.ones(N)
   mu[np.abs(data[:,0] - tc) < 0.5*width] = A - b

   # Student t distribution
   return N*scipy.special.gammaln(0.5*(nu+1.))\
        - N*scipy.special.gammaln(0.5*nu)\
        - np.sum(np.log(sig*np.sqrt(np.pi*nu)))\
        - 0.5*(nu + 1.)*np.sum(\
          np.log(1. + (data[:,1] - mu)**2/nu/sig**2))
```

With this modified likelihood function, a log-uniform prior for ν between 0.1 and 100, and a uniform prior for u_K between -1 and 1, we can calculate the consequences of this model for the transit example. The 2 extra parameters ν and K might be of interest, but unfortunately are not directly accessible since we elected to use $\log(\nu)$ and u_K as parameters instead, and derived ν and K from them. However, since these both had uniform priors, their marginal posteriors should be straightforward to interpret. These are plotted in Figure 1.9. The data has basically ruled out low values of $\log(\nu)$ as the posterior distribution favours high values. The Student-t distribution becomes

[7] If a variable has probability density $f(x)$ then the cumulative distribution function is $F(x) = \int_{-\infty}^{x} f(t)\,dt$.

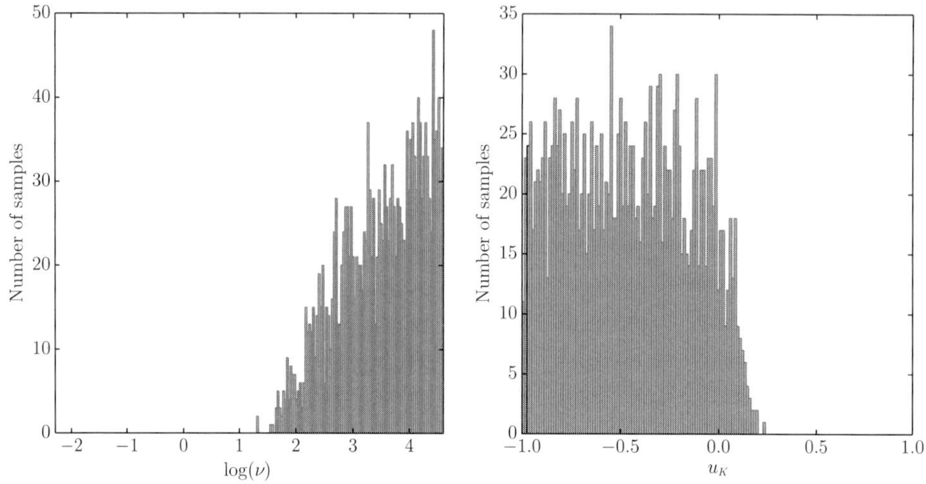

FIGURE 1.9 Marginal posterior distributions for $\log(\nu)$, which describes the shape of the Student-t distribution for the noise, and u_K, which determines the error bar inflation factor K.

approximately Gaussian when ν is high, so this result suggests the original Gaussian assumption was fine (even though the original Gaussian assumption is not technically even in this hypothesis space). For u_K, the data has ruled out values $\gtrsim 0.2$, implying that K is unlikely to be greater than about 1.2, so there is no real evidence that the error bars needed to be inflated.

Unlike with ν, with K our original assumption (equivalent to $K = 1$) is part of the parameter space, so we can compute its posterior probability. Its prior probability was 50%, and its posterior probability can be estimated using the fraction of MCMC samples for which $u_K < 0$:

```
np.mean(keep[:,5] < 0.)
```

The result is 92%, i.e. given this data and these assumptions, we should be quite confident that the error bars are of the appropriate size. This should be of no surprise, since I generated the data using a normal distribution, and the error bars in the data file were the same as the standard deviation of the normal distribution I used.

1.7 When the Results Are Sensitive to the Prior

The prior we used for K (Figure 1.8), with a spike at an especially plausible parameter value, is notorious for producing conclusions that depend sensitively on the shape of the prior. To see this for yourself, try keeping the prior probability for $K = 1$ constant at 50%, but make the scale length of the exponential part of the prior larger. You will find that the posterior probability for $K = 1$ decreases.

In the limit that the exponential scale length tends to infinity, the posterior probability for $K = 1$ will tend to 0, regardless of the dataset. This is not a mystery, but a logical consequence of the assumptions used. For example, if the scale length of the exponential were 1,000,000, then if $K > 1$ we would expect to see data scattered by an amount much greater than the error bars suggest. Since this was not observed, the data will favour $K = 1$. On the other hand, if the scale length of the exponential were 0.001, we would expect to see basically the same data whether $K = 1$ or not. In this situation, we would obtain a posterior probability very close to 50% (the same as the prior) for $K = 1$, unless we had a huge amount of data.

When the results of an analysis are very sensitive to the prior information assumed in the prior distribution and the sampling distribution, it is prudent to demonstrate this sensitivity to your readers. The problem may go away with more careful consideration of your priors, or it may just signal that the data cannot answer your question in a way that satisfies everyone.

1.8 What Is the Data?

It can sometimes be helpful to think carefully about exactly what your dataset contains, and whether it is all 'data' in the sense of Bayesian inference. In the transit example, we chose a sampling distribution for the y-values of the data (the flux measurements). Note that we never assigned a probability distribution for the *times* of the data points, even though we might think of that as part of the data (it is probably in the same file as the measurements, after all). In fact, because we did not assign a probability distribution for it, the times-stamps were not data at all, but part of the prior information I. Therefore, it is completely justifiable to use the timestamps when assigning the prior and the sampling distribution. It is not 'cheating', even slightly, to use the time information as we did to set the priors.

However, if we had used the y-values in the dataset to help choose the prior (perhaps for A), this would have been technically incorrect. Nevertheless, it is common practice to look at the data before assigning priors, and the real test of the legitimacy of this practice is whether it makes any difference to your results. In the transit example, if we had used a $U(-50, 50)$ prior instead of a $U(-100, 100)$ one, the posterior distribution would have been virtually the same. The only substantial difference would have been in the value of the marginal likelihood, had we calculated it (see Section 1.9 for more information about when this is important).

1.9 Model Selection and Nested Sampling

Nested sampling (NS) is a Monte Carlo algorithm introduced by Skilling (2006). It is not technically an MCMC algorithm, although MCMC can be used as part of the implementation. It has some advantages over related Monte Carlo methods such as parallel tempering (Hansmann, 1997; Gregory, 2005; Vousden et al., 2015), but also brings its own challenges. Several variants of NS exist, and most of them are somewhat complicated. Here, I present a simple version of the algorithm which is sufficient to solve a wide range of problems. This approach is similar to the one presented in the introductory textbook by Sivia and Skilling (2006).

Many more complex and sophisticated versions of NS exist, such as the popular MultiNest (Feroz, Hobson, and Bridges, 2009), my own Diffusive NS (Brewer, Pártay, and Csányi, 2011), and several others. These algorithms, while all based on the insights of Skilling (2006), are very different in detail (see the appendix).

Consider 2 different models, M_1 and M_2, which are mutually exclusive (they cannot both be true). Suppose M_1 has parameters θ_1, and M_2 has its own parameters θ_2. The methods described previously can be used to calculate the posterior distribution for M_1's parameters:

$$p(\theta_1|D, M_1) = \frac{p(\theta_1|M_1)p(D|\theta_1, M_1)}{p(D|M_1)} \tag{1.24}$$

You can also fit model M_2 to the data, i.e. get the posterior distribution for M_2's parameters:

$$p(\theta_2|D, M_2) = \frac{p(\theta_1|M_2)p(D|\theta_2, M_2)}{p(D|M_2)} \tag{1.25}$$

This is all very well, but you might want to know whether M_1 or M_2 is more plausible overall, given your data. That is, you want the posterior probabilities $P(M_1|D)$ and $P(M_2|D)$.

In this situation, it is usually easier to calculate the ratio of the 2 posterior probabilities, which is sometimes called the *posterior odds ratio*:

$$\frac{P(M_2|D)}{P(M_1|D)} = \frac{P(M_2)}{P(M_1)} \times \frac{P(D|M_2)}{P(D|M_1)} \tag{1.26}$$

As you might expect, the posterior odds for M_2 over M_1 depend on the prior odds: was M_2 more plausible than M_1 before taking into account the data? The other ratio is a ratio of likelihoods, sometimes called a Bayes factor; how probable was the data assuming M_2 versus assuming M_1? These are not the likelihoods for a specific value of the parameters, but instead likelihoods for the model as a whole. To distinguish this kind of likelihood from the standard kind (which is a function of the model parameters), the term *marginal likelihood* or *evidence* is used. The marginal likelihood for M_1 is:

$$p(D|M_1) = \int p(\theta_1|M_1)p(D|\theta_1, M_1)\, d\theta_1 \tag{1.27}$$

which you may recognise as the normalising constant in the denominator of Bayes' theorem in the context of getting the posterior for θ_1. For model selection, we need this as well as the marginal likelihood for M_2:

$$p(D|M_2) = \int p(\theta_2|M_2)p(D|\theta_2, M_2)\, d\theta_2. \tag{1.28}$$

Marginal likelihoods are integrals over the parameter space. They are expected values, not with respect to the posterior distribution, but rather the prior. Therefore, we cannot use standard MCMC methods (at least in any simple way) to calculate the marginal likelihood.[8]

A Monte Carlo approach that samples from the prior distribution also fails in most cases. The likelihood function $p(D|\theta_1, M_1)$ in Equation 1.27 is usually sharply peaked in a very small region of parameter space. The integral will be dominated by the high values of the likelihood in the tiny region, yet a Monte Carlo approach based on sampling the prior almost certainly will not give any samples in the important region.

1.10 An Easy Problem

Imagine a parameter-estimation problem with a single parameter X, a uniform prior distribution between 0 and 1, and a likelihood function which is a decreasing function of X (see Figure 1.10). In this situation the marginal likelihood would be a straightforward integral $Z = \int_0^1 L(X)\, dX$. If we could obtain some points X, and measure their likelihoods

[8] There is a method, called the 'harmonic mean estimator', for estimating the marginal likelihood from posterior samples. Bayesian statistician Radford Neal has described it as the 'worst Monte Carlo method ever'.

L, we could approximate the integral numerically, using the trapezoidal rule or some other numerical quadrature method. The key insight of NS is that any high-dimensional Bayesian problem can be mapped into a 1-dimensional problem like that shown in Figure 1.10. In this introductory discussion, I ignore some mathematical technicalities and focus more on the ideas behind the algorithm.

1.11 Making Hard Problems Easy

Consider a function that takes the entire parameter space and maps it to the real line between 0 and 1, such that the best-fit (highest-likelihood) point in the parameter space is mapped to $X = 0$, and the worst point in the space is mapped to $X = 1$. For all of the intermediate points, the X value is given by:

$$X(\theta) = \int \pi(\theta)\mathbb{1}\left(L(\theta) > L(\theta')\right) d\theta' \tag{1.29}$$

Since X only depends on θ through the likelihood function, it can also be considered a function of a likelihood value:

$$X(\ell) = \int \pi(\theta)\mathbb{1}\left(L(\theta) > \ell\right) d\theta. \tag{1.30}$$

An intuitive interpretation of X is the 'fraction of points in the space which beat this point', that is, X is a rank. Importantly, the prior distribution $\pi(\theta)$ for the parameters implies a uniform distribution $U(0, 1)$ for the ranks X. To see this, imagine measuring the height of all 900,000 residents of Tenerife, sorting them from tallest to shortest, and assigning a rank X to each. The tallest person would get a rank of 0, the shortest a rank of 1, and the person exactly in the middle a rank of 0.5. If you made a histogram of the

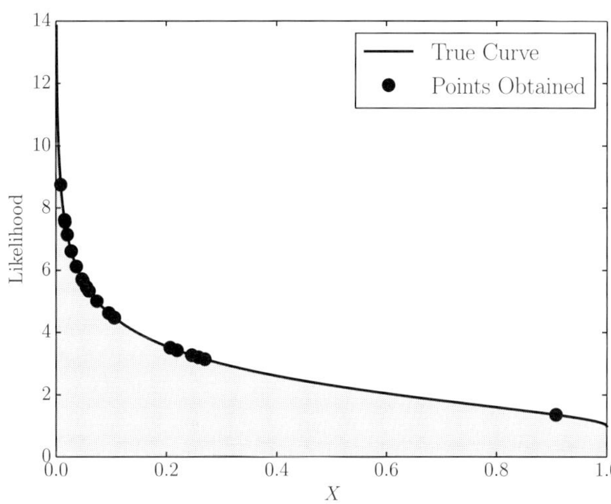

FIGURE 1.10 A simple, 1-dimensional parameter estimation problem where the prior is uniform between 0 and 1, and the decreasing curve shown is the likelihood function (which has the same shape as the posterior). The marginal likelihood is the integral of the likelihood function, and we could calculate this numerically if we had a few points along the curve. NS takes a high-dimensional space and uses it to compute a curve like this whose integral is the marginal likelihood Z.

ranks, it would be uniform since the range of X values (from 0 to 1) is divided evenly among the 900,000 people.

The NS algorithm itself works by starting with a population of N 'particles' (points in parameter space) drawn from the prior. In terms of X, these points will be uniformly distributed between 0 and 1. If we were to find the worst point (with the lowest likelihood, L_{worst}), this would correspond to the highest X value. We could also guess that the highest X value will not be close to $X = 0$, but instead will be close to $X = 1$ (exactly how close depends on N). The simplest prescription given by Skilling (2006) is to estimate $X_{\text{worst}} = \exp(-1/N)$ (more sophisticated possibilities exist but do not make much of a difference). We also know the likelihood of this point (we must have measured it in order to identify the worst point!), and can put it on a graph like the one in Figure 1.10.

To obtain more points, we generate a new point to replace the one just identified as the worst. This point is drawn from the prior, but with the restriction that its likelihood must exceed L_{worst}. In terms of X, the new point's location has a uniform distribution between 0 and $\exp(-1/N)$: just the same as the other $N - 1$ points. If we find the worst point now, we do the same thing but instead of looking at X values between 0 and 1 we look at X values between 0 and $\exp(-1/N)$. Therefore our estimate of the X value of the worst particle in the second iteration is $\exp(-1/N) \times \exp(-1/N) = \exp(-2/N)$. Since we know the likelihood, we can add another point to our graph and continue. The NS algorithm is summarised below.

(i) Generate N particles $\{\theta_1, \theta_2, \ldots, \theta_N\}$ from the prior, and calculate their likelihoods. Initialise a loop counter i at 1.

(ii) Find the particle with the lowest likelihood (call this likelihood L_{worst}). Estimate its X value as $\exp(-i/N)$. Save its properties (X_i, L_i), and its corresponding parameter values as well.

(iii) Generate a new particle to replace the one found in step (ii). The new particle should be drawn from the prior distribution, but its likelihood must be greater than L_{worst}.

(iv) Repeat steps (ii) and (iii) until enough iterations have been performed.

In step (ii), we assume that there is a unique 'worst' particle with the lowest likelihood. If you are working on a problem where likelihood ties are possible, you need to add an extra parameter to your model whose sole purpose is breaking ties in step (ii). See Murray (2007) for more details.

Step (iii) also hides a lot of complexity, and is the key point distinguishing different implementations of NS from each other. We have to be able to generate a point from a restricted version of the prior, proportional to $\pi(\theta)\mathbb{1}\left(L(\theta) > L_{\text{worst}}\right)$. Naive 'rejection sampling', i.e. generating from the prior until you get a point that satisfies the likelihood constraint, will not work because the region satisfying the constraint may be very small in volume. This volume is, in fact, exactly what X measures, and X decreases exponentially during NS. A simple and popular approach, which we will use, is to copy one of the surviving particles (which has a likelihood above L_{worst} by construction) and evolve it using MCMC as though we were trying to sample the prior, but rejecting any proposed move that would take the likelihood below L_{worst}. The main disadvantage of this approach is that the initial diversity of the N particles can become depleted quite quickly, leading to problems if there are multiple likelihood peaks.

In most problems, the region of the parameter space with high likelihood is quite small. Therefore, in practice, most of the marginal likelihood integral will be dominated by a very small range to the left of the plot in Figure 1.10. For this reason, logarithmic axes are more useful (Figure 1.11). In terms of the plots in Figure 1.11, NS steps towards the left, obtaining (X, L) pairs each time the worst point is found. In terms of $\log(X)$ the steps are of equal size $(1/N)$. Therefore, to cover a certain distance in terms of $\log(X)$,

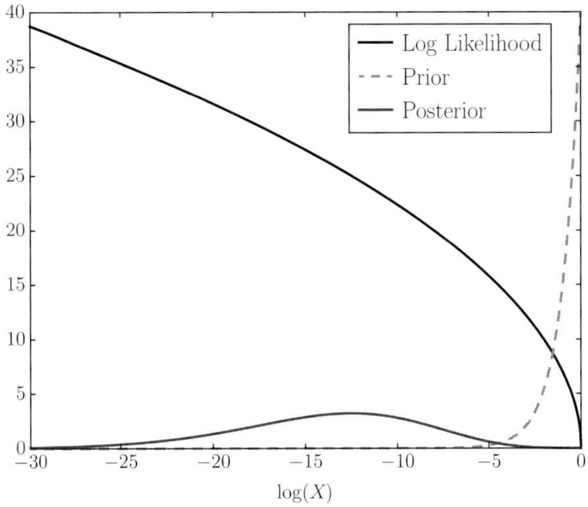

FIGURE 1.11 The same as Figure 1.10, but with logarithmic axes, since the important parts of parameter space tend to occupy a very small volume of parameter space. The black curve is the likelihood function, and the uniform prior for X corresponds to the dotted red exponential prior for $\log(X)$. The posterior (proportional to prior times likelihood) usually has a bell-shaped peak, but can be more complex.

the number of iterations you need is proportional to N. As you might expect, higher N leads to more accurate results but takes more CPU time.

Once the algorithm has been running for a while, the marginal likelihood can be obtained by numerically approximating the integral:

$$Z = \int_0^1 L(X)\, dX. \tag{1.31}$$

An additional output of NS is the 'information', also known as the Kullback-Leibler divergence of the posterior distribution from the prior distribution, which measures how compressed the posterior distribution is. Loosely speaking, the information is (minus the log of) the fraction of the prior volume occupied by the posterior. An information of 0 means the posterior is the same as the prior, and an information of 100 means the posterior occupies about e^{-100} times the prior volume. The definition is:

$$H = \int \frac{\pi(\theta)L(\theta)}{Z} \log\left[\frac{L(\theta)}{Z}\right] d\theta \tag{1.32}$$

$$= \int_0^1 \frac{L(X)}{Z} \log\left[\frac{L(X)}{Z}\right] dX \tag{1.33}$$

and can be computed numerically once Z has been calculated. The information also gives an estimate of the uncertainty in $\log(Z)$, given by $\sqrt{H/N}$ (Skilling, 2006).

As you might expect, NS can also be used to generate posterior samples; it is unlikely anyone would use it if it did not. Each discarded point (the worst point at each iteration) is assigned a prior 'width' based on the distance from its neighbours in terms of X, and a posterior weighting factor is given by the prior width times the likelihood. In the code provided with this chapter, a resampling technique is used to create equally weighted posterior samples which can be used like standard MCMC output.

Python code (minus bookkeeping) implementing basic NS is given below.

```
# Number of particles
N = 5

# Number of NS iterations
steps = 5*30

# MCMC steps per NS iteration
mcmc_steps = 10000

# Generate N particles from the prior
# and calculate their log likelihoods
particles = []
logp = np.empty(N)
logl = np.empty(N)
for i in range(0, N):
 x = from_prior()
 particles.append(x)
 logl[i] = log_likelihood(x)

# Main NS loop
for i in range(0, steps):
  # Find worst particle
  worst = np.nonzero(logl == logl.min())[0]

  # Save its details
  # ...

  # Copy survivor
  if N > 1:
    which = rng.randint(N)
    while which == worst:
      which = rng.randint(N)
    particles[worst] = copy.deepcopy(particles[which])

  # Likelihood threshold for new point
  threshold = copy.deepcopy(logl[worst])

  # Evolve within likelihood constraint using Metropolis
  for j in range(0, mcmc_steps):
    new = proposal(particles[worst])
    logp_new = log_prior(new)
    # Only evaluate likelihood if prior prob isn't zero
    logl_new = -np.Inf
    if logp_new != -np.Inf:
      logl_new = log_likelihood(new)
    loga = logp_new - logp[worst]
```

```
    if loga > 0.:
      loga = 0.

    # Accept
    if logl_new >= threshold and rng.rand() <= np.exp(loga):
      particles[worst] = new
      logp[worst] = logp_new
      logl[worst] = logl_new

  # Use the deterministic approximation
  # Estimate the X-values of the discarded worst points
  logX = -(np.arange(0, i+1) + 1.)/N
```

1.12 Nested Sampling on the Transit Example

To demonstrate NS, we can run it on the original version (Model 1) of the transit model, and the revised version with the t-distributed sampling distribution (Model 2), to calculate the 2 marginal likelihoods. I used $N = 5$ particles and 10,000 MCMC steps to generate a new particle in each iteration. For Model 1, I obtained $\log(Z_1) = -163.0 \pm 1.8$, and the graphical output is shown in Figure 1.12.

For Model 2, I got $\log(Z_2) = -165.0 \pm 1.9$. Thus, the data are approximately $e^2 \approx 10$ times as likely if the original model is true, compared to the modified model. At least, that is, if we ignore the numerical error bars. With 30 particles, I got $\log(Z_1) = -163.0 \pm 0.7$ for

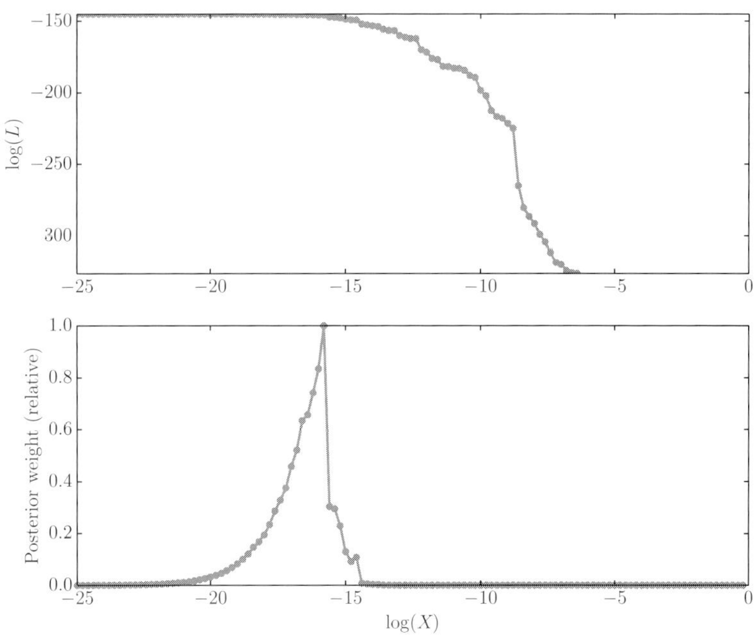

FIGURE 1.12 Numerical versions of the plots in Figure 1.11, produced using NS on the transit problem with the original priors. For an NS run to be satisfactorily completed, the posterior weights in the lower panel must peak and decline, as seen here. There is little point continuing the run further unless you suspect another peak may appear. This can happen in 'phase change' problems (Skilling, 2006).

the original model and $\log(Z) = -165.8 \pm 0.8$ for the modified model, so this conclusion seems robust.

If the Bayes factor (ratio of marginal likelihoods) favours Model 1 by a factor of 10, that does not necessarily imply that it is 10 times more plausible. We must remember the prior odds factor in Equation 1.26. It may be that we completely agree with the assumptions of Model 2, in which case the marginal likelihoods are irrelevant. Or we may consider the union of the 2 models to be sensible, but 50/50 prior probabilities to be inappropriate. Marginal likelihoods are not everything, but they are still important and worth calculating. As always, if a result is surprising, or depends on assumptions in a way you did not expect, treat it as a learning opportunity.

Appendix 1: Notation for Probability Distributions

There are 2 main types of notation used in probability theory. The first type is mathematically more correct and conceptually clear, yet results in equations that are very large, complicated, and difficult to read. The second type is a very compact notation which makes equations a lot shorter and easier to read, but could be misinterpreted if you are not used to it. I prefer the compact notation, and it is more popular in the Bayesian literature. To prevent any misunderstandings I describe both types of notation here.

Suppose we are about to flip a coin 10 times, and are interested in the quantity X, the number of times the result is heads. According to the standard assumptions, the possible values for X are the integers from 0 to 10, and the probability distribution for X is given by a binomial distribution:

$$\begin{aligned} P(X = x) &= \binom{10}{x}\left(\frac{1}{2}\right)^{x}\left(\frac{1}{2}\right)^{10-x} \\ &= \binom{10}{x}\left(\frac{1}{2}\right)^{10}. \end{aligned} \tag{1.34}$$

The equation gives us all the probabilities we want, i.e. $P(X = 0)$, $P(X = 1)$, and so on. The lowercase x is a dummy variable: like the index in a sum, it can be replaced by any other symbol and the equation still holds. For example, replacing x with a gives:

$$P(X = a) = \binom{10}{a}\left(\frac{1}{2}\right)^{10} \tag{1.35}$$

which has exactly the same meaning as Equation 1.34.

The compact notation is an alternative to the left-hand sides of Equations 1.34 and 1.35:

$$p(x) = \binom{10}{x}\left(\frac{1}{2}\right)^{10}. \tag{1.36}$$

Since X, the actual number of heads, is not written anywhere, we forget it exists and just use the lowercase x for that purpose, even though it was originally a dummy variable. The key to understanding this notation is to read $p(x)$ as 'the probability distribution for x', and to understand what the expression following it really means. If we have an expression involving $p(x)$ and $p(y)$, the 2 ps may not be the same function. If the set of possible x values is continuous (so $p(x)$ gives the probability *density* instead of a probability itself) then the concise notation does not change. Whether you are dealing with a probability density function or a probability mass function (the discrete case) must be understood from the context. For completeness, the full (non-compact) notation for the probability density for a variable X is $f_X(x)$.

The compact notation is especially helpful in Bayesian statistics because we frequently deal with conditional probability distributions. For example, if we have 3 variables, a, b, and c, the following is a true statement (it's an example of the *product rule* of probabilities):

$$p(a, b, c) = p(a)p(b|a)p(c|b, a). \tag{1.37}$$

Written in the non-compact notation, we have 3 variables, A, B, and C, and the corresponding equation would be

$$\begin{aligned} P(A = a|B = b, C = c) = {} & P(A = a)P(B = b|A = a) \\ & \times P(C = c|B = b, A = a) \end{aligned} \tag{1.38}$$

which is much harder to read and manipulate.

Appendix 2: A Rough Guide to Popular Bayesian Computation Packages

This is a brief list of popular Bayesian computation software packages that you might find useful. Far from an exhaustive list, it only includes packages that I know something about. I apologise if your favourite package is not listed here.

MultiNest (Feroz, Hobson, and Bridges, 2009): One of the most popular implementations of NS, MultiNest does not use MCMC, like the version presented in this chapter. It is designed to handle potentially multimodal posterior distributions in low ($\lesssim$30) dimensions. It is particularly useful in situations where the likelihood function is expensive to evaluate.

Emcee (Foreman-Mackey et al., 2012): Based on the affine-invariant ensemble sampling method introduced by Goodman and Weare (2010), Emcee is useful for getting quick, efficient results on low-dimensional ($\lesssim$30) unimodal posterior distributions. It is written in Python, and one of its main advantages is that the user only needs to write a function that evaluates the log of the posterior density; the proposal distribution is automatically generated by the algorithm and requires no tuning.

JAGS (Plummer, 2003): The main advantage of JAGS is that it uses the BUGS language, a neat way of specifying your modelling assumptions using a language similar to the '$\sim$' notation. This makes JAGS suitable for quickly implementing analyses of small to medium complexity without having to worry too much about MCMC algorithms themselves. JAGS can also be used easily from R (a programming language for statistics) through the `rjags` package, and as such is popular among statisticians. For a gentle introduction to JAGS, consult my lecture notes at `www.github.com/eggplantbren/STATS331`.

Stan (Stan Development Team, 2014): The modelling language of Stan is similar to that of JAGS, although the underlying MCMC methods used are completely different. Stan is based on the No-U-Turn Sampler (NUTS), a variant of the Hamiltonian MCMC method (Neal, 2011). This is a very efficient sampler for problems with a continuous parameter space and hierarchically defined priors. Like JAGS, Stan can be used from within R, and interfaces also exist for Python, Julia, and MATLAB.

DNest4 (Brewer, Pártay, and Csányi, 2011): Another version of NS quite different from MultiNest DNest uses MCMC, but rather than embedding MCMC in the standard NS framework as we did in this chapter, it uses MCMC to explore a target distribution which is inspired by NS. DNest is most useful for problems with large numbers of parameters and/or a complicated posterior distribution, but where the likelihood function can be evaluated quite quickly. It tends to outperform the version of NS presented here by a substantial margin. The main downside is that you need to write a fair bit of C++ code in order to implement a new model.

1.13 Acknowledgements

I would like to thank the organisers of the Winter School for their invitation, generosity, and hospitality, and the students and other lecturers for many interesting discussions.

The Astrostatistics Facebook group was very helpful with a MultiNest query, as was Thomas Lumley (University of Auckland), who helped me understand the finer mathematical points of the nested sampling 'mapping' trick. I am grateful to all the friends I have met through this subject, from whom I (hope I) have learned so much. Finally I would like to thank my wife, Lianne Zimmermann, for her support and understanding.

BIBLIOGRAPHY

Brewer, B. J., Pártay, L. B., and Csányi, G. 2011. Diffusive nested sampling. *Stat. Comput.*, 21(4), arXiv:0912.2380.

Brendon, B. J. and Foreman-Mackey, D. 2017. DNest4: Diffusive Nested Sampling in C++ and Python. *J. Stat. Softw.* arxiv: 1606.03757.

Feroz, F., Hobson, M. P., and Bridges, M. 2009. MultiNest: an efficient and robust Bayesian inference tool for cosmology and particle physics. *MNRAS*, 398, 1601–1614.

Foreman-Mackey, D., Hogg, D. W., Lang, D., and Goodman, J. 2012. emcee: the MCMC hammer. *Publ. Astron. Soc. Pac.*, arXiv:1202.3665.

Gelman, A., Carlin, J. B., Stern, H. S., et al. 2013. *Bayesian data analysis*, 3rd ed. Chapman & Hall/CRC, Boca Raton, Florida.

Goodman, J., and Weare, J. 2010. Ensemble samplers with affine invariance. *Comm. App. Math. Comp. Sci.*, 5, 65–80.

Gregory, P. 2005. *Bayesian logical data analysis for the physical sciences: a comparative approach with mathematica support.* Cambridge University Press, Cambridge.

Hansmann, U. H. E. 1997. Parallel tempering algorithm for conformational studies of biological molecules. *Chem. Phys. Lett.*, 281(1), 140–150.

Hogg, D. W., Bovy, J., and Lang, D. 2010. Data analysis recipes: fitting a model to data. arXiv:1008.4686.

Homan, M. D., and Gelman, A. 2014. The no-U-turn sampler: adaptively setting path lengths in Hamiltonian Monte Carlo. *J. Mach. Learn. Res.*, 15(1), 1593–1623.

Mackay, D. J. C. 2003. *Information theory, inference, and learning algorithms.* Cambridge University Press, Cambridge.

Murray, I. 2007. 'Advances in Markov chain Monte Carlo methods'. PhD thesis, University College London.

Neal, R. M. 2011. MCMC using Hamiltonian dynamics. **In** *Handbook of Markov chain Monte Carlo*, edited by Brooks, S., Gelman, A., Jones, G., and Meng, X.-L., 113–162. Chapman & Hall/CRC, Boca Raton, Florida.

Plummer, M. 2003. JAGS: a program for analysis of Bayesian graphical models using Gibbs sampling. *Proceedings of the 3rd International Workshop on Distributed Statistical Computing* (DSC 2003).

O'Hagan, A. and Forster, J. 2004. *Bayesian inference.* Arnold, London.

Sivia, D. S., and Skilling, J. 2006. *Data analysis: a Bayesian tutorial*, 2nd ed. Oxford University Press, Oxford.

Skilling, J. 2006. Nested sampling for general Bayesian computation. *Bayesian Anal.* 4, 833–860.

Stan Development Team. 2014. Stan: a C++ library for probability and sampling, Version 2.5.0, `http://mc-stan.org`.

Vousden, W., Farr, W. M., and Mandel, I. 2015. Dynamic temperature selection for parallel-tempering in Markov chain Monte Carlo simulations. *MNRAS*, arXiv:1501.05823.

2 Inverse Problems in Astronomy

J.-L. STARCK

2.1 Introduction

Many data-processing problems in astrophysics can be formalized as a linear inverse problem,

$$Y = HX \bullet N, \tag{2.1}$$

where Y is a set of noisy measurements, N is the unknown noise/error term, and the $\bullet$ operator translates the way the noise contaminates the data (e.g. + for additive noise). This noise/error can be either a stochastic-measurement noise induced by the sensor (e.g. additive Gaussian noise, Poisson noise, or mixed Gaussian and Poisson), or a deterministic perturbation due, for example, to an imperfect forward model. X is the solution to our problem, and H is a linear operator. Finding X is an inverse problem.

Typical cases of such inverse problems are:

- *Deconvolution*: If the optical system is linear and shift invariant, the relation between the data Y and the unknown X in the same coordinate frame is a convolution by a blurring kernel. H is the point spread function (PSF) of the optical system. We want to determine X knowing Y and H. This deconvolution problem has led to a large amount of work, the main difficulties being the existence of: (i) a cut-off frequency of the PSF and (ii) the additive noise.
- *Blind deconvolution*: This is similar to the deconvolution problem, except the PSF, i.e. matrix H, is unknown. Hence, we want to determine X and H knowing Y.
- *Myopic deconvolution*: This is an intermediate problem between deconvolution and blind deconvolution, where the PSF H is partially known.
- *Superresolution*: H is the convolution by a blurring kernel followed by a sub-sampling. The goal is to recover spatial frequency information outside the spatial bandwidth of the data-formation system.
- *Inpainting*: Missing data is a recurrent problem in astronomical data processing. H is a pixelwise multiplication by a binary mask – 1 for the pixels to be kept and 0 otherwise.
- *Radio-interferometry image reconstruction:* This is a specific case of the deconvolution problem. Indeed, the observed data, called visibilities, correspond to a subset of the Fourier component of the unknown X. The PSF, called *dirty beam*, is not compact in the direct space, and contains many 0s in the Fourier domain.
- *Blind source separation:* In the blind source separation setting, the instantaneous linear mixture model assumes that we are given m observations $Y = [y_1^T, \cdots, y_m^T]^T$ where each $(y_i)_{i=1,\cdots,m}$ is a row vector of size t; each measurement is the linear mixture of n source processes. As the measurements are m different mixtures, source-separation techniques aim at recovering the original sources $X = [x_1^T, \cdots, x_n^T]^T$ by taking advantage of some information contained in the way the signals are mixed in the observed data. The linear-mixture model is rewritten in matrix form, $Y = HX + N$, where Y is the $m \times t$ measurement matrix (i.e. observed data), X is the $n \times t$ source matrix, and H is the $m \times n$ mixing matrix. H defines the contribution of each source to each measurement. An $m \times t$ matrix N is added to account for

instrumental noise or model imperfections. Similar to the blind deconvolution, the H matrix is not known.
- *Image reconstruction:* An image is formed from projections.

When the inverse problem does not have a unique and stable solution, it is an *ill-posed problem*. A regularization is necessary to reduce the space of candidate solutions, and we need to incorporate some prior knowledge on the typical structure of the original signal X.

This prior information accounts typically for the 'smoothness' class of the solution and can range from the uniform smoothness assumption to more complex knowledge of the geometrical structures of the solution. A popular regularization in astronomy is the well-known Bayesian maximum entropy method (MEM), which is based on the principle that we want to select the simplest solution which fits the data. Sparsity is an alternative to describe the simplest solution, and has emerged as a powerful approach for regularization (Starck et al., 2010).

Sparse recovery involves four main ingredients which are the keys to its success. These ingredients are:

- **Data representation though the use of a dictionary:** Fixed dictionaries with fast analysis and synthesis operators, such as wavelets or curvelets (described in Sections 2.3 and 2.4), allow us to build very efficient algorithms, both in terms of computation time and quality of the recovered solutions. Such fast transform will be described in Section 2.3. We will see in Section 2.4.5 that these fixed dictionaries can be gathered together in order to build a larger dictionary, still using fast operators. To go beyond these fixed dictionaries, dictionary learning appears as a promising paradigm to learn dictionaries adapted to the analysed data. This will be discussed in Section 2.5.
- **Noise modelling:** Having the right dictionary is only part of the story and is not enough. In most inverse problems, one needs to disentangle the useful signal/information from the noise. This requires taking properly into account the noise behaviour in the observed data. Different kinds of noise models, such as Gaussian, Poisson, mixture of Gaussian and Poisson, correlated noise, etc., can derive the distribution of the coefficients in the chosen dictionary and therefore detect, with a given confidence level, significant coefficients that should be used to reconstruct the signal of interest. Section 2.3.3 shows how different kinds of noises can be considered in a wavelet analysis.
- **The variational problem and minimisation algorithm:** Piecing together the first ingredient and the second (i.e. regularization and data fidelity), solving the inverse problem at hand is solving a variational (optimization) problem. To solve the obtained optimization problem, a general class of optimization schemes can be used, namely proximal splitting algorithms that are fast and easy to implement, and enjoy provable convergence. An efficient proximal minimisation method is presented in Section 2.6.3.
- **The choice of parameter(s):** Variational formulation of inverse problems involves parameters that must be chosen. The scientist should not have to choose these parameters based on a tedious trial and error by running the algorithm many times with a range of parameters. Furthermore, one definitively does not want to make scientific conclusions based on an ad hoc parameter. We show in Section 2.6.4 how these parameters can be derived from noise modelling.

Specific inverse problems to be addressed in this chapter include denoising in Section 2.6.1 inpainting in Section 2.7, deconvolution in Section 2.6.3, and blind source separation in Section 2.8. Finally, an important aspect of sparsity is its relation to the compressed sensing sampling theory. This will be discussed in Section 2.2.4.

2.2 A Brief Tour of Sparsity

2.2.1 Definition

A signal $X = (X[1], \ldots, X[N])$ considered as a vector in $\mathbb{R}^N$, is said to be sparse if most of its entries are equal to 0. If k of the N samples are equal to 0, where $k \ll N$, then the signal is said to be k-sparse. In the case where only a few of the entries have large values and the rest are 0 or close to 0, the signal is said to be weakly sparse (or compressible). With a slight abuse of terminology, in the sequel, we will call compressible signals sparse. Generally signals are not sparse in direct space, but can be sparsified by transforming them to another domain. Think, for instance, of a purely sinusoidal signal which is 1-sparse in the Fourier domain, while it is clearly dense in the original one. In the *sparsity synthesis model*, a signal can be represented as the linear expansion

$$X = \Phi\alpha = \sum_{i=1}^{T} \phi_i \alpha[i], \tag{2.2}$$

where $\alpha[i]$ are the synthesis coefficients of X, $\Phi = (\phi_1, \ldots, \phi_T)$ is the dictionary, and ϕ_i are the atoms (elementary waveforms) of the dictionary Φ. In the language of linear algebra, the dictionary Φ is an $N \times T$ matrix whose columns are the atoms normalised, supposed here to be normalised to a unit ℓ_2-norm, i.e. $\forall i \in [1, T], \|\phi_i\|_2^2 = \sum_{n=1}^{N} |\phi_i[n]|^2 = 1$,[1] A function can be decomposed in many dictionaries, but the best dictionary is the one with the sparsest (most economical) representation of the signal. In practice, it is convenient to use dictionaries with fast implicit transform (Fourier transform, wavelet transform, etc.) which allow us to directly obtain the coefficients and reconstruct the signal from these coefficients using fast algorithms running in linear or almost linear time (unlike matrix-vector multiplications). The Fourier, wavelet, and discrete cosine transforms provide the most well-known dictionaries.

2.2.2 Sparsity as a Very Flexible Model to Represent Data

Most natural signals are not sparse but rather concentrated near a small set. Such signals are termed *compressible* or *weakly sparse* in the sense that the sorted magnitudes $|\alpha_{(i)}|$, i.e. $|\alpha_{(1)}| > |\alpha_{(2)}|, \ldots, > |\alpha_{(t)}|$, of the sequence of coefficients α decay quickly according to a power law, i.e. $|\alpha_{(i)}| \leq C i^{-1/r}$, $i = 1, \ldots, t$, where C is a constant. The larger r is, the faster the amplitudes of coefficients decay, and the more compressible the signal is. In turn, the non-linear ℓ_2 approximation error of α (and x) from its M largest entries in magnitude also decrease quickly. One can think, for instance, of the wavelet coefficients of a smooth signal away from isotropic singularities, or the curvelet coefficients of a piecewise regular image away from smooth contours. A comprehensive account on sparsity of signals and images can be found in Starck et al. (2010).

As an example, Figure 2.1 (top panels) shows a Saturn image and the same image where only the top 5% highest coefficients are kept (all other pixels are set to 0). The approximation error is very large, 95% of the energy (i.e. sum of the square of all pixel values) is lost. Figure 2.1 (bottom left) shows the top 5% highest coefficients of Saturn in the wavelet domain, while setting to 0 all other wavelet coefficients, and applying an inverse wavelet transform, we obtain the final image (bottom right). In this case, the approximation error is much smaller, and only 2% of the energy is lost. This means

[1] The l_p-norm of a vector X, $p \geq 1$, is defined as $\|X\|_p = \left(\sum_i |X[i]|^p\right)^{1/p}$, with the usual adaptation $\|X\|_\infty = \max_i X[i]$.

FIGURE 2.1 Example of sparse representation: Top left, Saturn image. Top right, same image but keeping only the top 5% highest coefficients (all other pixels are set to 0). Bottom left, top 5% highest coefficients in the wavelet domain. Bottom right, reconstructed image keeping only these coefficients in the wavelet domain.

that 5% of the coefficients in an appropriate space contains 98% of the energy of the Saturn image.

2.2.3 Sparse Regularization for Inverse Problems

In the following, for a vector z we denote $\|z\|_p^p = \sum_i |z_i|^p$ for $p \geq 0$. In particular, for $p \geq 1$, this is the p-th power of the ℓ_p norm, and for $p = 0$, we get the ℓ_0 pseudo-norm, which counts the number of non-0 entries in z. The ℓ_0 regularized problem amounts to minimising

$$\tilde{\alpha} \in \text{argmin}_\alpha \|Y - H\Phi\alpha\|_2^2 + \lambda\|\alpha\|_0, \tag{2.3}$$

where λ is a regularization parameter. A solution $\tilde{X}$ is reconstructed as $\tilde{X} = \Phi\tilde{\alpha}$. Clearly, the goal of Equation 2.3 is to minimise the number of non-0 coefficients describing the sought-after signal while ensuring that the forward model is faithful to the observations.

Solving Equation 2.3 is, however, known to be NP-hard. The ℓ_1 norm has been proposed as a tight convex relaxation of Equation 2.3, leading to the minimisation problem

$$\tilde{\alpha} \in \operatorname{argmin}_{\alpha} \|y - H\Phi\alpha\|_2^2 + \lambda\|\alpha\|_1, \tag{2.4}$$

where λ is again a regularization parameter different from that of Equation 2.3. Researchers spanning a wide range of disciplines have studied the structural properties of minimisers of Equation 2.4 and its equivalence with Equation 2.3. Equation 2.4 is computationally appealing and can be efficiently solved, and it has been proved that, under appropriate circumstances, Equation 2.4 produces exactly the same solutions as Equation 2.3 (see e.g. Donoho, 2006b).

2.2.4 Sparsity and Compressed Sensing

Compressed sensing (CS) (Donoho, 2006a; Candès and Tao, 2006) is a paradigm that allows for sampling a signal at a rate proportional to its information content rather than its bandwidth (think of sparsity as a measure of the information content) as in the classic Shannon sampling approach, where a signal needs to be sampled at least at twice its bandwidth (the so-called Nyquist rate). This tells us that a signal can be recovered from a small number of samples. CS requires the data Y of m pixels to be acquired through a random acquisition system H of size $m \times n$,

$$Y = HX, \tag{2.5}$$

and CS theory shows that, under certain conditions, one can exactly recover a k-sparse signal X of size n (a signal for which only k coefficients have values different from 0, out of n total coefficients, where $k < n$) from $m < n$ measurements.

Such a recovery is possible from undersampled data only if the sensing matrix H verifies the *restricted isometry property* (RIP) (see Candès and Tao, 2006, for more details). This property has the effect that each measurement Y_i contains some information about all of the pixels of X; in other words, the sensing operator H acts to spread the information contained in X across many measurements Y_i.

Under these 2 constraints – sparsity and a transformation meeting the RIP criterion – a signal can be recovered exactly even if the number of measurements m is much smaller than the number of unknown n.

The solution X to Equation 2.5 is obtained by minimising

$$\min_X \|X\|_1 \quad s.t. \;\; Y = HX, \tag{2.6}$$

where the ℓ_1 norm is defined by $\|X\|_1 = \sum_i |X_i|$. Many optimisation methods have been proposed to minimise this equation.

As mentioned earlier, signals in real life are generally not 'strictly' sparse, but are *compressible*; i.e. we can represent the signal in a basis or frame (Fourier, wavelets, curvelets, etc.) in which the curve obtained by plotting the obtained coefficients, sorted by their decreasing absolute values, exhibits a polynomial decay. Most natural signals and images are compressible in an appropriate basis.

We can therefore reformulate the CS equation above (Equation 2.6) to include the data transformation matrix Φ:

$$\min_\alpha \|\alpha\|_1 \quad s.t. \;\; Y = H\Phi\alpha, \tag{2.7}$$

where $X = \Phi^T \alpha$, and α are the coefficients of the transformed solution X in the dictionary Φ. Each column represents a vector (also called an *atom*), which ideally should be chosen to match the features contained in X.

CS requires the data to be acquired through a random acquisition system, which is not the case in general. Indeed, when considering CS in a given application, few matrices meet the RIP criterion. However, it has been shown that accurate recovery can be obtained as long as the mutual coherence between H and Φ, $\mu_{H,\Phi} = \max_{i,k} \left|\langle H_i, \Phi_k, \rangle\right|$, is low (Candès and Plan, 2010). The mutual coherence measures the degree of similarity between the sparsifying basis and the sensing operator.

Hence, in its relaxed definition, we consider a linear inverse problem $Y = H\Phi\alpha$ as being an instance of CS when the following fundamental points are verified:

- **Underdetermined problem:** We have fewer measurements (i.e. visibilities) than unknowns (i.e. pixel values of the reconstructed image).
- **Sparsity** of the signal: The signal to reconstruct can be represented with a small number of non-0 coefficients. For point-source observations, the solution is even strictly sparse (in the Dirac domain) since it is composed only of a list of spikes. For extended objects, sparsity can be obtained in another representation space, such as wavelets.
- **Incoherence** between the acquisition domain (i.e. Fourier space) and the domain of sparsity (e.g. wavelet space): Point sources, for instance, are localized in the pixel domain, but spread over a large domain of the visibility plane. Conversely, each visibility contains information about all sources in the field of view.

Most CS applications described in the literature are based on such a soft CS definition. In astronomy, CS has been studied in many applications, such as data transfer from satellites to Earth (Bobin et al., 2008; Barbey et al., 2011), 3-D weak lensing (Leonard et al., 2012), next-generation spectroscopic instrument design (Ramos et al., 2011), and aperture synthesis (Wiaux et al., 2009; Carrillo et al., 2012; Garsden et al., 2015).

2.2.5 Sparsity and the Bayesian Interpretation

In the Bayesian framework, a prior is imposed on the object of interest through a probability distribution. For instance, assume that coefficients α are i.i.d. Laplacian with the scale parameter τ, i.e. the density $P_\alpha(\alpha) \propto e^{-\tau\|\alpha\|_1}$, and the noise ε is 0-mean white Gaussian with variance σ^2, i.e. the conditional density $P_{Y|\alpha}(y) = (2\pi\sigma^2)^{-m/2} e^{-\|y-H\Phi\alpha\|_2^2/(2\sigma^2)}$. By traditional Bayesian arguments, the MAP estimator is obtained by maximizing the conditional posterior density $P_{\alpha|Y}(\alpha) \propto P_{Y|\alpha}(y)P_\alpha(\alpha)$, or equivalently by minimising its anti-log version

$$\min_{\alpha} \frac{1}{2\sigma^2} \|y - H\Phi\alpha\|_2^2 + \tau \|\alpha\|_1 . \tag{2.8}$$

This is exactly Equation 2.4 by identifying $\lambda = 2\sigma^2\tau$. This resemblance could be misleading, however (Starck et al., 2013a).

Should ℓ_1 Regularization Be the MAP?

In Bayesian cosmology, the following shortcut is often made: If a prior is at the basis of an algorithm, then to use this algorithm, the resulting coefficients must be distributed according to this prior. But it is a false logical chain in general, and high-dimensional phenomena completely invalidate it. A classic claim is that ℓ_1 regularization is *equivalent* to assuming that the solution is Laplacian and not Gaussian, which would be unsuitable for the case of cosmic microwave background (CMB) analysis. This argument, however, assumes that a MAP estimate follows the distribution of the prior. But it is now well established that MAP solutions substantially deviate from the prior model, and that the disparity between the prior and the effective distribution obtained from the true MAP estimate is a permanent contradiction in Bayesian MAP estimation (Nikolova, 2007). Even the supposedly correct ℓ_2 prior would yield an estimate (Wiener, which coincides with the MAP and posterior conditional mean) whose covariance is not that of the prior.

In addition, rigorously speaking, this MAP interpretation of ℓ_1 regularization is not the only possible interpretation. More precisely, it was shown in Gribonval et al. (2012) and Baraniuk et al. (2010) that solving a penalized least-squares regression problem with penalty $\psi(\alpha)$ (e.g. the ℓ_1 norm) should not necessarily be interpreted as assuming a Gibbsian prior $C\exp(-\psi(\alpha))$ and using the MAP estimator. In particular, for any prior P_α, the conditional mean can also be interpreted as a MAP with some prior $C\exp(-\psi(\alpha))$. Conversely, for certain penalties $\psi(\alpha)$ the solution of the penalized least-squares problem is indeed the conditional posterior mean, with a certain prior $P_\alpha(\alpha)$ which is generally different from $C\exp(-\psi(\alpha))$.

In summary, the MAP interpretation of such penalized least-squares regressions can be misleading, and using a MAP estimation, the solution does not necessarily follow the prior distribution, and an *incorrect* prior does not necessarily lead to a wrong solution.

Compressed Sensing: The Bayesian Interpretation Inadequacy

A beautiful example to illustrate this is the CS scenario (Donoho, 2006a; Candès and Tao, 2006), which tells us that a k-sparse, or compressible n-dimensional signal x can be recovered either exactly, or to a good approximation from much less random measurements m than the ambient dimension n, if m is sufficiently larger than the intrinsic dimension of x. Clearly, the underdetermined linear problem, $y = Hx$, where H is drawn from an appropriate random ensemble, with fewer equations than unknowns, *can* be solved exactly or approximately if the underlying object x is sparse or compressible. As mentioned earlier, this can be achieved by solving a computationally tractable ℓ_1-regularized convex optimization program.

If the underlying signal is exactly sparse, in a Bayesian framework, this would be a completely absurd way to solve the problem, since the Laplacian prior is very different from the actual properties of the original signal (i.e. k coefficients different from 0). In particular, what CS shows is that we can have prior A be completely true, but utterly impossible to use for computation time or any other reason, and we can use prior B instead and get the correct results. Therefore, from a Bayesian point of view, it is rather difficult to understand not only that the ℓ_1 norm is adequate, but also that it leads to the exact solution.

To conclude, Bayesian methodology offers an often-elegant framework that is extremely useful for many applications, but we should be careful not to be monolithic in the way we address a problem, and the prior itself cannot explain how a regularization penalty impacts the solution of an inverse problem. To understand it, we need to take into account the operator involved in the inverse problem, and this requires much deeper mathematical developments than a simple Bayesian interpretation. CS theory shows that, for some operators, beautiful geometrical phenomena allow us to recover perfectly the solution of an underdetermined inverse problem. Similar results were derived for a random sampling on the sphere (Rauhut and Ward, 2012).

2.3 The Wavelet Transform

The wavelet transform (WT) has been extensively used in astronomical data analysis during the last 10 years, and this holds for all astrophysical domains, from the study of the Sun through CMB analysis (Starck and Murtagh, 2006). X-ray and gamma-ray source catalogues are generally based on wavelets (Pacaud et al., 2006; Nolan et al., 2012). Using multiscale approaches such as the wavelet transform, an image can be decomposed into components at different scales, and the wavelet transform is therefore well adapted to the study of astronomical data (Starck and Murtagh, 2006). Several wavelet-transform algorithms are presented in the following. Since noise in physical sciences is not always Gaussian, modelling in wavelet space of many kinds of noise, such as Poisson noise, has been a key motivation for the use of wavelets in astrophysics (Schmitt et al., 2010).

2.3.1 Bi-Orthogonal Wavelet Transforms

The most commonly used wavelet transform algorithm is the decimated bi-orthogonal wavelet transform (OWT). Using the OWT, a signal s can be decomposed as follows:

$$s(l) = \sum_k c_{J,k}\phi_{J,l}(k) + \sum_k \sum_{j=1}^{J} \psi_{j,l}(k)\alpha_{j,k} \tag{2.9}$$

with $\phi_{j,l}(x) = 2^{-j}\phi(2^{-j}x - l)$ and $\psi_{j,l}(x) = 2^{-j}\psi(2^{-j}x - l)$, where ϕ and ψ are, respectively, the scaling and wavelet functions. J is the number of resolutions used in the decomposition, α_j is the wavelet coefficients (or details) at scale j, and c_J is a coarse or smooth version of the original signal s. The present indexing is such that $j = 1$ corresponds to the finest scale (high frequencies).

The 2-dimensional algorithm is based on separate variables leading to prioritizing of horizontal, vertical, and diagonal directions. The scaling function is defined by $\phi(x, y) = \phi(x)\phi(y)$, and the detail signal is obtained from 3 wavelets:

- vertical wavelet : $\psi^1(x, y) = \phi(x)\psi(y)$
- horizontal wavelet: $\psi^2(x, y) = \psi(x)\phi(y)$
- diagonal wavelet: $\psi^3(x, y) = \psi(x)\psi(y)$

which leads to three wavelet sub-images at each resolution level. A given wavelet band is therefore defined by its resolution level j ($j = 1 \ldots J$) and its direction number d ($d = 1 \ldots 3$, corresponding, respectively, to the horizontal, vertical, and diagonal bands).

2.3.2 The Starlet Transform

The isotropic undecimated wavelet transform (IUWT), or *starlet wavelet transform*, is known in the astronomical domain because it is well adapted to astronomical data where objects are more or less isotropic in most cases (Starck and Murtagh, 1994; Starck and Murtagh, 2006). For most astronomical images, the starlet dictionary is very well adapted.

The starlet wavelet transform (Starck et al., 2007) decomposes an $n \times n$ image c_0 into a coefficient set $\alpha = \{\alpha_1, \ldots, \alpha_J, c_J\}$, as a superposition of the form

$$c_0[k, l] = c_J[k, l] + \sum_{j=1}^{J} \alpha_j[k, l],$$

where c_J is a coarse or smooth version of the original image c_0 and α_j represents the details of c_0 at scale 2^{-j} (see Starck et al., 1998; Starck and Murtagh, 2002). Thus, the algorithm outputs $J + 1$ sub-band arrays of size $N \times N$. (The present indexing is such that $j = 1$ corresponds to the finest scale or high frequencies.)

FIGURE 2.2 Galaxy NGC 2997.

The decomposition is achieved using the filter bank ($h_{2D}, g_{2D} = \delta - h_{2D}, \tilde{h}_{2D} = \delta$, $\tilde{g}_{2D} = \delta$), where h_{2D} is the tensor product of 2 1-D filters h_{1D} and δ is the Dirac function. The passage from 1 resolution to the next is obtained using the *à trous* ('with holes') algorithm (Starck et al., 1998):

$$
\begin{aligned}
c_{j+1}[k,l] &= \sum_m \sum_n h_{1D}[m]\ h_{1D}[n]\ c_j[k+2^j m, l+2^j n], \\
\alpha_{j+1}[k,l] &= c_j[k,l] - c_{j+1}[k,l].
\end{aligned} \tag{2.10}
$$

Hence, we have a *multiscale pixel representation*, i.e. each pixel of the input image is associated with a set of pixels of the multiscale transform.

Figure 2.3 shows the starlet transform of the galaxy NGC 2997 displayed in Figure 2.2, 5 wavelet scales, and the final smoothed plane (lower right). The original image is given exactly by the sum of these 6 images.

The Starlet Reconstruction

The reconstruction is straightforward. A simple co-addition of all wavelet scales reproduces the original map: $c_0[k,l] = c_J[k,l] + \sum_{j=1}^{J} \alpha_j[k,l]$. But because the transform is not sub-sampled, there are many ways to reconstruct the original image from its wavelet transform (Starck et al., 2007). For a given wavelet filter bank (h, g), associated with a scaling function ϕ and a wavelet function ψ, any synthesis filter bank ($\tilde{h}$, $\tilde{g}$) which satisfies the following reconstruction condition

$$\hat{h}^*(\nu)\hat{\tilde{h}}(\nu) + \hat{g}^*(\nu)\hat{\tilde{g}}(\nu) = 1 \tag{2.11}$$

leads to exact reconstruction.

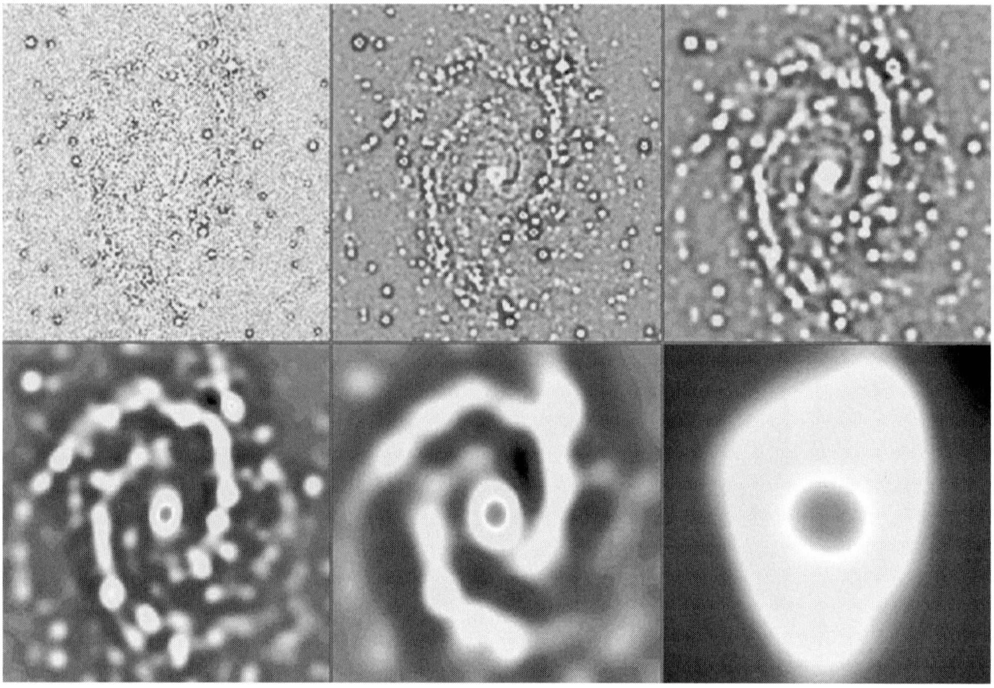

FIGURE 2.3 Wavelet transform of NGC 2997 by the IUWT. The co-addition of these 6 images reproduces exactly the original image.

Starlet Transform: Second Generation

A particular case is obtained when $\hat{\tilde{\phi}}_{1-\mathrm{D}} = \hat{\phi}_{1-\mathrm{D}}$ and $\hat{\psi}_{1-\mathrm{D}}(2\nu) = \frac{\hat{\phi}^2_{1-\mathrm{D}}(\nu)-\hat{\phi}^2_{1-\mathrm{D}}(2\nu)}{\hat{\phi}_{1-\mathrm{D}}(\nu)}$, which leads to a filter $g_{1-\mathrm{D}}$ equal to $\delta - h_{1-\mathrm{D}} \star h_{1-\mathrm{D}}$. In this case, the synthesis function $\tilde{\psi}_{1-\mathrm{D}}$ is defined by $\frac{1}{2}\tilde{\psi}_{1-\mathrm{D}}(\frac{t}{2}) = \phi_{1-\mathrm{D}}(t)$ and the filter $\tilde{g}_{1-\mathrm{D}} = \delta$ is the solution to Equation 2.11. We end up with a synthesis scheme where only the smooth part is convolved during the reconstruction.

Deriving h from a spline scaling function, for instance B_1 ($h_1 = [1, 2, 1]/4$) or B_3 ($h_3 = [1, 4, 6, 4, 1]/16$) (note that $h_3 = h_1 \star h_1$), we get the following filter bank:

$$
\begin{aligned}
h_{1-\mathrm{D}} &= h_3 = \tilde{h} = [1, 4, 6, 4, 1]/16 \\
g_{1-\mathrm{D}} &= \delta - h \star h = [-1, -8, -28, -56, 186, -56, -28, -8, -1]/256 \\
\tilde{g}_{1-\mathrm{D}} &= \delta \ .
\end{aligned}
\tag{2.12}
$$

As in the standard starlet transform, extension to 2-D is trivial. We just replace the convolution with $h_{1\mathrm{D}}$ by a convolution with the filter $h_{2\mathrm{D}}$, which is performed efficiently by using the separability.

With this filter bank, there is no convolution with the filter $\tilde{g}_{1-\mathrm{D}}$ during the reconstruction. Only the low-pass synthesis filter $\tilde{h}_{1-\mathrm{D}}$ is used. The reconstruction formula is

$$c_j[l] = (h^{(j)}_{1-\mathrm{D}} \star c_{j+1})[l] + \alpha_{j+1}[l], \tag{2.13}$$

and denoting $L^j = h^{(0)} \star \cdots \star h^{(j-1)}$ and $L^0 = \delta$, we have

$$c_0[l] = \left(L^J \star c_J\right)[l] + \sum_{j=1}^{J} \left(L^{j-1} \star \alpha_j\right)[l]. \tag{2.14}$$

Each wavelet scale is convolved with a low-pass filter.

As for the transformation, the 2-D extension consists just in replacing the convolution by $h_{1\mathrm{D}}$ with a convolution by $h_{2\mathrm{D}}$. In 2-D, similarly, the second-generation starlet transform leads to the representation of an image $X[k,l]$:

$$X[k,l] = \sum_{m,n} \phi^{(1)}_{j,k,l}(m,n) c_J[m,n] + \sum_{j=1}^{J} \sum_{m,n} \phi^{(2)}_{j,k,l}(m,n) \alpha_j[m,n], \tag{2.15}$$

where $\phi^{(1)}_{j,k,l}(m,n) = 2^{-2j}\tilde{\phi}_{1-\mathrm{D}}(2^{-j}(k-m))\tilde{\phi}_{1-\mathrm{D}}(2^{-j}(l-n))$, and $\phi^{(2)}_{j,k,l}(m,n) = 2^{-2j}\tilde{\psi}_{1-\mathrm{D}}(2^{-j}(k-m))\tilde{\psi}_{1-\mathrm{D}}(2^{-j}(l-n))$. $\phi^{(1)}$ and $\phi^{(2)}$ are positive, and α_j are 0- mean, 2-D wavelet coefficients. (More details can be found in Starck et al., 2007, 2010.)

2.3.3 Signal Detection in Wavelet Space

Observed data Y in the physical sciences are generally corrupted by noise, which is often additive and which follows in many cases a Gaussian distribution, a Poisson distribution, or a combination of both. It is important to detect the wavelet coefficients which are 'significant' or 'active', i.e. the wavelet coefficients which have an absolute value too large to be due to noise. We defined the multiresolution support M of an image Y by:

$$M_j[k,l] = \begin{cases} 1 & \text{if } \alpha_j[k,l] \text{ is significant} \\ 0 & \text{if } \alpha_j[k,l] \text{ is not significant} \end{cases} \tag{2.16}$$

where $\alpha_j[k,l]$ is the wavelet coefficient of Y at scale j and at position (k,l).

We now need to determine when a wavelet coefficient is significant. For Gaussian noise, it is easy to derive an estimation of the noise standard deviation σ_j at scale j from the noise standard deviation, which can be evaluated with good accuracy in an automated way (Starck and Murtagh, 1998). To detect the significant wavelet coefficients, it suffices to compare the wavelet coefficients $\alpha_j[k,l]$ to a threshold level t_j. t_j is generally taken to be equal to $K\sigma_j$, and K is chosen between 3 and 5. The value of 3 corresponds to a probability of false detection of 0.27%. If $\alpha_j[k,l]$ is small, then it is not significant and could be due to noise. If $\alpha_j[k,l]$ is large, it is significant:

$$\begin{array}{ll} \text{if } |\alpha_j[k,l]| \geq t_j & \text{then } \alpha_j[k,l] \text{ is significant} \\ \text{if } |\alpha_j[k,l]| < t_j & \text{then } \alpha_j[k,l] \text{ is not significant.} \end{array} \tag{2.17}$$

When the noise is not Gaussian, other strategies may be used:

- **Poisson noise:** If the noise in the data Y is Poisson, the transformation $\mathcal{A}(Y) = 2\sqrt{Y + \frac{3}{8}}$ (Anscombe, 1948) acts as if the data arose from a Gaussian white-noise model, with $\sigma = 1$, under the assumption that the mean value of Y is sufficiently large. However, this transform has some limits and it has been shown that it cannot be applied to data with less than 20 photons per pixel. So for X-ray or gamma-ray data, other solutions have to be chosen, which manage the case of a reduced number of events or photons under assumptions of Poisson statistics.

- **Gaussian + Poisson noise:** The generalization of variance stabilization Murtagh et al. (1995) is:

$$\mathcal{G}(Y[k,l]) = \frac{2}{\alpha}\sqrt{\alpha Y[k,l] + \frac{3}{8}\alpha^2 + \sigma^2 - \alpha g},$$

where α is the gain of the detector, and g and σ are the mean and the standard deviation of the read-out noise.

- **Poisson noise with few events using the MS-VST:** For images with very few photons, one solution consists in using the multiscale variance-stabilizing transform (MS-VST) (Zhang et al., 2008). The MS-VST combines the Anscombe transform and the IUWT to produce *stabilized* wavelet coefficients, i.e. coefficients corrupted by Gaussian noise with a standard deviation equal to 1. In this framework, wavelet coefficients are now calculated by:

$$\begin{array}{c}\text{IUWT}\\ + \\ \text{MS-VST}\end{array}\left\{\begin{array}{lcl} c_j & = & \sum_m \sum_n h_{1D}[m]h_{1D}[n] \\ & & c_{j-1}[k + 2^{j-1}m, l + 2^{j-1}n] \\ \alpha_j & = & \mathcal{A}_{j-1}(c_{j-1}) - \mathcal{A}_j(c_j) \end{array}\right. \tag{2.18}$$

where $\mathcal{A}_j$ is the MS-VST operator at scale j defined by:

$$\mathcal{A}_j(c_j) = b^{(j)}\sqrt{|c_j + e^{(j)}|}, \tag{2.19}$$

where the variance-stabilizing constants $b^{(j)}$ and $e^{(j)}$ only depend on the filter h_{1D} and the scale level j. They can all be pre-computed once for any given h_{1D} (Zhang et al., 2008). The multiresolution support is computed from the MS-VST coefficients, considering a Gaussian noise with a standard deviation equal to 1. This stabilization procedure is also invertible as we have:

$$c_0 = \mathcal{A}_0^{-1}\left[\mathcal{A}_J(a_J) + \sum_{j=1}^{J} \alpha_j\right]. \tag{2.20}$$

2.4 Multiscale Geometric Transforms

Wavelets rely on a dictionary of roughly isotropic elements occurring at all scales and locations. Despite the fact that they have had wide impact in image processing, they fail to efficiently represent objects with highly anisotropic elements such as lines or curvilinear structures (e.g. edges), independent of scale, because wavelets are nongeometrical and do not exploit the regularity of the edge curve. Following this reasoning, new constructions have been proposed, such as ridgelets (Candès and Donoho, 1999) and curvelets (Candès and Donoho, 2001, 2002; Starck et al., 2002). We review some of them in this section.

2.4.1 The Ridgelet Transform

The 2-dimensional continuous ridgelet transform in $\mathbb{R}^2$ can be defined as follows (Candès and Donoho, 1999). We pick a smooth univariate function $\psi : \mathbb{R} \to \mathbb{R}$ with sufficient decay and satisfying the admissibility condition

$$\int |\hat{\psi}(\xi)|^2/|\xi|^2 \, d\xi < \infty, \tag{2.21}$$

which holds if, say, ψ has a vanishing mean $\int \psi(t)dt = 0$. We suppose that ψ is normalised so that $\int |\hat{\psi}(\xi)|^2 \xi^{-2} d\xi = 1$.

For each $a > 0$, each $b \in \mathbb{R}$ and each $\theta \in [0, 2\pi]$, we define the bivariate *ridgelet* $\psi_{a,b,\theta} : \mathbb{R}^2 \to \mathbb{R}$ by

$$\psi_{a,b,\theta}(\mathbf{x}) = a^{-1/2} \cdot \psi((x_1 \cos\theta + x_2 \sin\theta - b)/a). \tag{2.22}$$

Given an integrable bivariate function $f(\mathbf{x})$, we define its ridgelet coefficients by:

$$\mathcal{R}_f(a, b, \theta) = \int \overline{\psi}_{a,b,\theta}(\mathbf{x}) f(\mathbf{x}) d\mathbf{x}.$$

We have the exact reconstruction formula

$$f(\mathbf{x}) = \int_0^{2\pi} \int_{-\infty}^{\infty} \int_0^{\infty} \mathcal{R}_f(a, b, \theta) \psi_{a,b,\theta}(\mathbf{x}) \frac{da}{a^3} db \frac{d\theta}{4\pi} \tag{2.23}$$

valid for functions which are both integrable and square integrable.

It has been shown (Candès and Donoho, 1999) that the ridgelet transform is precisely the application of a 1-dimensional wavelet transform to the slices of the Radon transform.

Local Ridgelet Transform

The ridgelet transform is optimal to find only lines of the size of the image. To detect line segments, a partitioning must be introduced. The image is decomposed into smoothly overlapping blocks of side-length B pixels in such a way that the overlap between 2 vertically adjacent blocks is a rectangular array of size $B \times B/2$; we use an overlap to avoid blocking artefacts. For an $n \times n$ image, we count $2n/B$ such blocks in each direction. The partitioning introduces redundancy, as a pixel belongs to 4 neighbouring blocks.

More details on the implementation of the digital ridgelet transform can be found in Starck et al. (2002, 2003). The ridgelet transform is therefore optimal to detect lines of a given size, which is the block size.

2.4.2 The First-Generation Curvelet Transform

The curvelet transform (Donoho and Duncan, 2000; Starck et al., 2003) opens the possibility of analysing an image with different block sizes, but with a single transform. The idea is to first decompose the image into a set of wavelet bands, and to analyse each band with a local ridgelet transform. The block size can be changed at each scale level. Roughly speaking, different levels of the multiscale ridgelet pyramids are used to represent different sub-bands of a filter bank output.

The side-length of the localising windows is doubled *at every other* dyadic sub-band, hence maintaining the fundamental property of the curvelet transform, that elements of length about $2^{-j/2}$ serve for the analysis and synthesis of the j-th sub-band $[2^j, 2^{j+1}]$. Note also that the coarse description of the image c_J is not processed. In Starck et al. (2003), a default block-size value $B_{\min} = 16$ pixels was used. This implementation of the curvelet transform is also redundant. The redundancy factor is equal to $16J + 1$ whenever J scales are employed. A given curvelet band is therefore defined by the resolution level j $(j = 1 \ldots J)$ related to the wavelet transform, and by the ridgelet scale r.

2.4.3 The Second-Generation Curvelet Transform

Despite these interesting properties, the first-generation curvelet (CurveletG1) construction presents some drawbacks. First, the construction involves a complicated 7-index structure among which we have parameters for scale, location, and orientation. Second, the spatial partitioning of the CurveletG1 transform uses overlapping windows to avoid blocking effects. This leads to an increase of the redundancy. The computational cost of the CurveletG1 algorithm may also be a limitation for large-scale data.

In contrast, the second-generation curvelet, CurveletG2 (Candès and Donoho, 2004; Candès et al., 2006), exhibits a much simpler and natural indexing structure with three parameters: scale, orientation (angle), and location – hence simplifying mathematical analysis. The CurveletG2 transform also implements a tight frame expansion (Candès and Donoho, 2004) and has a much lower redundancy. Unlike CurveletG1, the discrete CurveletG2 implementation will not use ridgelets, yielding a faster algorithm (Candès and Donoho, 2004; Candès et al., 2006).

CurveletG2 is defined at scale 2^{-j}, orientation θ_ℓ, and position $\mathbf{t}_{\mathbf{k}}^{j,\ell} = R_{\theta_\ell}^{-1}(2^{-j}k, 2^{-j/2}l)$ by translation and rotation of a mother curvelet φ_j as

$$\varphi_{j,\ell,\mathbf{k}}(\mathbf{t}) = \varphi_{j,\ell,\mathbf{k}}(t_1, t_2) = \varphi_j(R_{\theta_\ell}(\mathbf{t} - \mathbf{t}_{\mathbf{k}}^{j,\ell})), \tag{2.24}$$

where R_{θ_ℓ} is the rotation by θ_ℓ radians. θ_ℓ is the equi-spaced sequence of rotation angles $\theta_\ell = 2\pi 2^{-\lfloor j/2 \rfloor}\ell$, with integer ℓ such that $0 \leq \theta_\ell \leq 2\pi$ (note that the number of orientations varies as $1/\sqrt{\text{scale}}$). $\mathbf{k} = (k, l) \in \mathbb{Z}^2$ is the sequence of translation parameters. The waveform φ_j is defined by means of its Fourier transform $\hat{\varphi}_j(\boldsymbol{\nu})$, written in polar coordinates (r, θ) in the Fourier domain:

$$\hat{\varphi}_j(r, \theta) = 2^{-3j/4}\hat{\varpi}(2^{-j}r)\hat{v}\left(\frac{2^{\lfloor j/2 \rfloor}\theta}{2\pi}\right). \tag{2.25}$$

The support of $\hat{\varphi}_j$ is a polar *parabolic wedge* defined by the support of $\hat{\varpi}$ and $\hat{v}$, respectively the radial and angular windows (both smooth, nonnegative, and real valued), applied with scale-dependent window widths in each direction. $\hat{\varpi}$ and $\hat{v}$ must also satisfy the partition of unity property (Candès et al., 2006). See the frequency tiling in Figure 2.4(a).

In continuous frequency $\boldsymbol{\nu}$, the CurveletG2 coefficients of the 2-D function $f(\mathbf{t})$ are defined as the inner product

$$\alpha_{j,\ell,\mathbf{k}} := \langle f, \varphi_{j,\ell,\mathbf{k}} \rangle = \int_{\mathbb{R}^2} \hat{f}(\boldsymbol{\nu})\hat{\varphi}_j(R_{\theta_\ell}\boldsymbol{\nu})e^{i\mathbf{t}_{\mathbf{k}}^{j,\ell}\cdot\boldsymbol{\nu}}d\boldsymbol{\nu}. \tag{2.26}$$

This construction implies a few properties:

(i) CurveletG2 defines a tight frame of $L_2(\mathbb{R}^2)$

(ii) the effective length and width of these curvelets obey the parabolic scaling relation $(2^{-j} = \text{width}) = (\text{length} = 2^{-j/2})^2$

(iii) the curvelets exhibit an oscillating behaviour in the direction perpendicular to their orientation

Curvelets as just constructed are complex valued. It is easy to obtain real-valued curvelets by working on the symmetrized version $\hat{\varphi}_j(r, \theta) + \hat{\varphi}_j(r, \theta + \pi)$. More details can be found in Starck et al. (2010).

Figure 2.5 shows a few curvelets at different scales, orientations, and locations.

2.4.4 Contourlets and Shearlets

Other transforms have been proposed to build a decomposition with a frequency tiling similar to the curvelet representation. The contourlet transform (Do and Vetterli, 2003) consists in applying first a Laplacian pyramid (Burt and Adelson, 1983), followed by a direction filter bank (Bamberger and Smith, 1992) at each scale. The number of directions

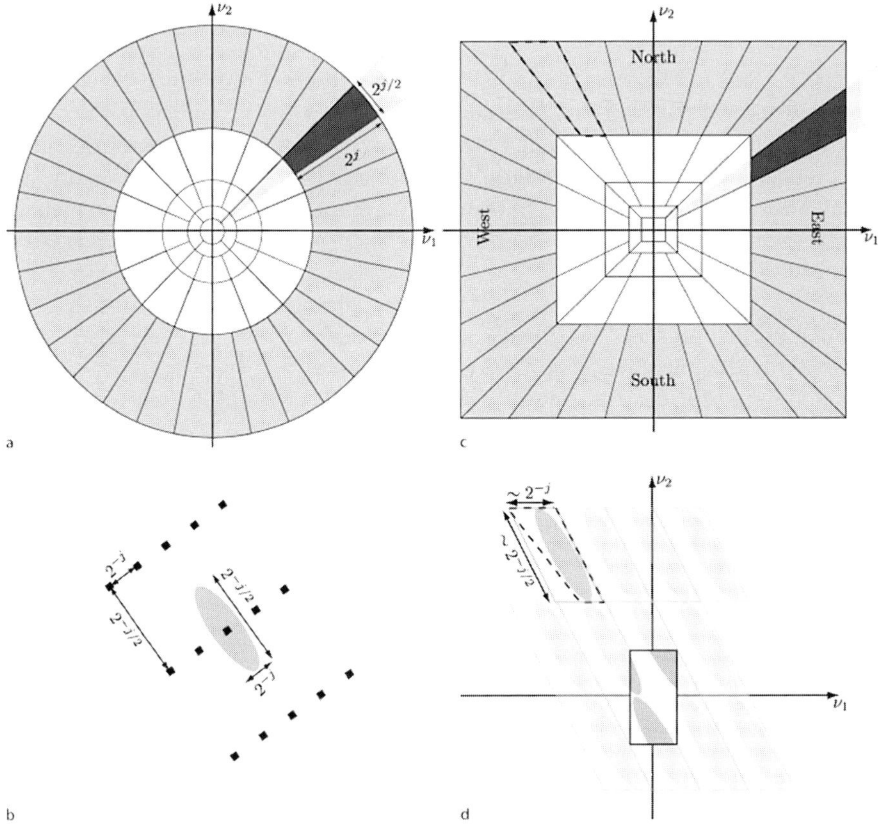

FIGURE 2.4 (a) Continuous curvelet frequency tiling. The grey area represents a wedge obtained as the product of the radial window (annulus shown in lighter colour) and the angular window (darker colour). (b) The Cartesian grid in space associated with the construction in (a) whose spacing also obeys the parabolic scaling by duality. (c) Discrete curvelet frequency tiling. The window $\hat{u}_{j,\ell}$ isolates the frequency near the trapezoidal wedge, such as the ones shown in grey. (d) The wrapping transformation. The dashed line shows the same trapezoidal wedge as in (b). The parallelogram contains this wedge and hence the support of the curvelet. After periodization, the wrapped Fourier samples can be collected in the rectangle centred at the origin.

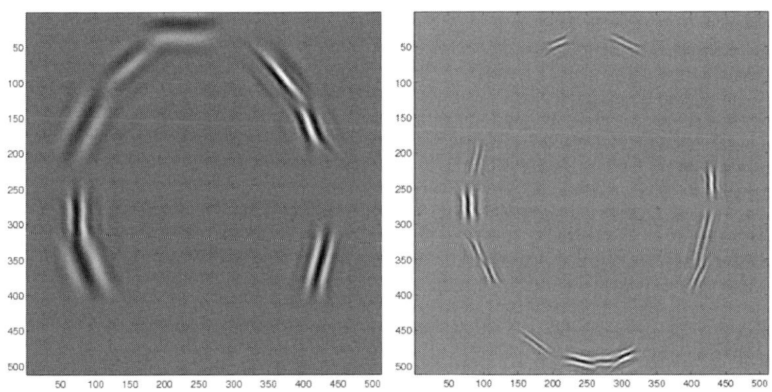

FIGURE 2.5 A few first-generation curvelets. Backprojections of a few curvelet coefficients at different positions and scales.

per scale has to be a power of 2, but we can have different numbers of direction through the scales. It is computationally efficient, all operations are done using filter banks, and the redundancy is small (33%), due to the Laplacian pyramid. However, this low redundancy has a cost: the loss of the translation-invariance property in the analysis, which limits its interest for restoration applications. For this reason, an undecimated contourlet transform was proposed (Da Cunha et al., 2006), where both the Laplacian filter bank and the directional filter bank are nonsubsampled. The redundancy is therefore much higher, since each contourlet band has the same size as the original image, but it was shown that denoising results were much better (Da Cunha et al., 2006). The contourlet transform can be seen as a discrete filter-bank version of the curvelet decomposition. A MATLAB toolbox that implements the contourlet transform and the nonsubsampled contourlet transform can be downloaded from MATLAB Central.[2]

Based on the concept of composite wavelets (Guo et al., 2004), the shearlet theory has been proposed (Labate et al., 2005; Guo et al., 2006; Guo and Labate, 2007; Kutyniok and Lim, 2011; Kutyniok et al., 2012a), and can be seen as a theoretical justification for contourlets (Easley et al., 2008). Several distinct numerical implementations of the discrete shearlet transform exist, based either on a Fourier implementation or a spatial domain implementation. The Fourier-based implementation developed in Easley et al. (2008) consists in the following:

- Apply the Laplacian pyramid scheme.
- Compute the pseudo-polar Fourier transform (PPFT) on each band of the Laplacian decomposition.
- Apply a directional band-pass filtering in the Fourier space.
- Apply an inverse 2-dimensional fast Fourier transform (FFT) or use the inverse PPFT from the previous step on each directional band.

Kutyniok et al. (2012b) propose another approach based on a weighted pseudo-polar transform, followed by a windowing and an inverse FFT for each direction. This transform is associated with band-limited tight shearlet frames, thereby allowing the adjoint frame operator for reconstruction (Davenport et al., 2012).

Several spatial-domain shearlet transforms were also developed (Easley et al., 2008; Kutyniok and Sauer, 2009; Lim, 2010; Han et al., 2011; Kutyniok et al., 2014). Easley et al. (2008) utilize directional filters, which are obtained as approximations of the inverse Fourier transforms of digitized band-limited window functions. In Lim (2010),[3] the use of separable shearlet generators allows a fast transform. The most efficient digitalization of the shearlet transform was derived in Lim (2013) and Kutyniok et al. (2014) by utilizing non-separable, compactly supported shearlet generators, which best approximate the classical band-limited generators.

2.4.5 Morphological Diversity

The morphological diversity concept was introduced in Starck et al. (2004) in order to model a signal as a finite linear mixture, each component of the mixture being sparse in a given dictionary. The idea is that a single transformation may not always represent an image well, especially if the image contains structures with different spatial morphologies. For instance, if an image is composed of edges and a locally oscillating texture, we can consider edges to be sparse in the curvelet domain while the oscillating texture is better

[2] `www.mathworks.com/matlabcentral`.

[3] Shearlet codes are available at `www.math.uh.edu/~dlabate` (Easley et al., 2008) and `www.ShearLab.org` (Kutyniok et al., 2012b; Lim, 2010; Kutyniok et al., 2014).

sparsified in the local Fourier domain. The Morphological Component Analysis method (MCA) (Starck et al., 2004) considers a new model where the observed signal Y is a mixture of K several components and can be described as

$$Y = \sum_{k=1}^{K} X_k, \tag{2.27}$$

where each component X_k can be described as $X_k = \Phi_k \alpha_k$ with a dictionary Φ_k and a sparse representation α_k. It is further assumed that, for any given component X_k, the sparsest decomposition over the proper dictionary Φ_k yields a highly sparse description, while its decomposition over the other dictionaries, $\mathbf{\Phi}_{k' \neq k}$, is highly non-sparse. Thus, the different $\mathbf{\Phi}_k$ can be seen as discriminating between the different components of the initial signal. The X_k components are the solutions of:

$$\min_{\{X_1,\ldots,\, X_n\}} \sum_{k=1}^{n} \|\Phi_k^t X_k\|_0 \quad \text{subject to} \quad Y = \sum_{k=1}^{n} X_k. \tag{2.28}$$

A detailed description of MCA is given in Starck et al. (2004).

Figure 2.6 (upper left) shows the Hubble Space Telescope A370 image. It contains many anisotropic features, such the gravitational arc and the arclets. The image has been decomposed using 3 transforms: the ridgelet transform, the curvelet transform, and the wavelet transform. Three images have then been reconstructed from the coefficients of the 3 basis. The upper right shows the coaddition of the ridgelet and curvelet reconstructed images. The wavelet reconstructed image is displayed in the lower left, and the coaddition of the 3 images can be seen in the lower right. The gravitational arc and the arclets are all represented in the ridgelet and the curvelet basis, while all isotropic features are better represented in the wavelet basis.[4]

MCA has been used for many astrophysical applications, such as stripping removal in infrared images (Starck et al., 2003), star-formation studies in Hershel data (André et al., 2010), the removal of point sources in the CMB (Bobin et al., 2014), and the detection of supernovae (Moller et al., 2015).

2.4.6 What Is the Best Dictionary?

The best data decomposition is the one which leads to the sparsest representation, i.e. few coefficients have a large magnitude, while most of them are close to 0 (Starck et al., 2010). Hence, for some astronomical data sets containing edges (planetary images, cosmic strings, etc.), curvelets should be preferred to wavelets. But for a signal composed of a sine, the Fourier dictionary is optimal from a sparsity standpoint since all information is contained in a single coefficient. Hence, the representation space that we use in our analysis can be seen as a prior we have on our observations. The MCA approach is more flexible, and can represent more complex images. An alternative approach is the dictionary learning that we present in this section.

2.5 Dictionary Learning

Fixed dictionaries, though they have very fast implicit analysis and synthesis operators, which make them attractive from a practical point of view, cannot guarantee sparse

[4] Further results are visible at `www.cosmostat.org/research/statistical-methods/mca`.

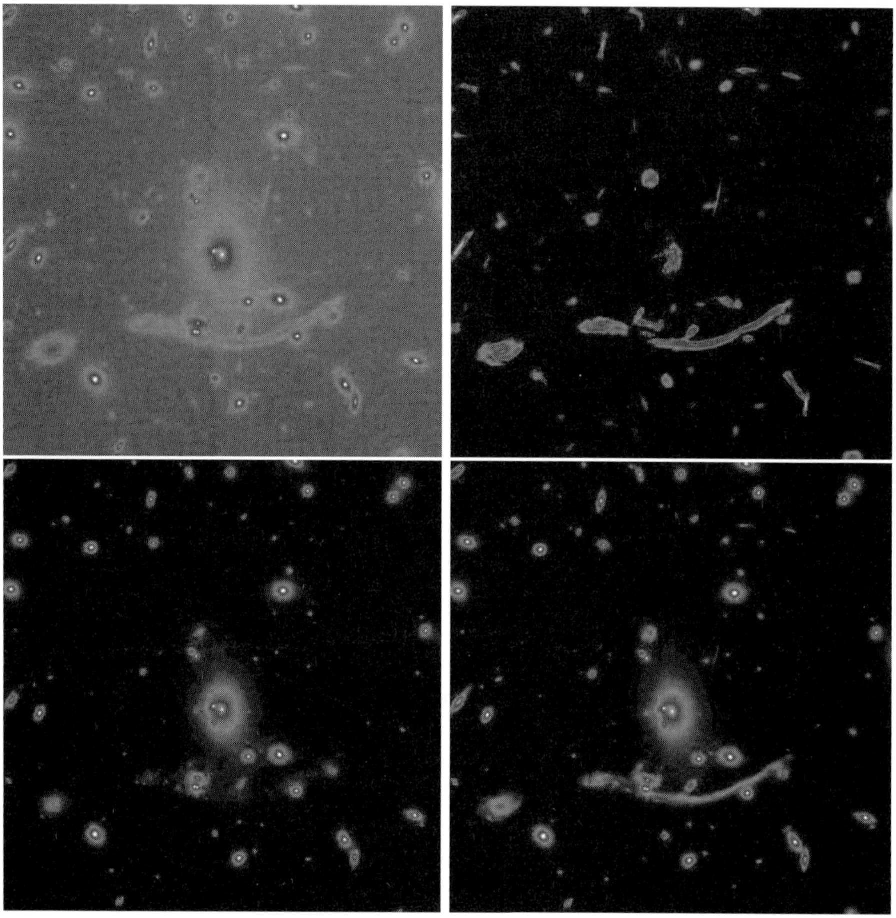

FIGURE 2.6 Top left, HST image of A370. Top right, coadded image from the reconstructions from the ridgelet and the curvelet coefficients. Bottom left, reconstruction from the wavelet coefficients. Bottom right, addition of the 3 reconstructed images.

representations of new classes of signals of interest that present more complex patterns and features. What can one do if the data cannot be sufficiently sparsely represented by any of these fixed (or combined) existing dictionaries? Or if we do not know the morphology of features contained in our data? Is there a way to make our data analysis more adaptive by optimizing for a dedicated dictionary? To answer these questions, a new field has emerged, called dictionary learning (DL) (Elad and Aharon, 2006; Mairal et al., 2010; Peyré et al., 2010). DL offers the possibility of learning an adaptive dictionary Φ directly from the data (or from a set of exemplars that we believe to represent the data well). DL is at the interface of machine learning and signal processing.

Assuming we have N signals x_i of P pixels that can be considered as a representative training set of the class of signal we are analysing and we consider a dictionary of T atoms of size P, then DL can be cast as the following optimization problem:

$$\min_{\Phi,\alpha} \frac{1}{2} \|X - \Phi\alpha\|^2 + \sum_{i=1}^{T} \lambda_i \|\alpha_i\|_p^p \quad s.t. \quad \Phi \in \mathcal{D}, \tag{2.29}$$

where $p \in [0, 1]$, X is an $N \times P$ matrix containing the training set (i.e. $X = [x_1, \ldots, x_N]$), Φ is the $N \times T$ dictionary matrix, and α is the $T \times P$ matrix whose i-th column is the synthesis coefficients vector α_i of the exemplar x_i in Φ, i.e. $x_i \approx \Phi\alpha_i$. The goal of DL is to jointly infer Φ and α from the sole knowledge of X. $\lambda_i > 0$ is the regularization parameter for vector α_i. Usually, $\lambda_i \equiv \lambda > 0$, for all i. The quadratic data fidelity accounts for the approximation error, and $\mathcal{D}$ is a non-empty, closed constraint set. An important constraint that $\mathcal{D}$ must deal with is to remove the scale ambiguity. Indeed, if $(\Phi^\star, \alpha^\star)$ is a minimiser of Equation 2.29, then for any s value $(s\Phi^\star, \alpha^\star/s)$ is also a minimiser with regularization parameter $|s|^p\lambda_i$. A standard constraint to remove this ambiguity is to enforce the atoms (columns of Φ) to be scaled to a norm equal to (respectively less than) 1. Additional constraints might be considered, for example, 0-mean atoms, orthogonality, or even physical constraints to accommodate for the specific sought-after applications (Beckouche et al., 2013).

To optimize the challenging Equation 2.29, we can adopt the following alternating minimisation strategy:

$$\text{Sparse coding: } \min_\alpha \tfrac{1}{2} \left\| X - \Phi^{(t)}\alpha \right\|^2 + \lambda \left\| \alpha_i \right\|_p^p \tag{2.30}$$

$$\text{Dictionary update: } \quad \min_{\Phi \in \mathcal{D}} \left\| X - \Phi\alpha^{(t+1)} \right\| \tag{2.31}$$

In Equation 2.30, the dictionary is fixed, and the problem is marginally minimised with respect to the coefficients. This corresponds to a standard sparse coding problem. Indeed, for $p = 1$, Equation 2.30 can be solved using the forward-backward method (Combettes and Wajs, 2005) or its accelerated version, fast iterative shrinkage-thresholding algorithm (FISTA) (Beck and Teboulle, 2009). For the dictionary update, the patch-based approach leads to different learning algorithms, such as method of optimal directions (Engan et al., 1999), projected gradient descent methods (Lin, 2007), or K-SVD (an algorithm for designing overcomplete dictionaries for sparse representation) (Aharon et al., 2006).

In practice, it is generally computationally problematic to deal with atoms of the same size as the data, and patch-based learning is generally preferred. This leads to patch-sized atoms, which entail a locality of the resulting algorithms. This locality can be turned back into a global treatment of larger images by appropriate tiling of the patches. The training set can also be directly derived by extracting patches at random positions in the data.

Figure 2.7 (left and right) show, respectively, a simulated cosmic string map ($1'' \times 1''$), and the different atoms that have been learned from the simulated image. It is interesting to note that many atoms do not look like wavelet or curvelet atoms.

FIGURE 2.7 Left, a simulated cosmic string map ($1'' \times 1''$), and right the learned dictionary.

If sparsity techniques are now widely applied in astrophysics, DL has almost never been used. Few refereed journal articles have been published in the astrophysical literature (Beckouche et al., 2013; Díaz-Hernández et al., 2014). In Beckouche et al. (2013), it was shown that DL clearly outperforms standard astronomical image denoising methods, such as those based on wavelets. However, the computation time is obviously a limitation for its use in important projects requiring the processing of a big data set.

2.6 Astronomical Image Restoration

2.6.1 Denoising

The denoising problem corresponds to the case where H is the identity (i.e. $Y = X + N$). The classical modus operandi is to first apply the transform operator to the noisy data Y, then apply a nonlinear estimation rule to the coefficients (each coefficient individually or as a group of coefficients), and finally compute the inverse transform to get an estimate $\tilde{X}$. In brief, a classic approach is to derive the threshold t from the noise modelling, as explained in Section 2.3.3. Many thresholding or shrinkage rules have been proposed. Among them, hard and soft thresholding are the most well known.

2.6.2 Hard and Soft Thresholding

Once coefficients $\alpha = \Phi^t Y$ have been calculated, *hard thresholding* (Starck and Bijaoui, 1994; Donoho et al., 1995) consists of setting to 0 all coefficients whose magnitude is less than a threshold t:

$$\tilde{\alpha}_k = \mathrm{HardThresh}_t(\alpha_k) = \begin{cases} \alpha_k & \text{if } |\alpha_k| \geq t, \\ 0 & \text{otherwise.} \end{cases} \tag{2.32}$$

Hard thresholding is a keep-or-kill procedure.

Soft thresholding (Weaver et al., 1991; Donoho, 1995) is defined as the kill-or-shrink rule:

$$\tilde{\alpha}_k = \begin{cases} \mathrm{sign}(\alpha_k)(|\alpha_k| - t) & \text{if } |\alpha_k| \geq t, \\ 0 & \text{otherwise.} \end{cases} \tag{2.33}$$

The coefficients above the threshold are shrunk towards the origin. This can be written in the compact form

$$\tilde{\alpha}_k = \mathrm{SoftThresh}_t(\alpha_k) = \mathrm{sign}(\alpha_k)(|\alpha_k| - t)_+, \tag{2.34}$$

where $(\cdot)_+ = \max(\cdot, 0)$.

Thresholding as a Minimisation Problem

Suppose that Φ is orthonormal. It can be proved that equations Equations 2.33 and 2.34 are the unique, closed-form solution to the following minimisation problems for a hard or soft thresholding with a threshold t:

$$\tilde{\alpha} = \underset{\alpha}{\mathrm{argmin}} \ \|Y - \Phi\alpha\|^2 + t^2 \|\alpha\|_0 \ \textbf{hard thresholding} \tag{2.35}$$

$$\tilde{\alpha} = \underset{\alpha}{\mathrm{argmin}} \ \tfrac{1}{2} \|Y - \Phi\alpha\|^2 + t \|\alpha\|_1 \ \textbf{soft thresholding} \tag{2.36}$$

Therefore, hard and soft thresholding have been used as building blocks to derive fast and efficient iterative thresholding techniques to minimise more complex inverse problems with sparsity constraints, such as inpainting and deconvolution.

For the OWT, hard thresholding is known to result in a larger variance estimate, while soft thresholding with the same threshold level creates undesirable bias because even large coefficients lying out of noise are shrunk. In practice, hard thresholding is generally preferred to soft thresholding.

2.6.3 Deconvolution

In a deconvolution problem, when the sensor is linear, H is the block Toeplitz matrix. An iterative thresholding deconvolution method, first proposed in Starck et al. (1995), consists in the following iterative scheme:

$$X^{(n+1)} = X^{(n)} + H^T \left(\mathbf{WDen}_{\Omega^{(n)}} \left(Y - HX^{(n)} \right) \right) \tag{2.37}$$

where **WDen** is an operator which performs a wavelet thresholding, i.e. applies the wavelet transform of the residual $R^{(n)}$ (i.e. $R^{(n)} = Y - HX^{(n)}$), thresholding some wavelet coefficients, and applies the inverse wavelet transform. Only coefficients that belong to the *multiresolution support* $\Omega^{(n)}$ (Starck et al., 1995) are kept, while the others are set to 0. At each iteration, the multiresolution support $\Omega^{(n)}$ is updated by selecting new coefficients in the wavelet transform of the residual which have an absolute value larger than a given threshold. The threshold is automatically derived assuming a given noise distribution such as Gaussian or Poisson noise.

More recently, it was shown (Figueiredo and Nowak, 2003; Daubechies et al., 2004; Combettes and Wajs, 2005) that a solution to Equation 2.4 can be obtained through the following iteration:

$$\alpha^{(n+1)} = \mathrm{SoftThresh}_\lambda \left(\alpha^{(n)} + \Phi^T H^T \left(Y - H\Phi\alpha^{(n)} \right) \right), \tag{2.38}$$

with $\|H\| = 1$. In the framework of monotone operator splitting theory, it is called a forward-backward (FB) splitting algorithm and is a classical gradient projection method for constrained convex optimization. It was shown that for frame dictionaries, it converges to the solution (Combettes and Wajs, 2005). In the last 10 years, many new optimization methods have been proposed to perform a sparse regularization in a more efficient way. For instance, the primal-dual splitting algorithm of Vũ (2013) can handle both sparse and positivity constraints, and leads to the following scheme:

(i) **Initialization:** Choose $\tau > 0$ and $\sigma > 0$ such that $1 - \tau\sigma \|\Phi\|^2 > \|H\|^2 / 2$; A sequence of relaxation parameters $\mu_n \in]0, 1]$; (x_0, u_0).

(ii) **Main iteration**
- $p_{n+1} = \left(x_n - \tau\Phi u_n + \tau H^T (y - Hx_n) \right)_+$;
- $q_{n+1} = (\mathbf{I} - \mathrm{SoftThresh}_\lambda)(u_n + \sigma\Phi(2p_{n+1} - x_n))$;
- $(x_{n+1}, u_{n+1}) = \mu_n(p_{n+1}, q_{n+1}) + (1 - \mu_n)(x_n, u_n)$.

The parameter μ_n is a relaxation that can be used to accelerate the algorithm. In the unrelaxed case, i.e. $\mu_n = 1$ for all n, we obtain $(x_{n+1}, u_{n+1}) = (p_{n+1}, q_{n+1})$.

Example

A simulated Hubble Space Telescope image of a distant cluster of galaxies is shown in Figure 2.8 (middle). The simulated data are shown Figure at left. Wavelet deconvolution solution is shown Figure at right. The method is stable for any kind of point-spread function, and any kind of noise modelling can be considered.

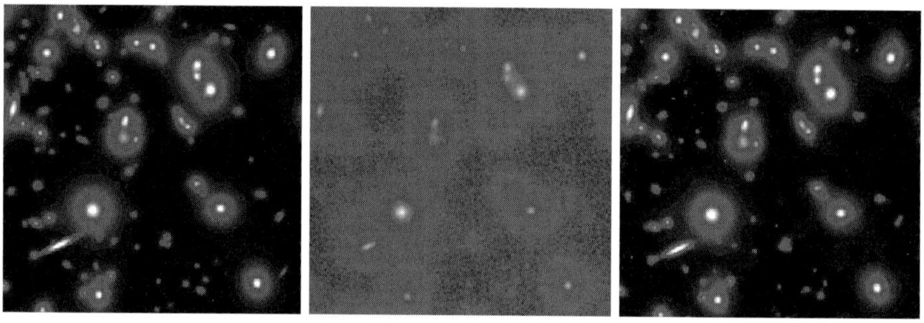

FIGURE 2.8 Simulated Hubble Space Telescope image of a distant cluster of galaxies. Left, original, unaberrated, and noise-free. Middle, input, aberrated, noise added. Right, wavelet-restoration wavelet.

2.6.4 Regularization Parameters

Once the dictionary and minimisation method are chosen, a final problem remains: choosing parameters to control the algorithm. Most minimisation methods, using l_0 and l_1, have a single thresholding step, where coefficients in the dictionary have to be soft or hard thresholded using a threshold value, which is a value λ common to projection in Φ. This parameter controls the trade-off between the fidelity to the observed visibility and the sparsity of the reconstructed solution. In real applications, however, the sought-after coefficients are likely to have very different scalings, and it would be wiser to adapt the regularization parameter to each coefficient. This corresponds to solving

$$\min_{\alpha} \frac{1}{2} \|y - H\Phi\alpha\|^2 + \sum_{i=1}^{T} \lambda_i |\alpha[i]| \tag{2.39}$$

where T is the size of the vector α (i.e. the number of coefficients).

Noise-Driven Strategy

Recall that, in the iterative schemes discussed earlier, the noise impacts the solution through the residual $R^{(n)} = Y - H\Phi\alpha^{(n)}$ and its backprojected coefficients $\alpha^{(R)} = \Phi^T H^T R^{(n)}$. If we see the regularization as a denoising step, then we can consider that λ_i should be related to the noise in $\alpha^{(R)}$, $\lambda_i = k\sigma_i$, where k is generally chosen between 3 and 5, and σ_i is the noise standard deviation on the coefficient $\alpha[i]$. If the noise is 0-mean additive with known covariance matrix Σ, then σ_i is the square root of the ith entry on the diagonal of the covariance matrix $\Phi^T H^T \Sigma H \Phi$. Alternatively, if we know how to simulate a set of N_r realistic noise realizations, we can calculate the coefficients $\alpha^{(r)} = \Phi^T H^T Z_r$ for each realization Z_r, and we have:

$$\sigma_i = \sqrt{\frac{1}{N_r} \sum_{r=1}^{N_r} \alpha^{(r)}[i]^2} .$$

Residual-Driven Strategy for Sub-Band Transforms

Lanusse et al. (2014) and Garsden et al. (2015) propose to fix the threshold only from the noise distribution at each wavelet scale j. A similar approach can be used for multiscale geometric transforms such as the curvelets.

Suppose that H is a convolution operator, Φ corresponds to a band-pass transform with J bands, and the noise N is stationary and Gaussian. Then $\alpha_j^{(N)} = (\Phi^T H^T N)_j$ in each band j is also Gaussian and stationary. We can therefore only consider 1 regularization parameter per band rather than 1 per coefficients, and Equation 2.39 becomes

$$\min_\alpha \frac{1}{2} \|y - H\Phi\alpha\|^2 + \sum_{j=1}^{J} \lambda_j \|\alpha_j\|_1, \tag{2.40}$$

where α_j are the coefficients in the band j and J is the number of bands. We could then use $\alpha_j^{(R^{(n)})}$ to estimate standard deviation σ_j. As $R^{(n)}$ may contain structure as well as noise, a more robust approach is to use the median of absolute deviation (MAD) estimator. Then we have

$$\sigma_j = \text{MAD}\left(\alpha_j^{(R^{(n)})}\right)/0.6745 = \text{median}\left(\left|\alpha_j^{(R^{(n)})} - \text{median}\left(\alpha_j^{(R^{(n)})}\right)\right|\right)/0.6745.$$

2.7 Inpainting

Missing data are a standard problem in astronomy. This can be due to bad pixels, or an image area we consider as problematic due to calibration or observational problems. These masked areas lead to difficulties for post-processing, especially to estimate statistical information, such as the power spectrum or bispectrum. The inpainting technique consists in filling the gaps. The typical image-inpainting problem can be defined as follows. Let X be the ideal complete signal, Y the observed incomplete data, and M the binary mask (i.e. $M_i = 1$ if we have information at pixel i, $M_i = 0$ otherwise). In short, we have: $Y = MX$. Inpainting consists in recovering X knowing Y and M. We thus want to minimise:

$$\min_X \|\Phi^T X\|_0 \quad \text{subject to} \quad Y = MX. \tag{2.41}$$

It was shown in Elad et al. (2005) that this optimization problem can be efficiently solved through an iterative thresholding algorithm (also see Fadili et al., 2009):

$$X^{(n+1)} = \Delta_{\Phi,\lambda_n}(X^{(n)} + Y - MX^{(n)}). \tag{2.42}$$

where the nonlinear operator $\Delta_{\Phi,\lambda}(Z)$ consists in (i) decomposing the signal Z in the dictionary Φ to derive the coefficients $\alpha = \Phi^T Z$; (ii) thresholding the coefficients: $\tilde{\alpha} = \rho(\alpha, \lambda)$, where the thresholding operator ρ can either be a hard thresholding or a soft thresholding; and (iii) reconstructing $\tilde{Z}$ from the thresholded coefficients $\tilde{\alpha}$.

The threshold parameter λ_n decreases with the iteration number and it plays a role similar to the cooling parameter of the simulated annealing techniques, i.e. it allows the solution to escape from local minima. More details on optimization in inpainting with sparsity can be found in Fadili et al. (2009) and Starck et al. (2010). The case where the dictionary is a union of subdictionaries $\Phi = \{\Phi_1, \ldots, \Phi_K\}$, where each Φ_i has a fast operator, has also been investigated in Elad et al. (2005) and Fadili et al. (2009).

Sparse inpainting has been used for different applications in astrophysics:

- High-order statistics on weak lensing and CMB data: For second-order statistics, such as power-spectrum estimation, robust methods exist and it is generally not necessary to use more sophisticated methods. For higher-order statistics (bispectrum, etc.), dealing with the mask of missing data is a challenging task. It was shown that sparse inpainting can fill the gap and attain a high accuracy on the statistical estimators for both weak lensing (Pires et al., 2009) and CMB data (Perotto et al., 2010).

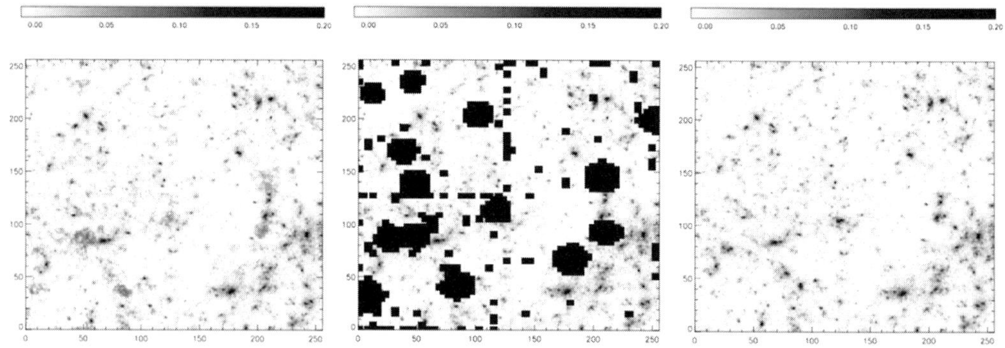

FIGURE 2.9 Left panel, simulated weak-lensing mass map. Middle panel, simulated mass map with a standard mask pattern. Right panel, inpainted mass map. The region shown is $1^\circ \times 1^\circ$.

- Asteroseismology: Sparse inpainting has been included in the official convection rotation and planetary transits (CoRoT) pipeline to correct the missing data in both asteroseismic and exoplanet channels (García et al., 2014; Pires et al., 2015).
- Large-scale studies on spherical maps for integrated Sachs-Wolfe effect signal reconstruction (Dupé et al., 2011) and CMB anomalies analysis (Starck et al., 2013b).

Example

The experiment illustrated in Figure 2.9 was conducted on a simulated weak-lensing mass map masked by a typical mask pattern Figure. The left panel shows the simulated mass map and the middle panel shows the masked map. The result of the inpainting method is shown in the right panel. Note that the gaps are undistinguishable by eye. More interesting, it has been shown that, using the inpainted map, we can reach an accuracy of about 1% for the power spectrum and 3% for the bispectrum (Pires et al., 2009).

2.8 Blind Source Separation

2.8.1 Sparsity and Blind Source Separation

In the blind source separation (BSS) setting, the instantaneous linear-mixture model assumes that we are given m observations $\{Y_1, \ldots, Y_m\}$, where each $\{Y_i\}_{i=1,\ldots,m}$ is a row vector of size t, and each measurement is the linear mixture of n source processes. As the measurements are m different mixtures, source-separation techniques aim at recovering the original sources $\mathbf{X} = [X_1^T, \ldots, X_n^T]^T$ by taking advantage of some information contained in the way the signals are mixed in the observed data. The linear mixture model is rewritten in matrix form

$$Y = HX + N, \tag{2.43}$$

where Y is the $m \times t$ measurement matrix (i.e. observed data), X is the $n \times t$ source matrix, and H is the $m \times n$ mixing matrix. H defines the contribution of each source to each measurement. An $m \times t$ matrix $\mathbf{N}$ is added to account for instrumental noise or model imperfections.

It has been shown that sparsity is a very robust regularization to solve BSS for both underdetermined (i.e. less observations than unknown, $m < n$) (Gribonval et al., 2008) and overdetermined mixtures ($m \geq n$) (Bobin et al., 2007, 2008). The generalized morphological component analysis (GMCA) algorithm (Bobin et al., 2007) finds the sparsest solution through the following iterative scheme:

$$X^{(k+1)} = \Delta_{\Phi,\lambda_k}(H^{+(k)}Y) \; ,$$
$$H^{(k+1)} = YX^{(k+1)^T}(X^{(k+1)}X^{(k+1)^T})^{-1}, \tag{2.44}$$

where $H^{+(k)}$ is the pseudo-inverse of the estimated mixing matrix $H^{(k)}$ at iteration k, λ_k is a decreasing threshold, and Δ_{Φ,λ_k} is the nonlinear operator which consists in decomposing each source X_i on the dictionary Φ, thresholding its coefficients, and reconstructing it. Finally, for hyperspectral data, it was advocated to impose sparsity constraints on the columns of the mixing matrix to enhance source recovery (Bobin et al., 2008). GMCA has shown to be very impressive to remove galactic and extragalactic foregrounds in order to detect the Epoch of Reionization 21-centimeter signal (Chapman et al., 2013, 2014).

2.8.2 Multiscale BSS

According to the mixture model underlying GMCA, all the observations are assumed to have the same resolution. However, this assumption is not always true. For instance, the Wilkinson microwave anisotropy probe (WMAP) (respectively Planck) full width at half maximum varies by a factor of about 5 (respectively 7) between the best resolution and the worst resolution. Assuming that the beam is invariant across the sky, the linear-mixture model should be substituted with the following:

$$\forall i = 1, \cdots, m; \quad Y_i = B_i \star \left(\sum_{s=1}^{n} H_{i,s} X_s \right) + N_i,$$

where B_i stands for the impulse response function of the PSF at channel i. This model can be expressed more globally by introducing a global multichannel convolution operator $\mathcal{B}$ defined so that its restriction to the channel i is equal to B_i:

$$X = \mathcal{B}(HX) + N. \tag{2.45}$$

A solution was proposed in Bobin et al. (2013) consisting in choosing a wavelet dictionary and adapting the wavelet decomposition for each channel such that the wavelet coefficients $\alpha_{1,j}, \ldots, \alpha_{M,j}$ of the M available channels at scale j have exactly the same resolution. This can be easily obtained by choosing a specific wavelet function for each channel i ($i = 1, \ldots, M$) such that:

$$\hat{\Psi}_{i,j}(u,v) = \frac{\hat{B}^{(\text{target})}(u,v)}{\hat{B}_i(u,v)} \hat{\psi}_j(u,v), \tag{2.46}$$

where $\hat{x}$ denotes the Fourier transform of x, B_i is the beam of the i-th channel, $B^{(\text{target})}$ is the Gaussian beam related to the target resolution, and ψ_j is the standard wavelet function at scale j.

Replacing the data matrix $\mathbf{X}$ by the matrix W formed of wavelet coefficients ($W_{i,*}$ contains the wavelet coefficients of the ith channel), and applying the pseudo inverse of H to W, we get $Z = H^+W$. The quantity $Z_{s,*}$ is now related to the wavelet coefficients of the s-th source. The estimated sources, at the resolution defined by $B^{(\mathrm{target})}$, can then be estimated via the following reconstruction formula:

$$X_s = \Phi^T Z_s, \tag{2.47}$$

where Φ is the dictionary related to the wavelet transform. In cases where the ratio $\hat{B}^{(\mathrm{target})}\hat{\psi}_j/\hat{B}_i$ is not defined for all channels, the solution is to use only a subset of the available channels. Hence the number of used channels in this multichannel GMCA (mGMCA) depends on the wavelet scale (Bobin et al., 2013). At the finest wavelet scales (i.e. highest frequencies), some channels may become inactive, while they will all be used at the largest wavelet scales.

2.8.3 Local Multiscale BSS

Both GMCA and mGMCA explicitly assume a model where the mixing matrix does not vary across pixels. This may be a strong limitation in some applications. For instance, for CMB map recovery, this model is not suited to capture the expected emissivity variation of galactic foregrounds across the sky. For this purpose, the mixture model has been extended to a multiscale local-mixture model (Bobin et al., 2013):

$$W_j[k,l] = (H_{j,k,l}\alpha_j)[k,l] + N_j[k,l], \tag{2.48}$$

where W_j is the matrix composed of the j-th wavelet scale of the data Y, $H_{j,k,l}$ is the mixing matrix at scale j and at location (k,l), α_j is the matrix composed of the j-th wavelet scale of the sources, and N_j is the noise at scale j.

A straightforward approach is to apply GMCA independently to decompose each wavelet scale into patches with a scale-dependent size, then perform GMCA on each patch independently. But fixing the patch size a priori is a very strong constraint, and the appropriate patch size should be space dependent as well and may vary from one region to another. A trade-off has to be made between small/large patches which would balance between statistical consistency (large patches) and adaptivity (small patches). The local GMCA (LGMCA) determines the optimal patch size from a quadtree decomposition (Bobin et al., 2013). Using a wavelet-based multiscale analysis and wavelet scales partitioning with adaptive patch sizes, LGMCA is able to capture properly both the beam evolution across channels and the variations across pixels of the emissivity of components. LGMCA is a sparse BSS technique well adapted for studying astrophysical phenomena with scale and space dependencies.[5]

2.8.4 CMB Sparse Recovery

Using both WMAP 9-year and Planck PR1 public data, a joint WMAP-Planck CMB map was reconstructed using LGMCA (Bobin et al., 2014). This allowed for the release of a clean, full-sky CMB map with very low foreground residuals (see Figure 2.10), especially on the galactic centre where estimation of the CMB is challenging. It was shown that this CMB map is also free of any significant thermal Sunyaev-Zel'dovich effect) residuals. This truly full-sky, clean estimate of the CMB map is obviously a good candidate for galactic studies. In addition, as it is free of any significant thermal SZ, it should also be suitable for kinetic SZ studies.[6]

[5] More details on LGMCA can be found at `www.cosmostat.org/research/statistical-methods/gmca`.

[6] The map and the software used to derive this CMB map are available at `www.cosmostat.org/research/cmb/planck_wpr1`.

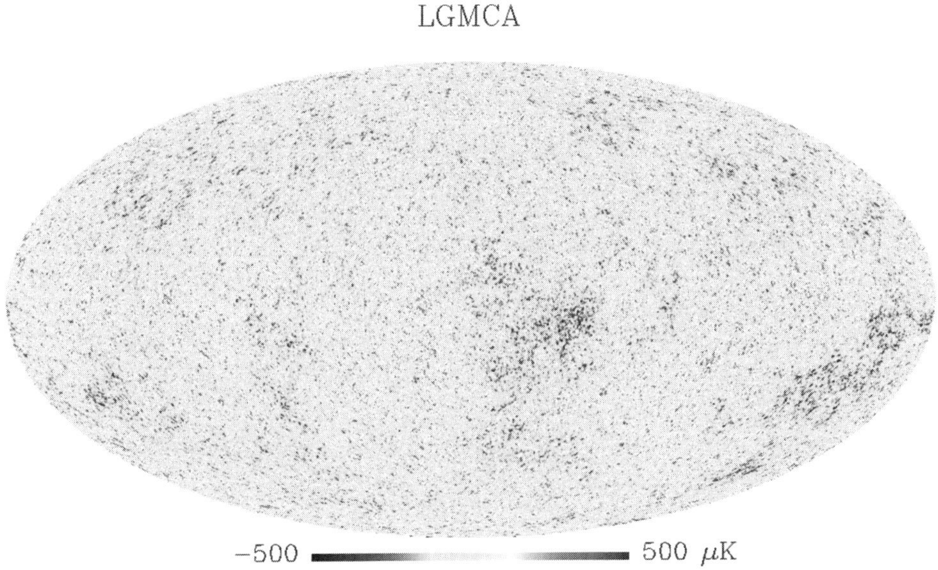

FIGURE 2.10 Sparse recovery of the CMB from a joint analysis of WMAP 9-year and Planck PR1 data.

2.9 Conclusion

Signal processing in the twentieth century was made mainly considering band-limited signals with an acquisition system based on the Shannon-Nyquist sampling theorem, and inverse problems were solved using least-squares estimators. The last 10 years have seen an emerging new paradigm based on sparse signals and the compressed sensing sampling theorem, where sparse regularization is used to solve inverse problems. We have presented in this chapter different sparse decomposition approaches and how to use them in applications such as denoising, deconvolution, inpainting, and blind source separation.

We conclude this chapter on a beautiful scientific result in astronomical imaging, where all the messages delivered above have been carefully taken into account. Figure 2.11 shows a reconstructed image from the Low-Frequency Array (LOFAR) radio-interferometry data of Cygnus A. LOFAR is a giant radiotelescope with a large antenna array comprising 25,000 elementary antennae grouped into 48 stations and distributed in different countries (the Netherlands, Germany, United Kingdom, France, Sweden, and soon, Poland). Cygnus A is one of the most powerful galaxies in the radio window. At a distance of 750 million light years from Earth, it is considered to have a central black hole of 2.5 billion solar masses generating 2 symmetric jets, visible in radio on each side of the black hole. Radio-interferometers provide measurements in the Fourier domain, but without covering the whole domain. Sparse recovery allows us to reconstruct the image with a resolution twice as good as standard methods (Garsden et al., 2015), showing hotspot features and a much sharper reconstruction of the structures inside the lobes. These high-resolution details cannot be seen when using the same data with standard methods. But Cygnus A was also observed with another instrument, the Very Large Array, in another frequency band allowing to have a resolution twice better. Figure 2.11 shows that both images clearly present the same quality of details. The resolved structures inside the lobes in the LOFAR image reconstructed using sparse recovery

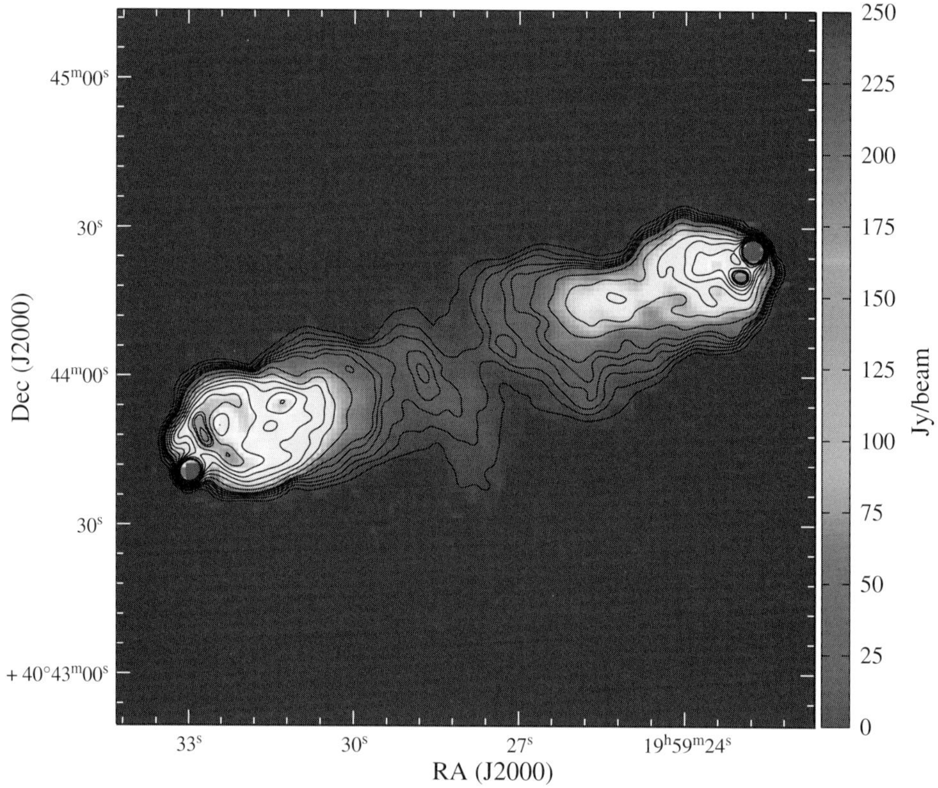

FIGURE 2.11 Colour scale: Reconstructed 512 × 512 image of Cygnus A at 151 MHz (with resolution 2.8″ and a pixel size of 1″). Contour levels from a 327.5 MHz Cyg A VLA image at 2.5″ angular resolution and a pixel size of 0.5″. Most of the recovered features in the image reconstructed using sparsity regularization correspond to real structures observed at higher frequencies.

(colour map) match the real structures observed at higher frequency and reconstructed with a standard method (contour), validating the capability of the method to reconstruct sources with high fidelity.

BIBLIOGRAPHY

Aharon, M., Elad, M., and Bruckstein, A. 2006. K-SVD: an algorithm for designing overcomplete dictionaries for sparse representation. *IEEE Trans. Signal Process.* **54**(11), 4311–4322.

André, P., Men'shchikov, A., Bontemps, S., et al. 2010. From filamentary clouds to prestellar cores to the stellar IMF: initial highlights from the Herschel Gould Belt Survey. *Astron. Astrophys.* **518**, L102.

Anscombe, F. 1948. The transformation of Poisson, binomial and negative-binomial data. *Biometrika* **15**, 246–254.

Bamberger, R. H., and Smith, M. J. T. 1992. A filter bank for the directional decomposition of images: theory and design. *IEEE Trans. Image Proc.* **40**, 882.

Baraniuk, R., Cevher, V., and Wakin, M. 2010. Low-dimensional models for dimensionality reduction and signal recovery: a geometric perspective, *Proc. IEEE.* **98**(6), 959.

Barbey, N., Sauvage, M., Starck, J.-L., et al. 2011. Feasibility and performances of compressed sensing and sparse map-making with Herschel/PACS data. *Astron. Astrophys.* **527**, A102.

Beck, A., and Teboulle, M. 2009. Fast iterative shrinkage-thresholding algorithm for linear inverse problems. *SIAM J. Imaging Sci.* **2**, 183–202.

Beckouche, S., Starck, J.-L., and Fadili, J. 2013. Astronomical image denoising using dictionary learning. *Astron. Astrophys.* **556**, A132.
Bobin, J., Starck, J.-L., Fadili, M. J., et al. 2007. Sparsity and morphological diversity in blind source separation. *IEEE T. Image Process.* **16**(11), 2662–2674.
Bobin, J., Starck, J.-L., Moudden, Y., et al. 2008. Blind source separation: the sparsity revolution. *Adv. Imag. Elect. Phys.* **152**, 221–306.
Bobin, J., Starck, J.-L., and Ottensamer, R. 2008. Compressed sensing in astronomy. *IEEE J. Sel. Top. Signa.* **2**, 718–726.
Bobin, J., Starck, J.-L., Sureau, F., et al. 2013, Sparse component separation for accurate cosmic microwave background estimation. *Astron. Astrophys.* **550**, A73.
Bobin, J., Sureau, F., Starck, J.-L., et al. 2014, Joint Planck and WMAP CMB map reconstruction, *Astron. Astrophys.* **563**, A105.
Burt, P., and Adelson, A. 1983. The Laplacian pyramid as a compact image code. *IEEE Trans. Commun.* **31**, 532–540.
Candès, E., Demanet, L., Donoho, D., et al. 2006. Fast discrete curvelet transforms. *Multiscale Model. Sim.* **5**(3), 861–899.
Candès, E., and Donoho, D. 1999, Ridgelets: the key to high dimensional intermittency? *Philos. T. R. Soc. Lond. A* **357**, 2495–2509.
Candès, E., and Donoho, D. 2001. Curvelets and curvilinear integrals. *J. Approx. Theory* **113**, 59–90.
Candès, E., and Donoho, D. 2002. Recovering edges in ill-posed inverse problems: optimality of curvelet frames. *Ann. Stat.* **30**, 784–842.
Candès, E., and Donoho, D. 2004. New tight frames of curvelets and optimal representations of objects with piecewise C^2 singularities. *Commun. Pure Appl. Math.* **57**(2), 219–266.
Candès, E., and Tao, T. 2006. Near optimal signal recovery from random projections: universal encoding strategies? *IEEE Trans. Inform. Theory* **52**, 5406–5425.
Candès, E. J., and Plan, Y. 2011. A probabilistic and RIPless theory of compressed sensing. *IEEE Trans. Inform. Theory* **57**(11), 7235–7254.
Carrillo, R. E., McEwen, J. D., and Wiaux, Y. 2012. Sparsity Averaging Reweighted Analysis (SARA): a novel algorithm for radio-interferometric imaging. *Mon. Not. R. Astron. Soc.* **426**, 1223.
Chapman, E., Abdalla, F. B., Bobin, J., et al. 2013. The scale of the problem: recovering images of reionization with Generalized Morphological Component Analysis. *Mon. Not. R. Astron. Soc.* **429**, 165.
Chapman, E., Zaroubi, S., Abdalla, F. B., et al. 2016. The effect of foreground mitigation strategy on EoR window recovery. *Mon. Not. R. Astron. Soc.* **458**(3), 2928–2939.
Combettes, P. L., and Wajs, V. R. 2005. Signal recovery by proximal forward-backward splitting. *Multiscale Model. Sim.* **4**(4), 1168–1200.
Da Cunha, A. L., Zhou, J., and Do, M. N. 2006. The nonsubsampled contourlet transform: theory, design, and applications. *IEEE Trans. Image Process.* **15**(10), 3089.
Daubechies, I., Defrise, M., and Mol, C. D. 2004. An iterative thresholding algorithm for linear inverse problems with a sparsity constraint. *Commun. Pure Appl. Math.* **57**, 1413–1541
Davenport, M., Duarte, M., Eldar, Y., et al. 2012. Introduction to compressed sensing. In *Compressed sensing: theory and applications*, edited by Eldar, Y., and Kutyniok, G., 1–68. Cambridge University Press, Cambridge.
Díaz-Hernández, R., Peregrina-Barreto, H., Altamirano-Robles, L., et al. 2014. Automatic stellar spectral classification via sparse representations and dictionary learning. *Exp. Astron.* **38**, 193.
Do, M. N., and Vetterli, M. 2003. Contourlets. In *Beyond wavelets*, edited by Stoeckler, J., and Welland, G. V., –. Academic Press, New York.
Donoho, D. 1995. De-noising by soft-thresholding. *IEEE Trans. Inform. Theory* **41**(3), 613–627.
Donoho, D. 2006a. Compressed sensing. *IEEE Trans. Inform. Theory.* **52**(4), 1289–1306.
Donoho, D. 2006b. For most large underdetermined systems of linear equations, the minimal ℓ_1 solution is also the sparsest solution. *Commun. Pure Appl. Math.* **59**(7), 907–934.
Donoho, D., and Duncan, M. 2000. Digital curvelet transform: strategy, implementation and experiments. Edited by Szu, H., Vetterli, M., Campbell, W., et al., [–] *Proc. SPIE* 4056, *Wavelet Appl. VII*, 12–29.
Donoho, D., Johnstone, I., Kerkyacharian, G., et al. 1995. Wavelet shrinkage: asymptopia? *J. Roy. Stat. Soc. B* **57**, 301–369.

Dupé, F.-X., Rassat, A., Starck, J.-L., et al. 2011. Measuring the integrated Sachs-Wolfe effect. *Astron. Astrophys.* **534**, A51.

Easley, G., Labate, D., and Lim, W.-Q. 2008. Sparse directional image representation using the discrete shearlet transform. *Appl. Comput. Harmon. Anal.* **25**, 25.

Elad, M., and Aharon, M. 2006. Image denoising via sparse and redundant representations over learned dictionaries. *IEEE Trans. Image Process.* **15**(12), 3736.

Elad, M., Starck, J.-L., Donoho, D., et al. 2005. Simultaneous cartoon and texture image inpainting using morphological component analysis (MCA). *Appl. Comput. Harmon. A* **19**, 340–358.

Engan, K., Aase, S. O., and Husoy, J. H. 1999. Method of optimal directions for frame design. *Int. Conf. Acoust Speeh.*, 2443–2446.

Fadili, M. J., Starck, J.-L., and Murtagh, F. 2009. Inpainting and zooming using sparse representations. *Comput. J.* **52**, 64–79.

Figueiredo, M., and Nowak, R. 2003. An EM algorithm for wavelet-based image restoration. *IEEE Trans. Image Process.* **12**(8), 906–916.

García, R. A., Mathur, S., Pires, S., et al. 2014. Impact on asteroseismic analyses of regular gaps in Kepler data. *Astron. Astrophys.* **568**, A10.

Garsden, H., Girard, J. N., Starck, J. L., et al. 2015. LOFAR sparse image reconstruction. *Astron. Astrophys.* **575**, A90.

Gribonval, R., Cevher, V., and Davies, M. E. 2012. Compressible distributions for high-dimensional statistics. *IEEE Trans. Inform. Theory* **58**(8), 5016–5034.

Gribonval, R., Rauhut, H., Schnass, K., et al. 2008. Atoms of all channels, unite! Average case analysis of multi-channel sparse recovery using greedy algorithms. *J. Fourier Anal. Appl.* **14**(5), 655–687.

Guo, K., Kutyniok, G., and Labate, D. 2006. Sparse multidimensional representations using anisotropic dilation and shear operators. In *Wavelets and splines*, edited by Chen, G., and Lai, M. J., 189–201. Nashboro Press, Nashville, TN, USA.

Guo, K., and Labate, D. 2007, Optimally sparse multidimensional representation using shearlets. *SIAM J. Math. Anal.* **39**(1), 298.

Guo, K., Labate, D., Lim, W.-Q., et al. 2004. Wavelets with composite dilations. *Electron. Res. Announc.* **10**(9), 78.

Han, B., Kutyniok, G., and Shen, Z. 2011. Adaptive multiresolution analysis structures and shearlet systems. *SIAM J. Numer. Anal.* **49**, 1921.

Kutyniok, G., and Labate, D., eds. 2012a. *Shearlets: multiscale analysis for multivariate data.* Springer Science & Business Media, Dordrecht, Netherlands.

Kutyniok, G., and Lim, W.-Q. 2011. Compactly supported shearlets are optimally sparse. *J. Approx. Theory* **163**(11), 1564.

Kutyniok, G., Lim, W.-Q., and Reisenhofer, R. 2014. Shearlab 3D: faithful digital shearlet transforms based on compactly supported shearlets. *ACM Trans. Math. Software*, **42**(1).

Kutyniok, G., and Sauer, T. 2009. Adaptive directional subdivision schemes and shearlet multiresolution analysis. *SIAM J. Math. Anal.* **41**, 1436.

Kutyniok, G., Shahram, M., and Zhuang, X. 2012b. Shearlab: a rational design of a digital parabolic scaling algorithm. *SIAM J. Imaging Sci.* **5**, 1291.

Labate, D., Lim, W.-Q., Kutyniok, G., et al. 2005. Sparse multidimensional representation using shearlets. *Proc. SPIE 4056, Wavelet Appl. VII*, 254–262.

Lanusse, F., Paykari, P., Starck, J.-L., et al. 2014. PRISM: recovery of the primordial spectrum from Planck data. *Astron. Astrophys.* **571**, L1.

Leonard, A., Dupé, F.-X., and Starck, J.-L. 2012. A compressed sensing approach to 3D weak lensing. *Astron. Astrophys.* **539**, A85.

Lim, W.-Q. 2010. The discrete shearlet transform: a new directional transform and compactly supported shearlet frames. *IEEE Trans. Image Proc.* **19**, 1166.

Lim, W.-Q. 2013. Nonseparable shearlet transform. *IEEE Trans. Image Proc.* **22**(5), 2056–2065.

Lin, C. 2007. Projected gradient methods for nonnegative matrix factorization. *Neural Comput.* **19**(10), 2756.

Mairal, J., Bach, F., Ponce, J., et al. 2010. Online learning for matrix factorization and sparse coding. *J. Mach. Learn. Res.* **11**, 19.

Moller, A., Ruhlmann-Kleider, V., Lanusse, F., et al. 2015. SNIa detection in the SNLS photometric analysis using Morphological Component Analysis. arXiv:1501.02110v4.

Murtagh, F., Starck, J.-L., and Bijaoui, A. 1995. Image restoration with noise suppression using a multiresolution support. *Astron. Astrophys. Sup.* **112**, 179–189.

Nikolova, M. 2007. Model distortions in Bayesian map reconstruction, *Inverse Probl. Imag.* **1**(2), 399.

Nolan, P. L., Abdo, A. A., Ackermann, M., et al. 2012. Fermi large area telescope second source catalog. *Astrophys. J. Suppl.* **199**, 31.

Pacaud, F., Pierre, M., Refregier, A., et al. 2006. The XMM large-scale structure survey: the X-ray pipeline and survey selection function. *Mon. Not. R. Astron. Soc.* **372**, 578.

Perotto, L., Bobin, J., Plaszczynski, S., et al. 2010. Reconstruction of the cosmic microwave background lensing for Planck. *Astron. Astrophys.* **519**, A4.

Peyré, G., Fadili, J., and Starck, J. L. 2010. Learning the morphological diversity. *SIAM J. Imaging Sci.* **3**(3), 646.

Pires, S., Mathur, S., García, R. A., et al. 2015. Gap interpolation by inpainting methods: application to ground and space-based asteroseismic data. *Astron. Astrophys.* **574**, A18.

Pires, S., Starck, J.-L., Amara, A., et al. 2009. Cosmological model discrimination with weak lensing. *Astron. Astrophys.* **505**, 969.

Ramos, E. P. R. G., Vio, R., and Andreani, P. 2011. Detection of new point sources in WMAP cosmic microwave background maps at high galactic latitude. A new technique to extract point sources from CMB maps. *Astron. Astrophys.* **528**, A75.

Rauhut, H., and Ward, R. 2012. Sparse Legendre expansions via ℓ_1 minimization. *J. Approx. Theory* **164**(5).

Schmitt, J., Starck, J. L., Casandjian, J. M., et al. 2010. Poisson denoising on the sphere: application to the Fermi gamma ray space telescope. *Astron. Astrophys.* **517**, A26.

Starck, J.-L., and Bijaoui, A. 1994. Filtering and deconvolution by the wavelet transform. *Signal Process.* **35**, 195–211.

Starck, J.-L., Bijaoui, A., and Murtagh, F. 1995. Multiresolution support applied to image filtering and deconvolution. *CVGIP-Graph. Model. Image.* **57**, 420–431.

Starck, J.-L., Candès, E., and Donoho, D. 2002. The curvelet transform for image denoising. *IEEE Trans. Image Process.* **11**(6), 131–141.

Starck, J.-L., Candès, E., and Donoho, D. 2003. Astronomical image representation by the curvelet transform. *Astron. Astrophys.* **398**, 785–800.

Starck, J.-L., Donoho, D. L., Fadili, M. J., et al. 2013a. Sparsity and the Bayesian perspective. *Astron. Astrophys.* **552**, A133.

Starck, J.-L., Elad, M., and Donoho, D. 2004. Redundant multiscale transforms and their application for morphological component analysis. *Adv. Imag. Elect. Phys.* **132**.

Starck, J.-L., Fadili, J., and Murtagh, F. 2007, The undecimated wavelet decomposition and its reconstruction. *IEEE Trans. Image Process.* **16**, 297–309.

Starck, J.-L., Fadili, M. J., and Rassat, A. 2013b. Low-ℓ CMB analysis and inpainting. *Astron. Astrophys.* **550**, A15.

Starck, J.-L., and Murtagh, F. 1994. Image restoration with noise suppression using the wavelet transform. *Astron. Astrophys.* **288**, 343–348.

Starck, J.-L., and Murtagh, F. 1998. Automatic noise estimation from the multiresolution support. *Publ. Astron. Soc. Pac.* **110**, 193–199.

Starck, J.-L., and Murtagh, F. 2002. *Astronomical Image and Data Analysis.* Springer-Verlag, Berlin, Germany.

Starck, J.-L., and Murtagh, F. 2006. *Astronomical Image and Data Analysis*, 2nd ed. Springer-Verlag, Berlin, Germany.

Starck, J.-L., Murtagh, F., and Bijaoui, A. 1998. *Image processing and data analysis: the multiscale approach.* Cambridge University Press, Cambridge.

Starck, J.-L., Murtagh, F., and Fadili, M. 2010. *Sparse image and signal processing.* Cambridge University Press, Cambridge.

Vũ, B. 2013. A splitting algorithm for dual monotone inclusions involving cocoercive operators. *Adv. Comput. Math.* **38**(3), 667.

Weaver, J. B., Yansun, X., Healy, D. M., et al. 1991. Filtering noise from images with wavelet transforms. *Magn. Reson. Med.* **21**(2), 288–295.

Wiaux, Y., Jacques, L., Puy, G., et al. 2009. Compressed sensing imaging techniques for radio interferometry, *Mon. Not. R. Astron. Soc.* **395**, 1733.

Zhang, B., Fadili, M. J., and Starck, J.-L. 2008. Wavelets, ridgelets and curvelets for Poisson noise removal. *IEEE Trans. Image Process.* **17**(7), 1093–1108.

3 Bayesian Inference in Extrasolar Planet Searches

PHIL GREGORY

3.1 Introduction

A remarkable array of ground-based and space-based astronomical tools are providing astronomers access to other solar systems. Close to 2,000 planets have been discovered to date, starting from the pioneering work of Campbell et al. (1998), Wolszczan and Frail (1992), Mayor and Queloz (1995), and Marcy and Butler (1996). Figure 3.1 illustrates the pace of discovery[1] up to October 2014.

A wide range of techniques, including radial velocity, transiting, gravitational microlensing, timing, and direct imaging, have contributed to exoplanet discoveries. Because a typical star is approximately a billion times brighter than a planet, only a small fraction of the planets have been detected by direct imaging. As of this writing, we know of 1,832 planets in 1,145 planetary systems; 469 planets are known to be in multiple-planet systems, the largest of which has 7 planets (Lovis et al., 2011). Many additional candidates from NASA's Kepler transit detection mission are awaiting confirmation. One analysis of the Kepler data (Petigura, Mercy, and Howard, 2013) estimates that the occurrence rate of Earth-size planets (radii 1–2 times that of the Earth) in the habitable zone (where liquid water could exist) of sun-like stars is $22 \pm 8\%$.

These observers' successes have spurred a significant effort to improve the statistical tools for analysing data in this field (e.g. Loredo and Chernoff, 2003; Cumming, 2004; Loredo, 2004; Gregory, 2005a; Ford, 2005, 2006; Ford and Gregory, 2007; Zechmeister and Kürster, 2009; Cumming and Dragomir, 2010; Dawson and Fabrycky, 2010; Feroz, Balan, and Hobson, 2011). The majority of the planets have been detected through transits or the reflex motion of the star caused by the gravitational tug of unseen planets, using precision radial velocity (RV) measurements. In both cases the key problem is isolating a periodic signal in a time series, which is typically very sparse in the case of RV measurements. Increasing attention is also being focussed on identifying stellar-activity-induced RV signals (Aigrain, Pont, and Zucker, 2012; Baluev, 2013; Tuomi et al., 2013; Haywood et al., 2014).

Classical periodograms like Lomb-Scargle (Lomb, 1976; Scargle, 1982) have played an important role but the field has also spurred the development of new Bayesian periodograms to make use of the valuable prior information concerning the nature of the signal. Much of this work has highlighted a Bayesian Markov chain Monte Carlo (MCMC) approach as a way to better understand parameter uncertainties and degeneracies and to compute model probabilities. MCMC algorithms provide a powerful means for efficiently computing the required Bayesian integrals in many dimensions (e.g. an 8-planet model has 42 unknown parameters).

In this chapter I introduce a new, general purpose tool for Bayesian data analysis called fusion MCMC (FMCMC), used for nonlinear model fitting and regression analysis. When supplied with a multi-planet Kepler model it becomes a multi-planet Kepler periodogram. It is the outgrowth of an earlier attempt to achieve an automated MCMC algorithm

[1] Data from the Extrasolar Planet Encyclopedia, Exoplanet.eu. Schneider et al. (2011).

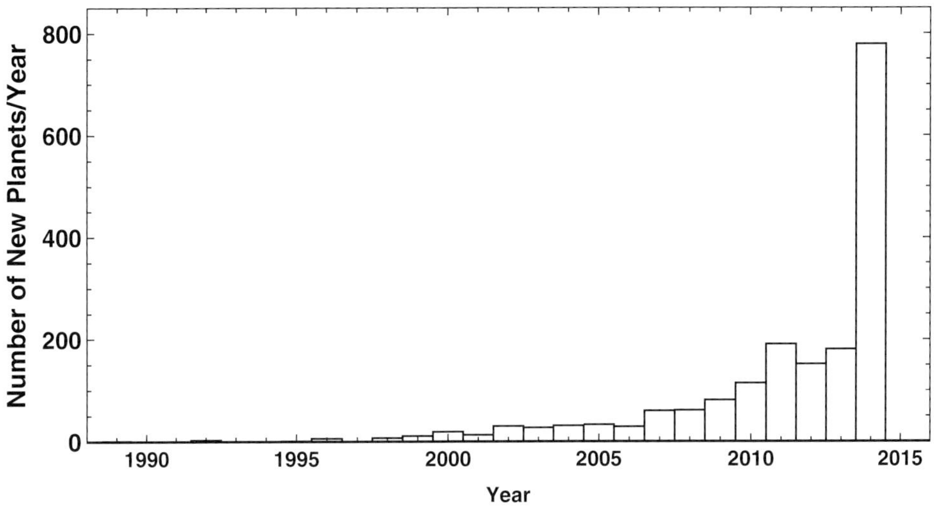

FIGURE 3.1 The pace of exoplanet discoveries (October 2014).

discussed in *Bayesian Logical Data Analysis for the Physical Sciences* (Gregory, 2005b, section 12.8).

FMCMC is a special version of the metropolis MCMC algorithm that incorporates parallel tempering, simulated annealing, and genetic-crossover operations. Each of these features facilitates the detection of a global minimum in chi-squared in a highly multimodal environment. By combining all three, the algorithm greatly increases the probability of realizing this goal.

FMCMC is controlled by a unique adaptive control system that automates the tuning of the MCMC proposal distributions for efficient exploration of the model parameter space, even when the parameters are highly correlated. This controlled, statistical-fusion approach has the potential to integrate other relevant statistical tools as desired. The FMCMC algorithm has been implemented in Mathematica using parallelized code to take advantage of multiple core computers.

Section 3.2 provides a brief, self-contained introduction to Bayesian inference applied to the Kepler exoplanet problem leading to the MCMC challenge. The different components of FMCMC are described in Section 3.3. In Section 3.4, the performance of the algorithm is illustrated with some exoplanet examples. Section 3.5 deals with the challenges of Bayesian model comparison and describes a new method for computing marginal likelihoods called nested restricted Monte Carlo. Its performance is compared with two other marginal likelihood estimators that depend on the posterior MCMC samples. Appendix 1 provides important details behind the choices of priors employed in this work.

3.2 Bayesian Primer

What Is Bayesian Probability Theory (BPT)?

$$\text{BPT} = \text{extended logic}$$

Deductive logic is based on axiomatic knowledge. In science we never know whether any theory of nature is absolutely true because our reasoning is based on incomplete information. Our conclusions are at best probabilities. Any extension of logic to deal with situations of incomplete information (realm of inductive logic) requires a theory of probability.

A new perception of probability has arisen in recognition that the mathematical rules of probability are not merely rules for manipulating random variables. They are now recognised as valid principles of logic for conducting inference about any hypothesis of interest. This view, of 'probability theory as logic', was championed in the late twentieth century by E. T. Jaynes in his book,[2] Probability Theory: The Logic of Science, (Jaynes, 2003). It is also commonly referred to as BPT in recognition of the work of the eighteenth-century English clergyman and mathematician Thomas Bayes.

Logic is concerned with the truth of propositions. A proposition asserts that something is true. Below are some examples of propositions.

- $A \equiv$ Theory X is correct.
- $\bar{A} \equiv$ Theory X is not correct.
- $D \equiv$ The measured redshift of the galaxy is 0.15 ± 0.02.
- $B \equiv$ The star has 5 planets.
- $A \equiv$ The orbital frequency is between f and $f + df$.

In the last example, f is a continuous parameter giving rise to a continuous hypothesis space. Negation of a proposition is indicated by a bar over the top, i.e. $\bar{A}$.

The propositions that we want to reason about in science are commonly referred to as hypotheses or models. We need to consider compound propositions like A, B which asserts that propositions A and B are both true, conditional on the truth of another proposition C. This is written $A, B|C$.

Rules for Manipulating Probabilities and Bayes' Theorem

There are only 2 rules for manipulating probabilities.

$$\text{Sum rule: } p(A|C) + p(\bar{A}|C) = 1 \tag{3.1}$$

$$\begin{aligned}\text{Product rule: } p(A, B|C) &= p(A|C) \times p(B|A, C) \\ &= p(B|C) \times p(A|B, C)\end{aligned} \tag{3.2}$$

Bayes' theorem is obtained by rearranging the two right-hand sides of the product rule.

$$\text{Bayes' theorem: } p(A|B, C) = \frac{p(A|C) \times p(B|A, C)}{p(B|C)} \tag{3.3}$$

Figure 3.2 shows Bayes' theorem in its more usual form for data analysis purposes. In a well-posed Bayesian problem the prior information, I, specifies the hypothesis space of current interest (range of models actively under consideration) and the procedure for computing the likelihood. The starting point is always Bayes' theorem. In any given problem the expressions for the prior and likelihood may be quite complex and require repeated applications of the sum and product rules to obtain an expression that can be solved.

As a theory of extended logic, BPT can be used to find optimal answers to well-posed scientific questions for a given state of knowledge, in contrast to a numerical recipe approach.

[2] The book was published 5 years after Jaynes's death through the efforts of a former graduate student, Dr G. L. Bretthorst.

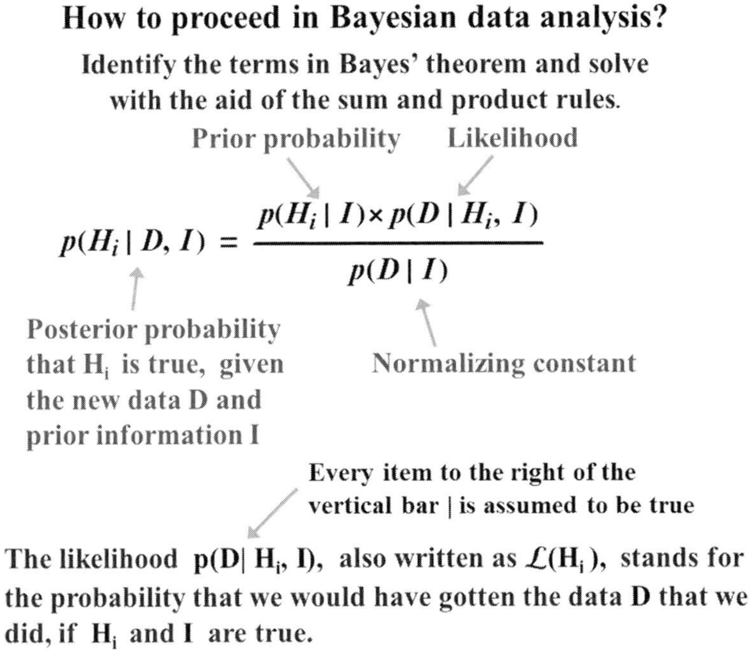

FIGURE 3.2 How to proceed in Bayesian data analysis.

3.2.1 Two Common Inference Problems

(i) **Model comparison (discrete hypothesis space):** Which of 2 or more models (hypotheses) is most probable given our current state of knowledge? Examples:
- Hypothesis or model M_0 asserts that the star has no planets.
- Hypothesis M_1 asserts that the star has 1 planet.
- Hypothesis M_i asserts that the star has i planets.

(ii) **Parameter estimation (continuous hypothesis):** Assuming the truth of M_1, solve for the probability-density distribution for each of the model parameters based on our current state of knowledge. Example:
- Hypothesis P asserts that the orbital period is between P and $P + dP$.

For a continuous hypothesis space the same symbol P is often employed in two different ways. When it appears as an argument of a probability, e.g. $p(P|D, I)$, it acts as a proposition (obeying the rules of Boolean algebra) and asserts that the true value of the parameter lies in the infinitesimal numerical range P to $P + dP$. In other situations it acts as an ordinary algebraic variable standing for possible numerical values.

Figure 3.3 illustrates the significance of the extended logic provided by BPT. Deductive logic is just a special case of BPT in the idealised limit of complete information.[3]

Calculation of a Simple Likelihood $p(D|M, I)$

Let d_i represent the ith measured data value. We model d_i by

$$d_i = f_i(X) + e_i, \tag{3.4}$$

[3] For a demonstration of this see Gregory, 2005b, section 2.5.4.

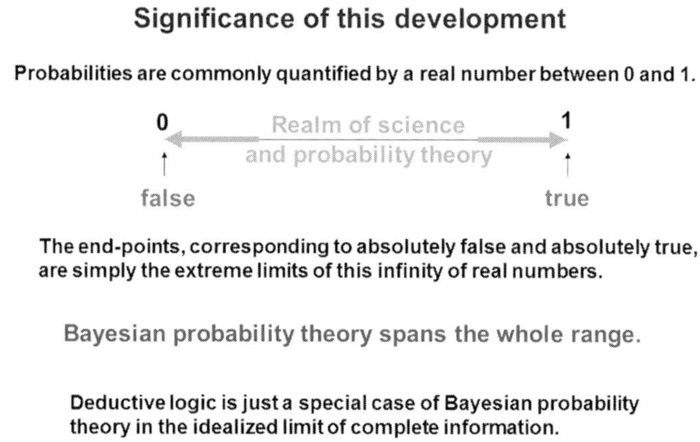

FIGURE 3.3 Significance of the extended logic provided by BPT.

where X represents the set of model parameters and e_i represents our knowledge of the measurement error, which can be different for each data point (heteroscedastic data).

If the prior information I indicates that distribution of the measurement errors are independent Gaussians, then

$$\begin{aligned} p(D_i|M,X,I) &= \frac{1}{\sigma_i\sqrt{2\pi}}\ Exp[-\frac{e_i^2}{2\sigma_i^2}] \\ &= \frac{1}{\sigma_i\sqrt{2\pi}}\ Exp\left[-\frac{(d_i - f_i(X))^2}{2\sigma_i^2}\right]. \end{aligned} \tag{3.5}$$

For independent data the likelihood for the entire data set $D = D_1, D_2, \ldots, D_N$ is the product of N Gaussians.

$$\begin{aligned} p(D|M,X,I) &= (2\pi)^{-N/2}\left\{\prod_{i=1}^{N}\sigma_i^{-1}\right\}\ Exp\left[-0.5\sum_{i=1}^{N}\frac{(d_i - f_i(X))^2}{\sigma_i^2}\right] \\ &= (2\pi)^{-N/2}\left\{\prod_{i=1}^{N}\sigma_i^{-1}\right\}\ Exp\left[-0.5\ \chi^2\right], \end{aligned} \tag{3.6}$$

where the summation within the square brackets is the familiar χ^2 statistic that is minimised in the method of least squares. Thus maximising the likelihood corresponds to minimising χ^2.

Recall: Bayesian posterior $\propto$ prior $\times$ likelihood.

Thus only for a uniform prior will a least-squares analysis yield the same result as the Bayesian *maximum a posterior* solution.

In the exoplanet problem the prior range for the unknown orbital period P is very large, from 1/2 day to 1,000 years (upper limit set by perturbations from neighbouring stars). Suppose we assume a uniform prior probability density for the P parameter. According to Equation 3.7, this would imply that we believed that it was 10^4 times more probable that the true period was in the upper decade (10^4–10^5 d) of the prior range than in the decade from 1 to 10 d.

$$\frac{\int_{10^4}^{10^5} p(P|M,I)dP}{\int_{1}^{10} p(P|M,I)dP} = 10^4 \tag{3.7}$$

Usually, expressing great uncertainty in some quantity corresponds more closely to a statement of scale invariance or equal probability per decade (uniform on ln$P$). A scale invariant prior has this property. An analysis of the occurrence rate of transiting planets versus orbital period found that the occurrence rate is constant, within 15%, between 12.5 and 100 d (Petigura, Mercy, and Howard, 2013).

$$p(\ln P|M,I)\ d\ln P = \frac{d\ \ln P}{\ln(P_{max}/P_{min})} \tag{3.8}$$

3.2.2 Marginalisation: An Important Bayesian Tool

Suppose our model parameter set, X, consists of 2 continuous parameters θ and ϕ. In parameter estimation, we are often interested in the implications of our current state of knowledge data D and prior information I for each parameter separately, independent of the values of the other parameters.[4] We can write

$$p(\theta|D,I) = \int d\phi\ p(\theta,\phi|D,I). \tag{3.9}$$

This can be expanded using Bayes' theorem. If our prior information for ϕ is independent of θ, this yields

$$p(\theta|D,I) \propto p(\theta|I) \int d\phi\ p(\phi|I)p(D|\theta,\phi,I). \tag{3.10}$$

This gives the marginal[5] posterior distribution $p(\theta|D,I)$, in terms of the weighted average of the likelihood function, $p(D|\theta,\phi,I)$, weighted by $p(\phi|I)$, the prior probability density function for ϕ. This operation marginalises out the ϕ parameter. For exoplanet detection, the need to fit at least an 8-planet model to the data requires integration in 42 dimensions. The integral in Equation 3.9 can sometimes be evaluated analytically, which can greatly reduce the computational aspects of the problem, especially when many parameters are involved. If the joint posterior in θ,ϕ is non-Gaussian then the marginal distribution $p(\theta|D,I)$ can look very different from the projection of the joint posterior onto the θ axis.[6] This is because the marginal, $p(\theta|D,I)$, for any particular choice of θ, is proportional to the integral over ϕ, which depends on the width of the distribution in the ϕ coordinate as well as the height.

In Bayesian model comparison, we are interested in the most probable model, independent of the model parameters (i.e. marginalise out all parameters). This is illustrated in the equation below for model M_2:

$$p(M_2|D,I) = \int_{\Delta X} dX\ p(M_2,X|D,I), \tag{3.11}$$

where ΔX designates the appropriate range of integration for the set of model parameters designated by X, as specified by our prior information, I.

[4] As shown in Gregory, 2005b, section 1.5.

[5] Since a parameter of a model is not a random variable, the frequentist statistical approach is denied the concept of the probability of a parameter.

[6] This is clearly demonstrated in Gregory, 2005b, figure 11.3.

Integration and MCMC

It should be clear from the above that a full Bayesian analysis involves integrating over model parameter spaces. Integration is more difficult than minimisation. However, the Bayesian solution provides the most accurate information about the parameter errors and correlations without the need for any lengthy additional calculations, i.e. Monte Carlo simulations. Fortunately the MCMC algorithms (Metropolis et al., 1953) provide a powerful means for efficiently computing integrals in many dimensions to within a constant factor. This factor is not required for parameter estimation.

The output at each iteration of the MCMC is a sample from the model parameter space of the desired joint posterior distribution (the *target distribution*). All MCMC algorithms generate the desired samples by constructing a kind of random walk in the model parameter space. The random walk is accomplished using a Markov chain, whereby the new sample, X_{t+1}, depends on previous sample X_t according to an entity called the *transition probability* or *transition kernel*, $p(X_{t+1}|X_t)$. The transition kernel is assumed to be time independent. Each sample is correlated with nearby samples.[7] The remarkable property of $p(X_{t+1}|X_t)$ is that after an initial burn-in period (which is discarded) it generates samples of X with a probability density equal to the desired joint posterior probability distribution of the parameters.

The marginal posterior probability density function (PDF) for any single parameter is given by a histogram of that component of the sample for all post burn-in iterations. Because the density of MCMC samples is proportional to the joint posterior probability distribution, it does not waste time exploring regions where the joint posterior density is very small in contrast to straight Monte Carlo integration.

In general the target distribution is complex and difficult to draw samples from. Instead new samples are drawn from a distribution which is easy to sample from, like a multivariate normal with mean equal to the current X_t. Figure 3.4 shows the operation of a Metropolis-Hastings MCMC algorithm.[8] In this example the same Gaussian proposal distribution is used for both parameters. For details on why the Metropolis-Hastings algorithm works, see my book (Gregory, 2005b, section 12.3).

Figure 3.5 shows the behaviour of the Metropolis-Hastings algorithm for a simple toy target distribution consisting of two 2-dimensional Gaussians. A single Gaussian proposal distribution (with a different choice of σ in each panel) was employed for both parameters, X_1, X_2. Each simulation started from the same location (top left) and used the same number of iterations. For $\sigma = 0.1$ the acceptance rate is very high at 95% and the samples are strongly correlated. For $\sigma = 1$ the acceptance rate is 63% and the correlations are much weaker. One can barely detect the burn-in samples in this case. For $\sigma = 10$ the acceptance rate is only 4%. In this case most samples were rejected resulting in many repeats of the current sample. Simulation (b) is close to ideal, while the other two would have to be run much longer to achieve reasonable sampling of the underlying target distribution. Based on empirical studies, Roberts, Gelman, and Gilks (1997) recommend calibrating the acceptance rate to about 25% for high-dimensional models and to about 50% for models of 1 or 2 dimensions.

[7] Care must be taken if independent samples are desired (typically by thinning the resulting chain of samples by only taking every nth value, e.g. every hundredth value).

[8] Gibbs sampling is a special case of the Metropolis-Hastings algorithm which is frequently employed. Gibbs sampling is applicable when the joint distribution is not known explicitly or is difficult to sample from directly, but the conditional distribution of each variable is known and is relatively easy to sample from. One advantage is that it does not require the tuning of proposal distributions.

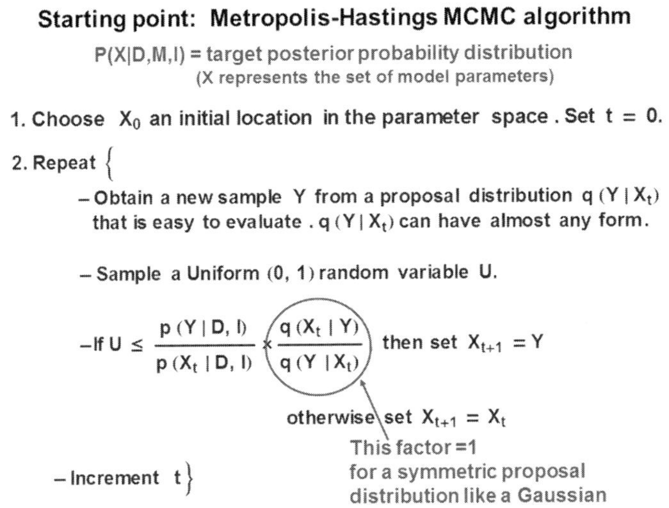

FIGURE 3.4 The Metropolis-Hastings algorithm. In this example the same Gaussian proposal distribution is used for both parameters.

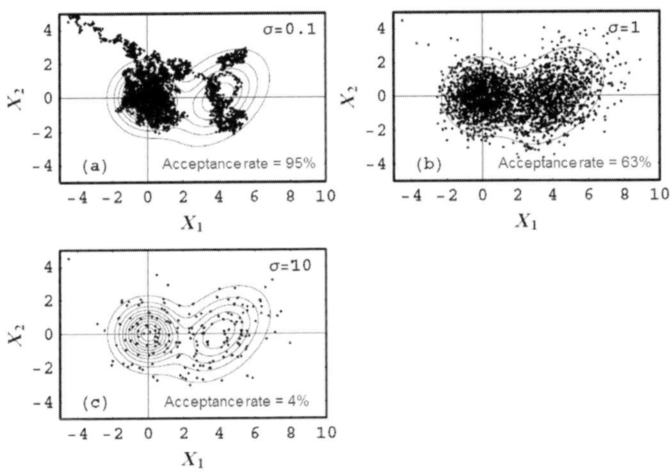

FIGURE 3.5 Toy Metropolis-Hastings MCMC simulations for a range of proposal distributions σ, showing how the efficiency depends on tuning this proposal distribution. It can be difficult to tune the efficiency when sampling many parameters. In this example, the target probability distribution consists of two 2-dimensional Gaussians indicated by the contours.

3.3 FMCMC

Frequently, MCMC algorithms have been augmented with an additional tool such as parallel tempering, simulated annealing, or differential evolution, depending on the complexity of the problem. The exoplanet detection problem is particularly challenging because of the large search range in period space coupled with the sparse sampling in time which gives rise to many peaks in the target probability distribution. The goal of FMCMC (Gregory, 2011b) has been to fuse together the advantages of all of the above tools together with a genetic crossover operation in a single, automated MCMC algorithm to facilitate the detection of a global minimum in χ^2 (maximum posterior probability in the Bayesian context).

To achieve this, a unique multistage adaptive control system was developed that automates the tuning of the proposal distributions for efficient exploration of the model parameter space even when the parameters are highly correlated. The FMCMC algorithm is currently implemented in Mathematica using parallelized code and run on an 8-core PC. When implemented with a multiplanet Kepler model,[9] it is able to identify any significant periodic signal component in the data that satisfies Kepler's laws and function as a multiplanet Kepler periodogram.[10]

The adaptive FMCMC is intended as a general Bayesian nonlinear model fitting program. After specifying the model, M_i, the data, D, and priors, I, Bayes' theorem dictates the target joint probability distribution for the model parameters, which is given by

$$p(X|D, M_i, I) = C\ p(X|M_i, I) \times p(D|M_i, X, I) \tag{3.12}$$

where C is the normalisation constant, which is not required for parameter-estimation purposes and X represents the set of model parameters. The term $p(X|M_i, I)$ is the prior probability distribution of X, prior to the consideration of the current data D. The term $p(D|, M_i, I)$ is called the likelihood and is the probability that we would have obtained the measured data D for this particular choice of parameter vector X, model M_i, and prior information I. At the very least, the prior information I must specify the class of alternative models being considered (hypothesis space of interest) and the relationship between the models and the data (how to compute the likelihood). In some simple cases the log of the likelihood is simply proportional to the familiar χ^2 statistic.[11]

3.3.1 Parallel Tempering

An important feature that prevents FMCMC from becoming stuck in a local probability maximum is *parallel tempering* (Geyer, 1991), and re-invented under the name *exchange Monte Carlo* (Hukushima and Nemoto, 1996). Multiple MCMC chains are run in parallel. The joint distribution for the parameters of model M_i, for a particular chain, is given by

$$\pi(X|D, M_i, I, \beta) \propto p(X|M_i, I) \times p(D|X, M_i, I)^{\beta}. \tag{3.13}$$

Each MCMC chain corresponds to a different β, with the value of β ranging from 0 to 1. When the exponent $\beta = 1$, the term on the left-hand side of the equation is the target joint probability distribution for the model parameters, $p(X|D, M_i, I)$. For $\beta \ll 1$, the distribution is much flatter.

In Equation 3.13, an exponent $\beta = 0$ yields a joint distribution equal to the prior. The reciprocal of β is analogous to a temperature: the higher the temperature the broader the distribution. For parameter estimation purposes 8 chains were employed. A representative set of β values is shown in Figure 3.6. At an interval of 10 to 40 iterations, a pair of adjacent chains on the tempering ladder is chosen at random and a proposal made to swap their parameter states. A Monte Carlo acceptance rule determines the probability for the proposed swap to occur (e.g. equation 12.12 in Gregory, 2005b). This swap allows for an exchange of information across the population of parallel simulations. In low β (higher-temperature) simulations, radically different configurations can arise, whereas in higher β (lower-temperature) states, a configuration is given the chance to refine itself.

[9] For multiple planet models, there is no analytic expression for the exact RV perturbation. In many cases, the RV perturbation can be well modelled as the sum of multiple, independent Keplerian orbits, which is what has been assumed in this work.

[10] Following the pioneering work on Bayesian periodograms by Jaynes (1987) and Bretthorst (1988).

[11] For further details of the likelihood function for this type of problem see Gregory (2005a).

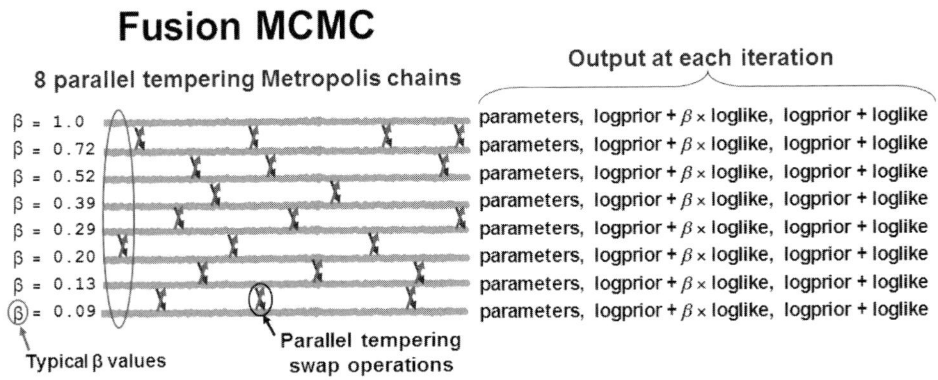

$$\pi(\boldsymbol{X}|D, M_i, I, \beta) \propto p(\boldsymbol{X}|M_i, I) \times p(D|\boldsymbol{X}, M_i, I)^{\beta}$$

FIGURE 3.6 Parallel tempering schematic. Eight parallel chains with $\beta = 1 \ldots 0$ are employed to avoid becoming trapped in a local probability maximum. $\beta = 1$ corresponds to our desired target probability distribution. The others correspond to progressively flatter distributions. At intervals, a pair of adjacent chains is chosen at random and a proposal made to swap their parameter states. The swap allows for an exchange of information across the ladder of chains.

The lower β chains can be likened to a series of scouts that explore the parameter terrain on different scales. The final samples are drawn from the $\beta = 1$ chain, which corresponds to the desired target probability distribution. The choice of β values can be checked by computing the swap acceptance rate. When they are too far apart the swap rate drops to very low values. In this work a typical swap acceptance rate of $\approx 30\%$ was employed but rates in a broad range from 0.15 to 0.5 were deemed acceptable as they did not exhibit any clear differences in performance. For a swap acceptance rate of 30%, jumps to adjacent chains occur at an interval of $\sim$230–920 iterations, while information from more distant chains diffuses much more slowly. Recently, Atchade et al. (2010) have shown that, under certain conditions, the optimal swap acceptance rate is 0.234.

3.3.2 FMCMC Adaptive Control System

At each iteration, a single joint proposal to jump to a new location in the parameter space is generated from independent Gaussian proposal distributions (centred on the current parameter location), 1 for each parameter. In general, the values of σ for these Gaussian proposal distributions are different because the parameters can be very different entities. If the chosen values of σ are too small, successive samples will be highly correlated and require many iterations to obtain an equilibrium set of samples. If the values of σ are too large, the proposed samples will very rarely be accepted. The process of choosing a set of useful proposal values of σ when dealing with a large number of different parameters can be time consuming. In parallel tempering MCMC, this problem is compounded because of the need for a separate set of Gaussian proposal distributions for each tempering chain. This process is automated by an innovative statistical control system (Gregory, 2007b; Gregory and Fischer, 2010) in which the error signal is proportional to the difference between the current joint parameter acceptance rate and a target acceptance rate,[12] λ

[12] (Roberts, Gelman, and Gilks, 1997)

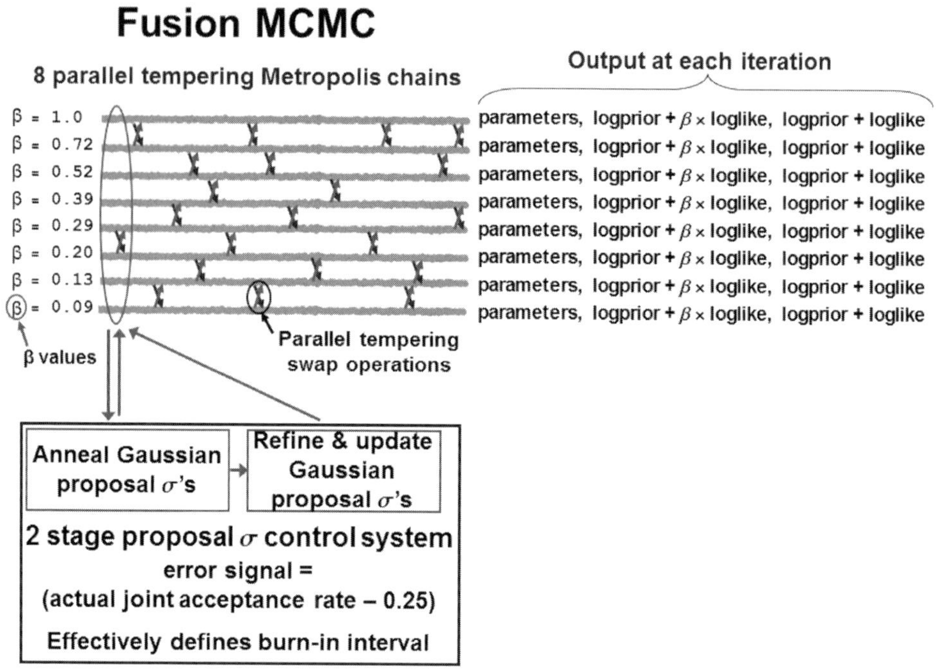

FIGURE 3.7 First two stages of the adaptive control system. The control system depicted shows how to automate the selection of an efficient set of σ values for the independent Gaussian proposal distributions.

(typically $\lambda \sim 0.25$). A schematic of the first two stages of the adaptive control system (CS) is shown in Figure 3.7.[13,14]

The adaptive capability of the control system can be appreciated from an examination of Figure 3.8. The upper-left portion of the figure depicts the FMCMC iterations from the 8 parallel chains, each corresponding to a different tempering level β as indicated on the extreme left. One of the outputs obtained from each chain at every iteration (shown at the far right) is the log prior + log likelihood. This information is continuously fed to the CS which constantly updates the most probable parameter combination regardless of which chain the parameter set occurred in. This is passed to the 'peak parameter set' block of the CS. Its job is to decide whether a significantly more probable parameter set has emerged since the last execution of the second-stage CS. If so, the second-stage CS is re-run using the new, more probable parameter set, which is the basic adaptive feature of the existing CS.[15] Figure 3.8 illustrates how the second stage of the control system is restarted if a significantly more probable parameter set is detected regardless of which chain it occurs in. This also causes the burn-in phase to be extended.

[13] The interval between tempering swap operations is typically much smaller than suggested by this schematic.
[14] Further details on the operation of the CS are given in the supplement to Gregory (2005b, appendix A). This supplement is available in the free resources section of the book website.
[15] Mathematica code that implements a recent version of FMCMC is available on the Cambridge University Press website for my textbook, *Bayesian logical data analysis for the physical sciences* (Gregory, 2005b), in the free resources section. There you will also find additional book examples with a Mathematica 8 tutorial. Non-Mathematica users can download a free Wolfram CDF Player to view the resource material.

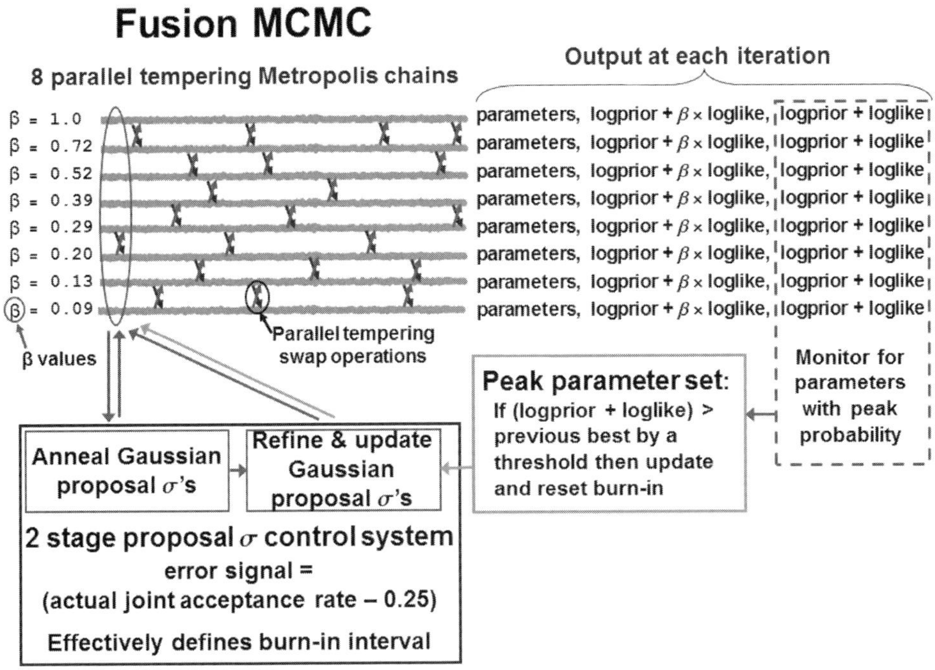

FIGURE 3.8 Schematic illustrating how the second stage of the control system is restarted if a significantly more probable parameter set is detected.

The control system also includes a genetic algorithm block, which is shown in the bottom right of Figure 3.9. The current parameter set can be treated as a set of genes. In the present version, 1 gene consists of the parameter set that specifies 1 orbit. On this basis, a 3-planet model has 3 genes. At any iteration there exists within the CS the most probable parameter set to date $X_{\rm max}$, and the most probable parameter set of the 8 chains $X_{\rm cur}$. At regular intervals (user specified) each gene from $X_{\rm cur}$ is swapped for the corresponding gene in $X_{\rm max}$. If either substitution leads to a higher probability it is retained and $X_{\rm max}$ is updated. The effectiveness of this operation was tested by comparing the number of times the gene crossover operation gave rise to a new value of $X_{\rm max}$ compared to the number of new $X_{\rm max}$ arising from the normal parallel tempering MCMC operations. The gene crossover operations proved to be very effective, and gave rise to new $X_{\rm max}$ values $\approx$1.7 times more often than MCMC operations. It turns out that individual gene swaps from $X_{\rm cur}$ to $X_{\rm max}$ are much more effective (in one test by a factor of 17) than the other way around (reverse swaps). Since it costs just as much time to compute the probability for a swap either way, we no longer carry out the reverse swaps. Instead, we extend this operation to swaps from $X_{\rm cur2}$, the parameters of the second-most probable current chain, to $X_{\rm max}$. This gave rise to new values of $\mathrm{X}_{\rm max}$ at a rate $\sim$ 70% that of swaps from $X_{\rm cur}$ to $X_{\rm max}$. Crossover operations at a random point in the entire parameter set did not prove as effective except in the single-planet case where there is only one gene.

3.3.3 Automatic Simulated Annealing

The annealing of the proposal σ values occurs while the MCMC is homing in on any significant peaks in the target probability distribution. Concurrent with this, another aspect of the annealing operation takes place whenever the Markov chain is started from

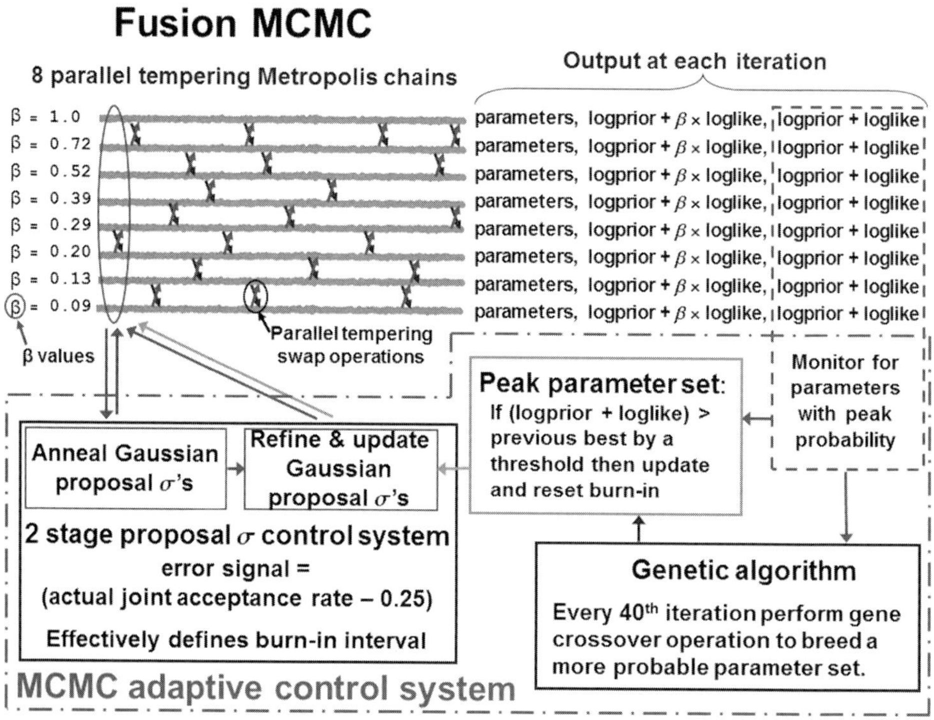

FIGURE 3.9 This schematic shows how the genetic crossover operation is integrated into the adaptive control system.

a location in parameter space that is far from the best fit values. This automatically arises because all the models considered incorporate an extra additive noise (Gregory, 2005a), whose probability distribution is independent and identically distributed Gaussian with zero mean and an unknown standard deviation s. When the χ^2 of the fit is very large, the Bayesian Markov chain automatically inflates s to include anything in the data that cannot be accounted for by the model with the current set of parameters and the known measurement errors. This results in smoothing out the detailed structure in the χ^2 surface and, as pointed out by Ford (2006), allows the Markov chain to explore the large-scale structure in parameter space more quickly. This is illustrated in Figure 3.10, which shows a simulated toy posterior PDF for a single parameter model with (dashed) and without (solid) an extra noise term s. Figure 3.11 shows the behaviour of $\text{Log}_{10}[\text{Prior} \times \text{Likelihood}]$ and s versus MCMC iteration for some real data. In the early stages s is inflated to around 38 m s^{-1} and then decays to a value of $\approx$4 m s^{-1} over the first 9,000 iterations as $\text{Log}_{10}[\text{Prior} \times \text{Likelihood}]$ reaches a maximum. This is similar to simulated annealing, but does not require choosing a cooling scheme.

3.3.4 Highly Correlated Parameters

For some models the data is such that the resulting estimates of the model parameters are highly correlated and the MCMC exploration of the parameter space can be very inefficient. Figure 3.12 shows an example of two highly correlated parameters and possible ways of dealing with this issue, which includes a transformation to a more orthogonal parameter set. It would be highly desirable to employ a method that automatically samples correlated parameters efficiently. One potential solution in the literature is differential

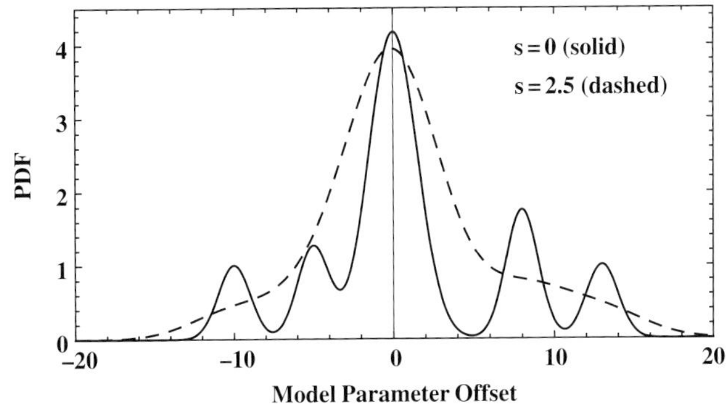

FIGURE 3.10 A simulated toy posterior probability distribution (PDF) for a single parameter model with (dashed) and without (solid) an extra noise term s.

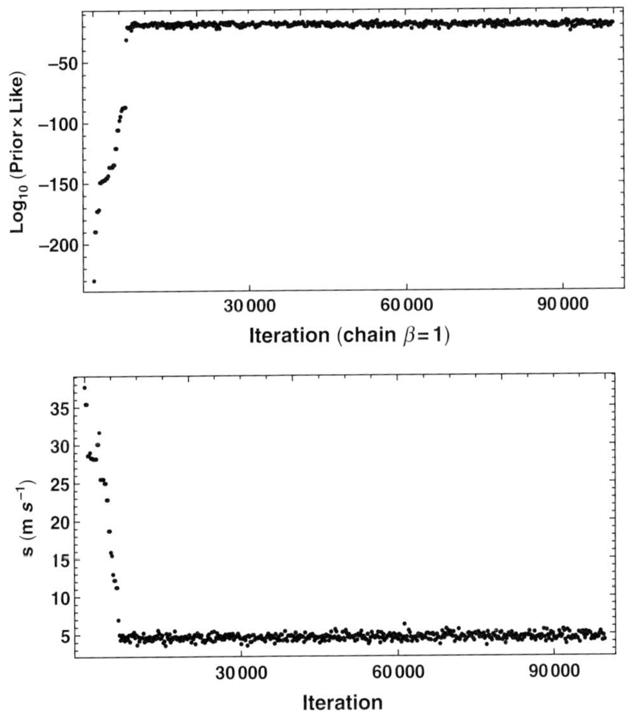

FIGURE 3.11 The upper panel is a plot of the Log_{10}[Prior × Likelihood] versus MCMC iteration. The lower panel is a similar plot for the extra noise term s. Initially s is inflated and then rapidly decays to a much lower level as the best fit parameter values are approached.

evolution Markov chain (DE-MC) (Ter Braak, 2006). DE-MC is a population MCMC algorithm, in which multiple chains are run in parallel, typically from 15 to 40. DE-MC solves an important problem in MCMC, namely that of choosing an appropriate scale and orientation for the jumping distribution.

For the FMCMC algorithm, I developed and tested a new method (Gregory, 2011a), in the spirit of DE, that automatically achieves efficient MCMC sampling in highly correlated parameter spaces without the need for additional chains. The block in the lower-left panel of Figure 3.13 automates the selection of efficient proposal distributions

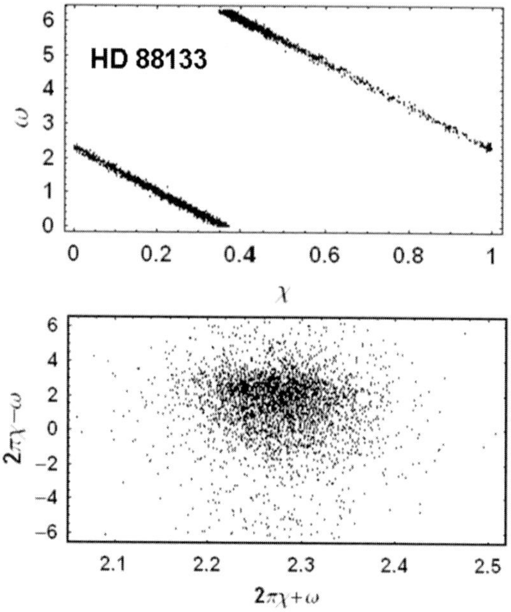

FIGURE 3.12 An example of two highly correlated parameters and possible ways of dealing with this issue, which includes a transformation to a more orthogonal parameter set. The top figure shows an exoplanet example. For low eccentricity orbits, the parameters ω and χ are not separately well determined. This shows up as a strong correlation between ω and χ. The bottom figure shows how the combination $2\pi\chi + \omega$ is well determined for all eccentricities. Although $2\pi\chi - \omega$ is not well determined for low eccentricities, it is at least orthogonal to $2\pi\chi + \omega$ as shown. Another option is to apply an algorithm that learns about the parameter correlations during burn-in and generates proposals with these statistical correlations.

when working with model parameters that are independent or transformed to new independent parameters. New parameter values are jointly proposed based on independent Gaussian proposal distributions ('I' scheme), one for each parameter. Initially, only this 'I' proposal system is used and it is clear that if there are strong correlations between any parameters the σ values of the independent Gaussian proposals need to be very small for any proposal to be accepted and consequently convergence will be slow. However, the accepted 'I' proposals generally cluster along the correlation path. In the optional third stage of the control system every second [16] accepted 'I' proposal is appended to a correlated sample buffer. There is a separate buffer for each parallel tempering level. Only the 300 most recent additions to the buffer are retained. A 'C' proposal is generated from the difference between a pair of randomly selected samples drawn from the correlated sample buffer for that tempering level after multiplication by a constant. The value of this constant (for each tempering level) is computed automatically (Gregory, 2011a) by another control system module which ensures that the 'C' proposal acceptance rate is close to 25%. With very little computational overhead, the 'C' proposals provide the scale and direction for efficient jumps in a correlated parameter space.

The final proposal distribution is a random selection of 'I' and 'C' proposals such that each is employed 50% of the time. The combination ensures that the whole parameter

[16] Thinning by a factor of 10 has already occurred, meaning only every tenth iteration is recorded.

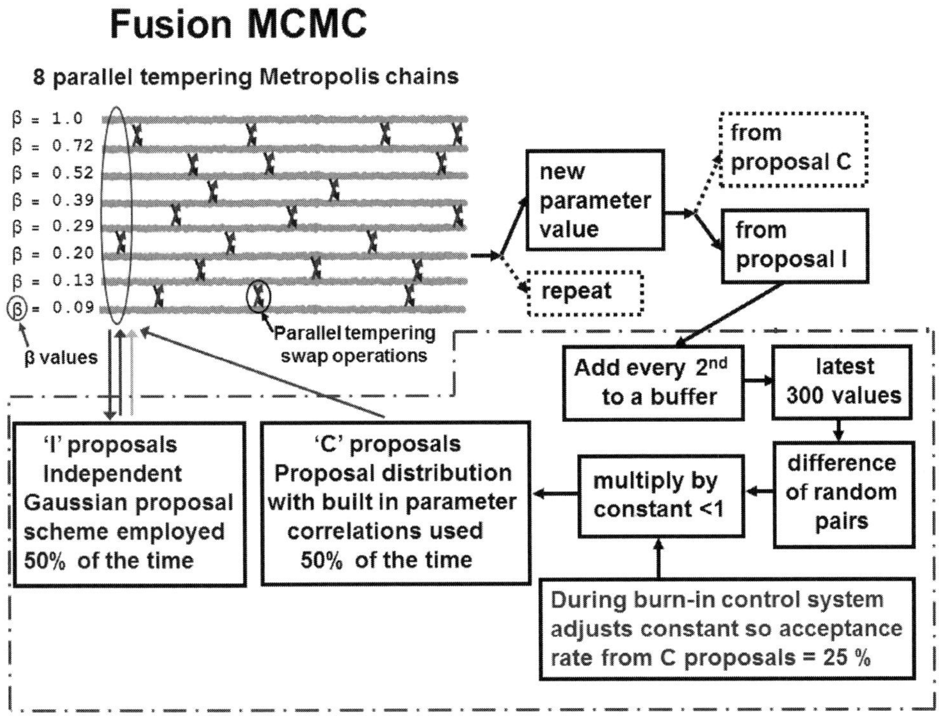

FIGURE 3.13 This schematic illustrates the automatic proposal scheme for handling correlated ('C') parameters.

space can be reached and that the FMCMC chain is aperiodic. The parallel tempering feature operates as before to avoid becoming trapped in a local probability maximum.

Because the 'C' proposals reflect the parameter correlations, large jumps are possible, allowing for much more efficient movement in parameter space than can be achieved by the 'I' proposals alone. Once the first two stages of the control system have been turned off, the third stage continues until a minimum of an additional 300 accepted 'I' proposals have been added to the buffer and the 'C' proposal acceptance rate is within the range ≥ 0.22 and ≤ 0.28. At this point further additions to the buffer are terminated and this sets a lower bound on the burn-in period.

Tests of the 'C' Proposal Scheme

I carried out two tests of the 'C' proposal scheme using (a) simulated exoplanet astrometry data, and (b) a sample of real RV data (Gregory, 2011a). In the latter test I analysed a sample of 17 HD 88133 precision RV measurements (Fischer et al., 2005) using a single-planet model in three different ways. Figure 3.14 shows a comparison of the resulting post-burn-in marginal distributions for two correlated parameters χ and ω, together with a comparison of the autocorrelation functions. The black trace corresponds to a search in χ and ω using only 'I' proposals. The light grey trace corresponds to a search in χ and ω with 'C' proposals turned on. The dark grey trace corresponds to a search in the transformed orthogonal coordinates $\psi = 2\pi\chi + \omega$ and $\phi = 2\pi\chi - \omega$ using only 'I' proposals. It is clear that a search in χ and ω with 'C' proposals turned on achieves the same excellent results as a search in the transformed orthogonal coordinates ψ and ϕ using only 'I' proposals.

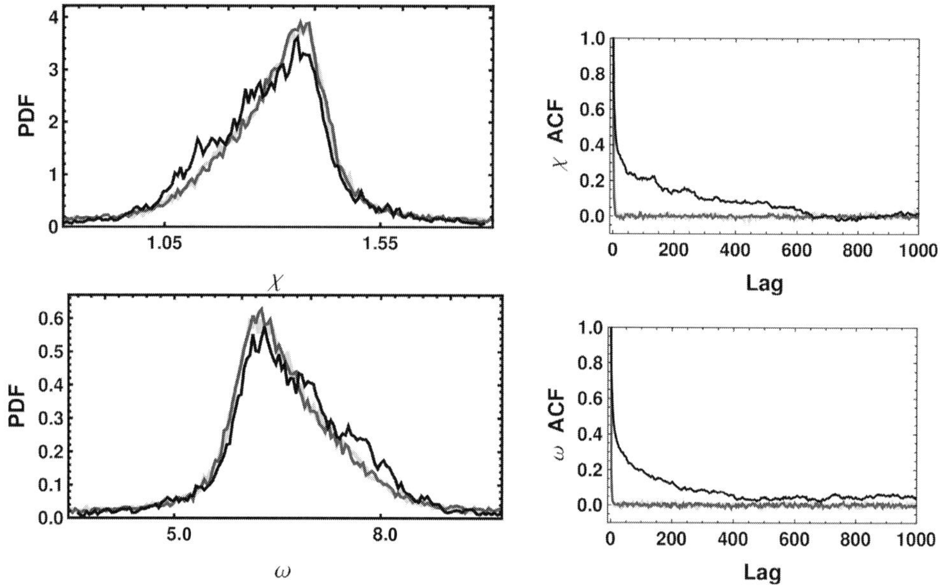

FIGURE 3.14 The two panels on the left show a comparison of the post-burn-in marginal distributions for χ and ω. The two panels on the right show a comparison of their MCMC autocorrelation functions. The solid black trace corresponds to a search in χ and ω using only 'I' proposals. The light grey trace corresponds to a search in χ and ω with 'C' proposals turned on. The dark grey trace corresponds to a search in the transformed orthogonal coordinates $\psi = 2\pi\chi + \omega$ and $\phi = 2\pi\chi - \omega$ using only 'I' proposals.

3.4 Exoplanet Applications

As previously mentioned, the FMCMC algorithm is designed to be a general tool for nonlinear model fitting. When implemented with a multi-planet Kepler model it is able to identify any significant periodic signal component in the data that satisfies Kepler's laws and is able to function as a multiplanet Kepler periodogram. This approach leads to the detection of planetary candidates. One reason to think of them as planetary candidates is because it is known that stellar activity (spots and larger-scale magnetically active regions) can lead to RV artefact signals (e.g. Queloz et al., 2001; Robertson et al., 2014). A great deal of attention is now focussed on correlating stellar activity signals with those found in RV data. Also it is necessary to carry out N-body simulations to establish the long-term stability of the remaining candidate planets.

In this section we describe the model fitting equations and the selection of priors for the model parameters. For a 1-planet model the predicted RV is given by

$$f(t_i) = V + K[\cos\{\theta(t_i + \chi P) + \omega\} + e\cos\omega] \tag{3.14}$$

and involves the 6 unknown parameters:

- V = a constant velocity
- K = velocity semi-amplitude $= \frac{2\pi}{P}\frac{a\ \sin i}{\sqrt{1-e^2}}$, where a = semi-major axis and i = inclination
- P = the orbital period
- e = the orbital eccentricity
- ω = the longitude of periastron

- χ = the fraction of an orbit, prior to the start of data taking, at which periastron occurred, thus, χP = the number of days prior to $t_i = 0$ that the star was at periastron, for an orbital period of P days
- $\theta(t_i + \chi P)$ = the true anomaly – the angle of the star in its orbit relative to periastron at time t_i

We utilise this form of the equation because we obtain the dependence of θ on t_i by solving the conservation of angular momentum equation

$$\frac{d\theta}{dt} - \frac{2\pi[1 + e\cos\theta(t_i + \chi\, P)]^2}{P(1 - e^2)^{3/2}} = 0. \tag{3.15}$$

Our algorithm is implemented in Mathematica and it proves faster for Mathematica to solve this differential equation than solving the equations relating the true anomaly to the mean anomaly via the eccentric anomaly. Mathematica generates an accurate interpolating function between t and θ so the differential equation does not need to be solved separately for each t_i. Evaluating the interpolating function for each t_i is fast compared to solving the differential equation.[17]

We employ a re-parameterization of χ and ω to improve the MCMC convergence speed motivated by the work of Ford (2006). The two new parameters are $\psi = 2\pi\chi + \omega$ and $\phi = 2\pi\chi - \omega$. Parameter ψ is well determined for all eccentricities. Although ϕ is not well determined for low eccentricities, it is at least orthogonal to the ψ parameter. We use a uniform prior for ψ in the interval 0 to 4π and uniform prior for ϕ in the interval -2π to $+2\pi$. This insures that a prior that is wraparound continuous in (χ, ω) maps into a wraparound continuous distribution in (ψ, ϕ). To account for the Jacobian of this re-parameterization it is necessary to multiply the Bayesian integrals by a factor of $(4\pi)^{-nplan}$, where $nplan$ = the number of planets in the model. By utilising the orthogonal combination (ψ, ϕ), there is less need to make use of the 'C' proposal scheme outlined in Section 3.3.4 but to allow for other possible correlations (e.g. a planet with a period greater than the data duration) it is safest to always make use of the 'C' proposal scheme as well.

3.4.1 Exoplanet Priors

In a Bayesian analysis we specify a suitable prior for each parameter. These are tabulated in Table 3.1. For the current problem, the prior given in Equation 3.13 is the product of the individual parameter priors. Detailed arguments for the choice of each prior are given in Appendix 1 (see also Gregory, 2007a; Gregory and Fischer, 2010).

As mentioned in Section 3.3.3, all of the models considered in this chapter incorporate an extra noise parameter s that can allow for any additional noise beyond the known measurement uncertainties.[18] We assume the noise variance is finite and adopt a Gaussian distribution with a variance s^2. Thus, the combination of the known errors and extra noise has a Gaussian distribution with variance $= \sigma_i^2 + s^2$, where σ_i is the standard deviation of the known noise for ith data point. For example, suppose that the star actually has 2 planets but the model assumes only one is present. In regard to the single-planet model, the velocity variations induced by the unknown second planet act like an

[17] Details on how Equation 3.15 is implemented are given in Gregory (2011b).

[18] In the absence of detailed knowledge of the sampling distribution for the extra noise, we pick a Gaussian because for any given finite noise variance it is the distribution with the largest uncertainty as measured by the entropy, i.e. the maximum entropy distribution Jaynes (1957) and Gregory (2005b, section 8.7.4).

TABLE 3.1. Prior parameter probability distributions

Parameter	Prior	Lower Bound	Upper Bound
Orbital frequency	$p(\ln f_1, \ln f_2, \ldots \ln f_n \mid M_n, I) = \frac{n!}{[\ln(f_H/f_L)]^n}$ (n = number of planets)	1/0.5 d	1/1000 yr
Velocity K_i (m s^{-1})	Modified scale invariant [a] $\frac{(K+K_0)^{-1}}{\ln\left[1+\frac{K_{\rm max}}{K_0}\left(\frac{P_{\rm min}}{P}\right)^{1/3}\frac{1}{\sqrt{1-e^2}}\right]}$	0 ($K_0 = 1$)	$K_{\rm max}\left(\frac{P_{\rm min}}{P}\right)^{1/3}\frac{1}{\sqrt{1-e^2}}$ $K_{\rm max} = 2129$
V (m s^{-1})	Uniform	$-K_{\rm max}$	$K_{\rm max}$
e Eccentricity	$3.1(1-e)^{2.1}$	0	0.99
χ Orbit fraction	Uniform	0	1
ω Longitude of periastron	Uniform	0	2π
s Extra noise (m s^{-1})	$\frac{(s+s_0)^{-1}}{\ln\left(1+\frac{s_{\rm max}}{s_0}\right)}$	0 ($s_0 = 1$–10)	$K_{\rm max}$

[a] Since the prior lower limits for K and s include 0, we used a modified scale invariant prior of the form

$$p(X|M,I) = \frac{1}{X + X_0}\,\frac{1}{\ln\left(1 + \frac{X_{\rm max}}{X_0}\right)}. \tag{3.16}$$

For $X \ll X_0$, $p(X|M,I)$ behaves like a uniform prior and for $X \gg X_0$ it behaves like a scale invariant prior.

The $\ln\left(1 + \frac{X_{\rm max}}{X_0}\right)$ term in the denominator ensures that the prior is normalized in the interval 0 to $X_{\rm max}$.

additional unknown noise term. Other factors like star spots and chromospheric activity also contribute to this extra-velocity noise term, which is often referred to as *stellar jitter*. In general, nature is more complicated than our model and known noise terms. Marginalizing s has the desirable effect of treating anything in the data that cannot be explained by the model and known measurement errors as noise, leading to conservative estimates of orbital parameters.[19] If there is no extra noise then the posterior probability distribution for s peaks at $s = 0$. The upper limit on s was set equal to $K_{\rm max}$. This is much larger than the estimates of stellar jitter for individual stars based on statistical correlations with observables (e.g. Saar and Donahue, 1997; Saar, Butler, and Marcy, 1998; Wright, 2005).

In our Bayesian analysis, s serves two purposes. First it allows for an automatic simulated annealing operation as described in Section 3.3.3 and for this purpose it is desirable to have a much larger range. The final s value after the annealing is complete provides a crude measure of the residuals that cannot be accounted for by the model

[19] See sections 9.2.3 and 9.2.4 of Gregory (2005b) for a tutorial demonstration of this point.

and known measurement uncertainties. Of course, the true residuals exhibit correlations if there are additional planets present not specified by the current model. In addition, stellar activity RV artefacts can lead to correlated noise and a number of attempts are being explored to jointly model the planetary signals and stellar activity diagnostics (e.g. Queloz et al., 2009; Aigrain, Pont, and Zucker, 2012; Tuomi et al., 2013; Haywood et al., 2014). These correlations are not accounted for by this simple additional noise term. We use the same prior range for s for all the models ranging from the 0-planet case to the many-planet case. We employed a modified scale invariant prior for s with a knee s_0 in the range 1–10 m s^{-1}, according to Equation 3.15.

3.4.2 HD 208487 Example

In 2007, using an automatic Bayesian multiplanet Kepler periodogram, I found evidence for a second planetary candidate with a period of ~900 d in HD 208487 (Gregory, 2007a). We use this as an example data set to illustrate a number of issues that can arise in the analysis using an FMCMC-powered, multiplanet Kepler periodogram. Figure 3.15 shows sample FMCMC traces for the two-planet fit to the 35 RV measurements (Tinney et al., 2005) for HD 208487 based on our latest version of the FMCMC algorithm and employing the updated eccentricity prior. The top-left panel is a display of the Log$_{10}$ [Prior × Likelihood] versus FMCMC iteration number. In total, 5×10^5 iterations were executed and only every tenth value saved. It is clear from this trace that the burn-in period is very short. In this example the control system ceased tuning 'I' and 'C' proposal distributions at iteration 3,220. The top-right panel shows the trace for the extra noise parameter. During the automatic annealing operation it dropped from a high around 18 m s^{-1} to an equilibrium value of around 1 m s^{-1} within the burn-in period. As explained in Appendix 1, it is more efficient to allow the individual orbital frequency parameters to roam over the entire frequency space and re-label afterwards so that parameters associated with the lower frequency are always identified with planet 1 and vice versa. In this approach nothing constrains f_1 to always be below f_2 so that degenerate parameter peaks often occur. This behaviour can be seen clearly in the panels for P_1 and P_2, where there are frequent transitions between the 130 and 900 d periods. The lower 4 panels show corresponding transitions occurring for the other orbital parameters. The traces shown in Figure 3.15 are before relabelling and those in Figure 3.16 after.

Figure 3.17 illustrates a variety of 2-planet periodogram plots for the HD 208487 data. The top left shows the evolution of the 2-period parameters (after relabelling) from their starting values, marked by the 2 dots that occur before the 0 on the iteration axis. With the default scale, invariant, orbital-frequency prior only 2 periods were detected. The top-right panel shows a sample of the 2-period parameter values versus a normalised value of Log$_{10}$[Prior × Likelihood]. The bottom left shows a plot of eccentricity versus period and the bottom right K versus eccentricity. It is clear from the latter plot that eccentricity values are heavily concentrated around a value of 0.2 for the 130 d signal. The distribution of eccentricity for the secondary period is broader but the greatest concentration is towards low values. The MAP values are shown by the filled black circles. The combination of the lower 2 panels indicates that the eccentricity of the secondary period is lowest towards the 800 d end of the period range.

Figure 3.18 shows a plot of K versus eccentricity for a 1-planet fit to the HD 208487 data for comparison. Clearly the single-planet fit finds a larger K value for the dominant 130 d period. The MAP solution is shown by the filled black circle. Even for the single-planet fit there is a preference for an eccentricity of ~0.2. Panel (a) of Figure 3.19 shows the RV data by Tinney et al. (2005). Panels (b) and (c) show the 2-planet fit

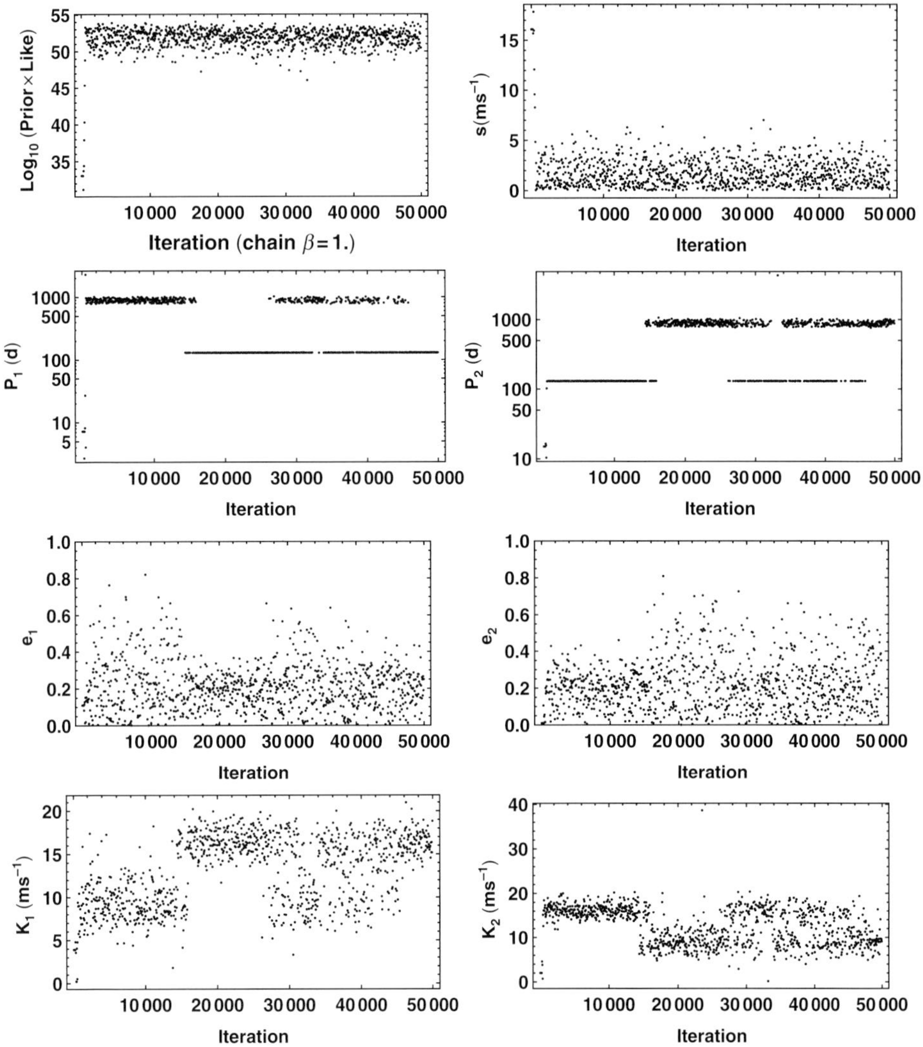

FIGURE 3.15 Sample FMCMC traces for a 2-planet fit to HD 208487 RV data before relabelling.

to the data and the fit residuals, respectively. Figure 3.20 shows the FMCMC marginal distributions for a subset of the 2-planet-fit parameters for HD 208487. The dominant 130 d peak prefers a modest eccentricity of ∼0.2. The secondary period of 900 d exhibits a broader eccentricity range with a preference for low values. The marginal for the extra noise parameter peaks near 0.

Independent analysis of this data by Wright (2007) found two possible solutions for the second period at 27 and 1,000 d. Restricting the period range for the second period in our analysis to a range that excluded the 900 d period confirmed a feature near their shorter period value. We subsequently investigated the effect of redoing our analysis with a frequency prior $p(f|M, I) \propto 1/\sqrt{f}$ to give more weight to shorter-period signals. This resulted in the parallel tempering jumping between 28 and 900 d periods for the secondary period. The periodogram plots for the frequency prior $p(f|M, I) \propto 1/\sqrt{f}$ are shown in Figure 3.21. Because there are now three periods present, the dominant signal

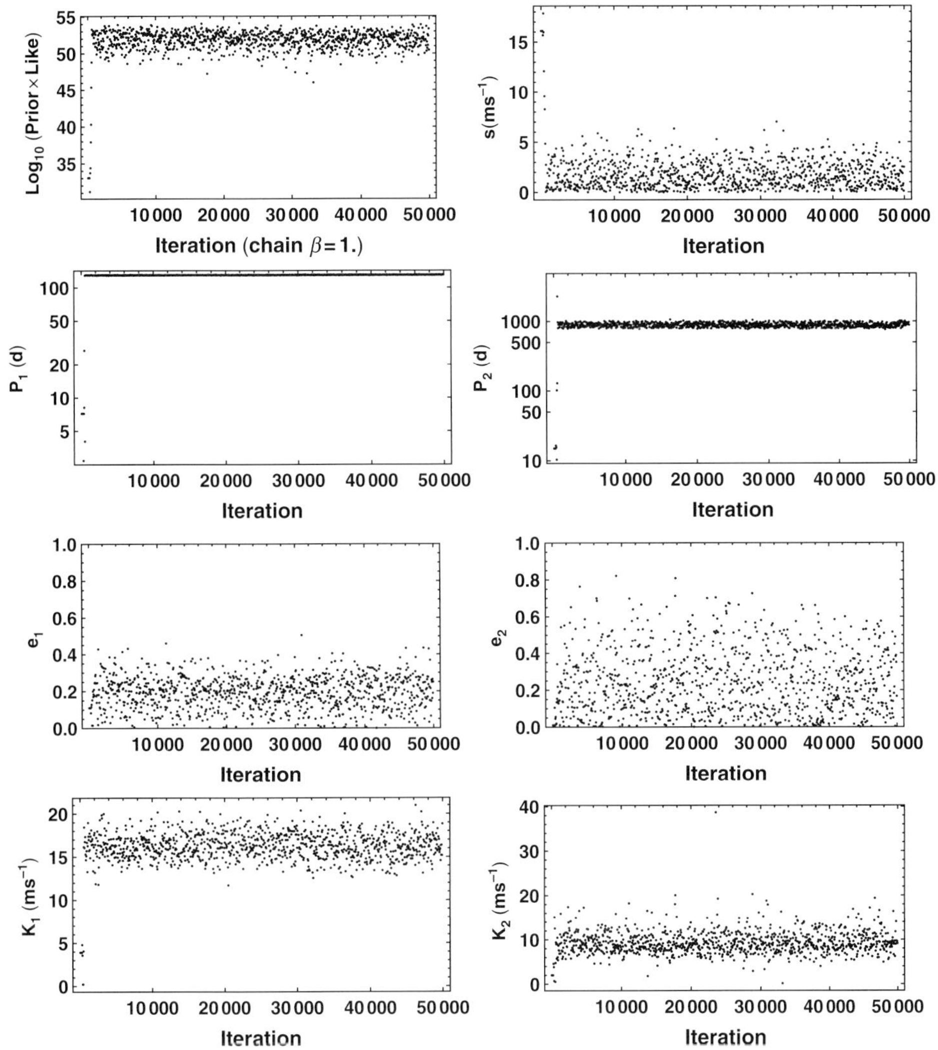

FIGURE 3.16 Sample FMCMC traces for a 2-planet fit to HD 208487 RV data after relabelling.

($P = 130$ d) does not have a unique shade as it pairs first with the longer-period signal ($p = 900$ d) and then with the shorter period ($p = 28$ d). We now find it useful to employ both choices of priors during the search for interesting candidate planetary systems. It is only in the model-comparison phase that we strictly employ a scale-invariant frequency prior to allocate the relative probabilities of two different 2-planet models with different combinations of periods.

In the right panel of Figure 3.21, the black stars are samples of the 28 d period and the upper black stars are the corresponding samples of the dominant 130 d period. Similarly, the large boxes are samples of the 900 d period and the small boxes are the corresponding samples for the 130 d period. The MAP values are shown by the filled black circles. Both the 28 and 900 d samples have their highest concentrations at low values of eccentricity, where the average K values for the 28 and 900 d periods are $\sim$8 m s^{-1} and $\sim$9 m s^{-1}, respectively.

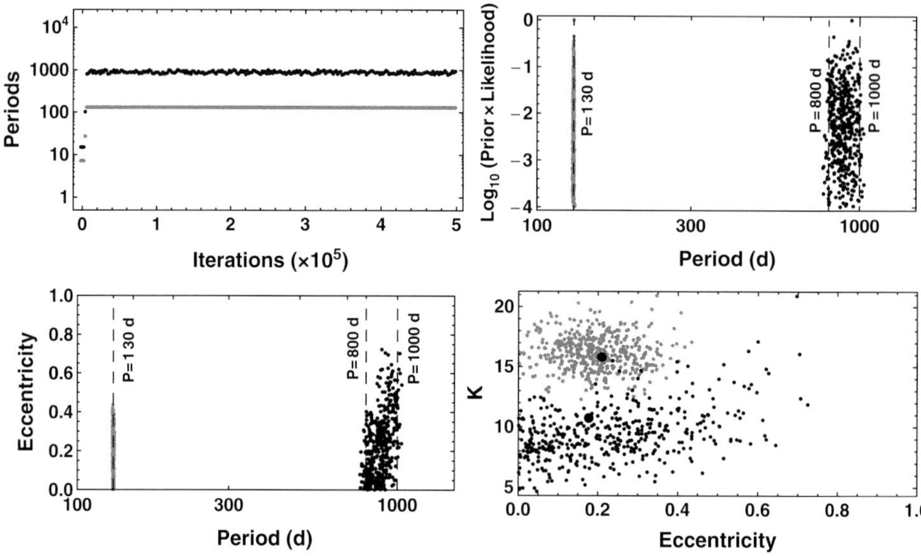

FIGURE 3.17 A variety of 2-planet periodogram plots for HD 208487.

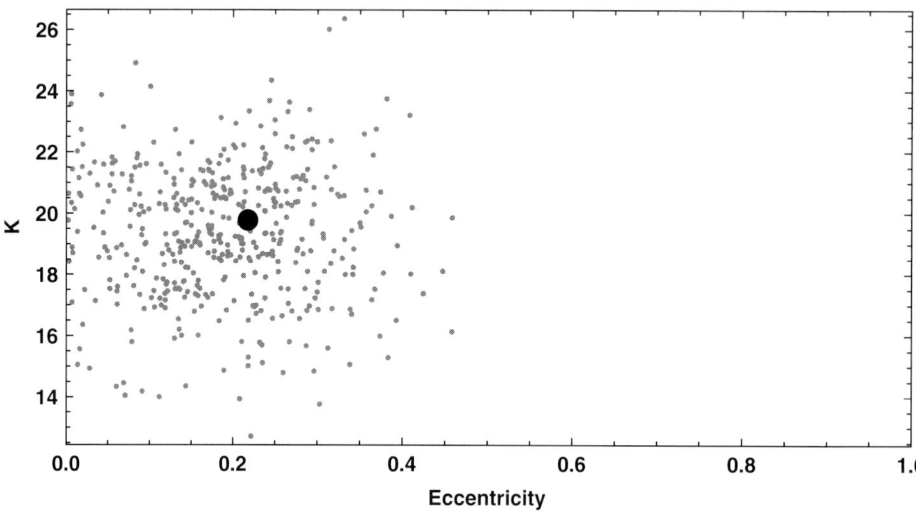

FIGURE 3.18 A K versus eccentricity plot for a 1-planet fit for HD 208487.

The extra noise parameter for the 2-planet fit peaks at 0, which indicates that there is no additional signal to be accounted for. For consistency purposes, a three-planet model was run using a frequency prior $\propto 1/\sqrt{f}$. Both the 130 d and 900 d signals were clearly detected. A wide range of third period options were observed but did not include a clear detection of the 28 d signal.

3.4.3 Aliases

Dawson and Fabrycky (2010) drew attention to the importance of aliases in the analysis of RV data even for nonuniform sampling. Although the sampling is nonuniform when the star is observable, the restrictions imposed by the need to observe between sunset and

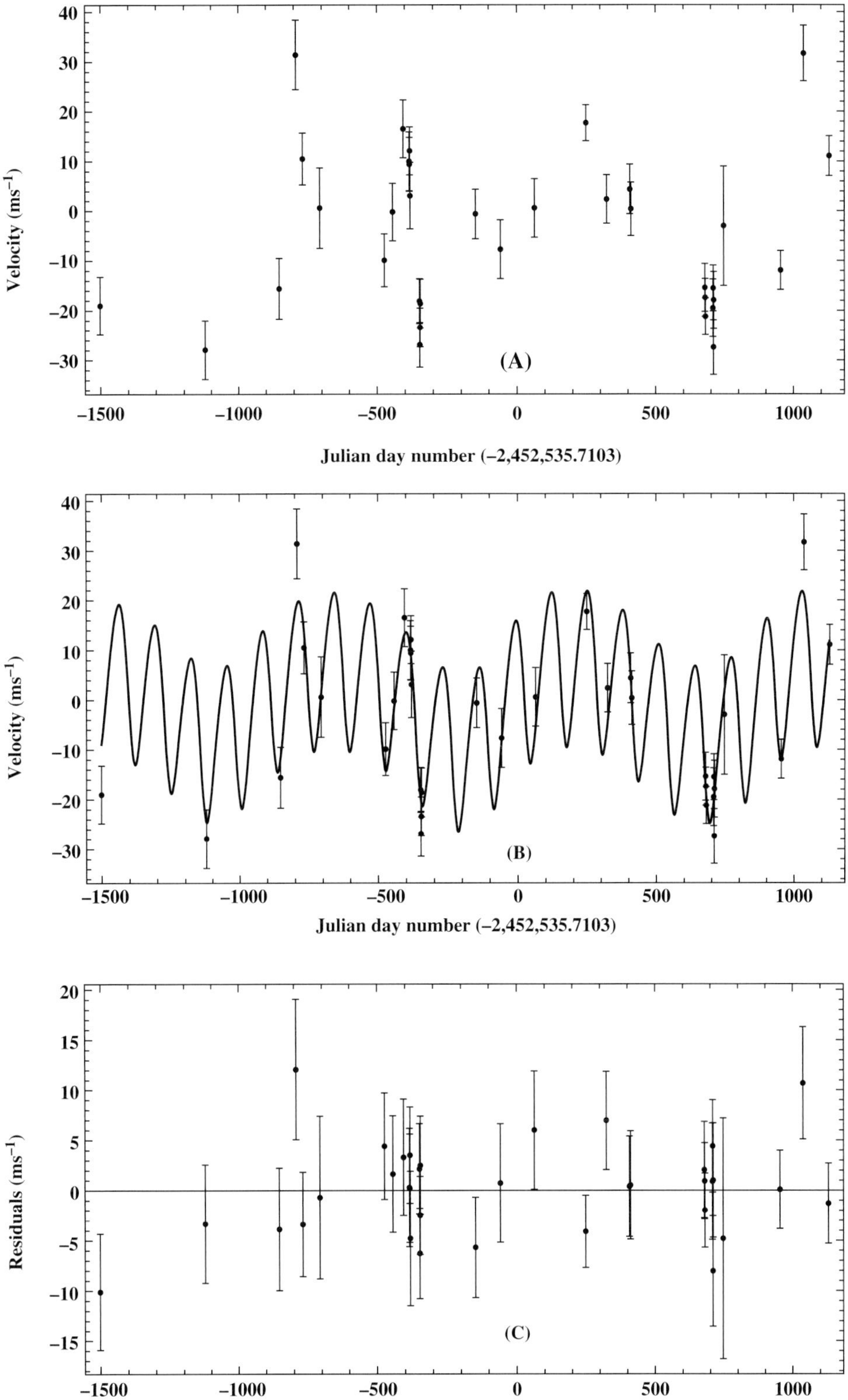

FIGURE 3.19 Panel (A) shows the RV data by Tinney et al. (2005). Panel (B) and (C) show the 2-planet fit to the data and the fit residuals, respectively.

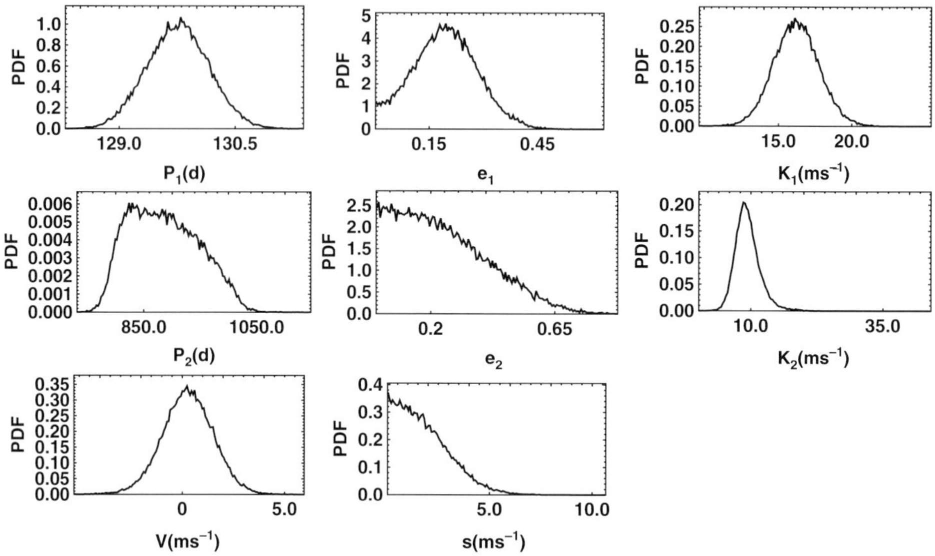

FIGURE 3.20 A plot of a subset of the FMCMCparameter marginal distributions for the 2-planet fit for the HD 208487 data.

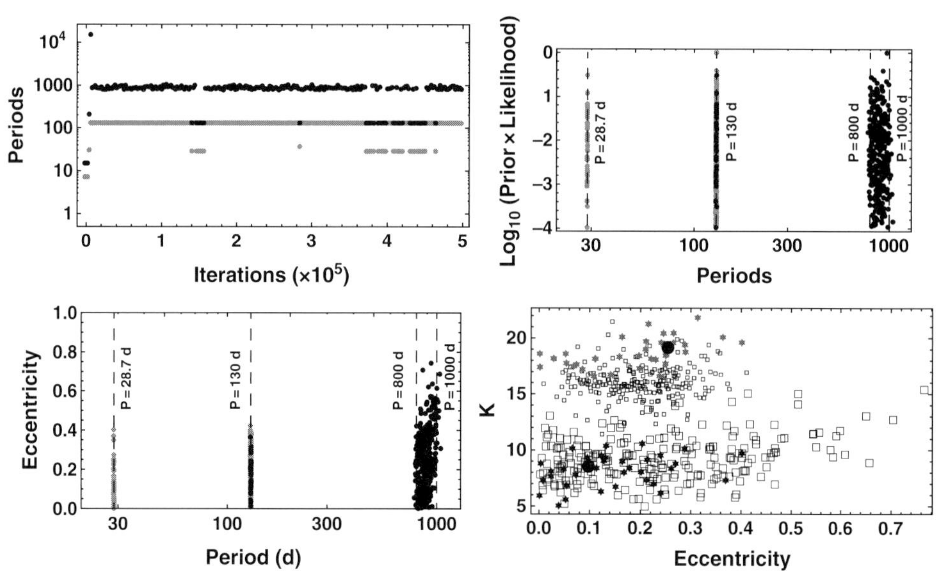

FIGURE 3.21 Two-planet periodogram plots for HD 208487 using a frequency prior $\propto 1/\sqrt{f}$.

sunrise, and when the star is above the horizon, means that there are periodic intervals of time where no sample is possible. These periodicities give rise to aliases, which we investigate in this section. Figure 3.22 shows the location of the HD 208487 RV samples (Tinney et al., 2005) modulo time of day and time of year using $t = JD - 2,451,224.19$ for convenience.

Deeming (1975 and 1976) showed that a discrete Fourier transform (F_N) can be defined for arbitrary sampling of a deterministic time series (including a periodic function) and

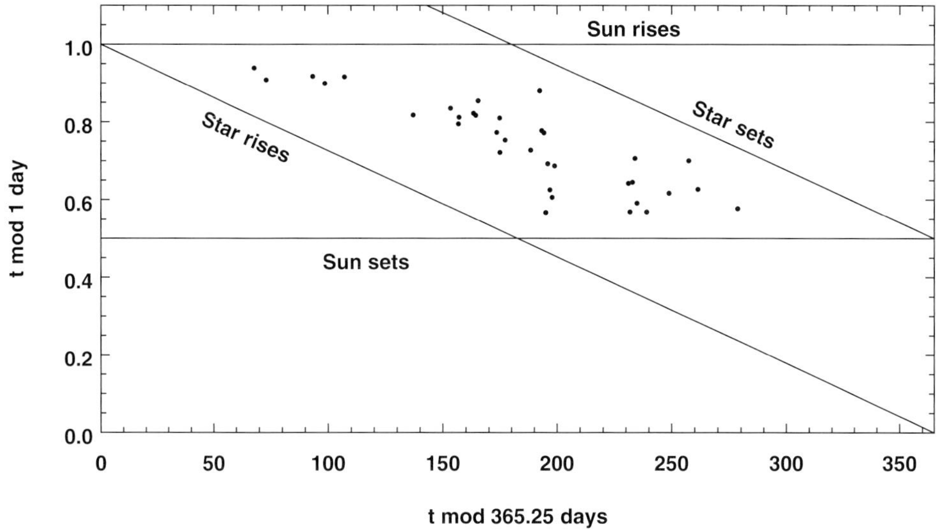

FIGURE 3.22 Times of HD 208487 RV observations folded modulo the time of day and time of year. We used $t = JD - 2,451,224.19$ for convenience.

that F_N is equal to the convolution[20] of the true Fourier transform (FT) with a spectral window function

$$F_N(f) = F(f) * w(f) \equiv \int_{-\infty}^{\infty} F(f - f')w(f')df', \tag{3.17}$$

where $F(f)$, and true FT of the continuous time series is given by

$$F(f) = \int_{-\infty}^{\infty} f(t)e^{i2\pi ft}. \tag{3.18}$$

If $f(t)$ is a pure cosine function of frequency f_0, then $F(f)$ will be a pair of Dirac delta functions $\delta(f \pm f_0)$. $w(f)$ is the spectral window function given by

$$w(f) = \sum_{k=1}^{N} e^{i2\pi ft_k}. \tag{3.19}$$

It is also evident that $w(-f) = w^*(f)$. In the limit of continuous sampling in the time interval $(-T/2, +T/2)$, $w(f)$ tends to the Dirac delta function $\delta(0)$ as $t \to \infty$. In general, the sampling will not be continuous so $w(f)$ may be significantly different from 0 at frequencies other than $f = 0$. For N evenly sampled data, unless the only physically possible frequencies are less than the Nyquist frequency $= N/(2T)$, the frequencies cannot be unambiguously determined. An important point is that $w(f)$ can be calculated directly from the data spacing alone.

It is common practice to use a normalised spectral window function

$$W(f) = N^{-1} \sum_{k=1}^{N} e^{i2\pi ft_k}, \tag{3.20}$$

[20] Such a convolution is nicely illustrated in Gregory, 2005b, figures 13.6 and B.2.

normalised so $W(0) = 1$. In this case, the convolution Equation 3.17 becomes

$$\frac{1}{N}F_N(f) = F(f) * W(f) \equiv \int_{-\infty}^{\infty} F(f - f')W(f')df'. \quad (3.21)$$

In general both $F(f)$ and $W(f)$ are complex so it is usual to examine the amplitude[21] of $W(f)$. Also, Dawson and Fabrycky (2010) show how the phase information is sometimes useful to distinguish between an alias and a physical frequency. Figure 3.23 shows a plot of the amplitude of the spectral window function, $W(f)$. Clearly with such sparse sampling there are many peaks in $W(f)$. The plots show a well-defined peak at $f = 0$ with a width of order T^{-1} as well as strong peaks at 1 sidereal day, 1 solar day, the synodic month (29.53 d), and $32.13 = 1/(1/29.53 - 1/365.25)$ d.

Most exoplanet periodogram analysis makes use of the data weighted by $wt_k = 1/\sigma_k^2$ so it is more appropriate to employ a spectral window function of the form

$$W(f) = wt_{\text{sum}}^{-1} \sum_{k=1}^{N} wt_k \; e^{i2\pi f t_k}, \quad (3.23)$$

where $wt_{\text{sum}} = \sum_{k=1}^{N} wt_k$. The dominant spectral features at 1 day, the synodic month, and one year remain.

Now that we understand the $W(f)$ for HD 208487, we can see how the 28 d period could be an alias of the 900 d period and vice versa. If the true spectrum contains a physical signal of frequency f, then when convolved with $W(f)$ we expect to see aliases in the observed spectrum at $f \pm f_w$. If $f_w > f$, we still see a positive frequency alias at $|f - f_w|$ but to see this you need to recall that $W(f)$ has negative frequency peaks that satisfy $W(-f) = W^*(f)$. Also, the true spectrum of a periodic signal likewise has positive and negative frequencies when we use the common exponential notation as employed in the Fourier transform defined by Equation 3.17.

Suppose the 28 d period is the true signal. The 68% credible region boundaries of the 28 d peak extend from 28.58 to 28.70 d. One of the aliases produced by the synodic month feature (29.53 d) in $W(f)$ would be expected to give rise to an alias somewhere in the range $1/(1/28.58–1/29.53) = 890$ d to $1/(1/28.70–1/29.53) = 1,026$ d. This range nicely overlaps the 68% credible region of 804–940 d of the 900 d peak. Similarly, if the 900 d signal is the true signal, its synodic-month alias should be found in the 68% credible range 28.48–28.63 d, which it does.

3.4.4 Which Is the Alias?

In this section we attempt to answer the question of which of the two secondary Kepler solutions at 28 and 900 d is a real physical signal. Below we consider some criteria that have proven useful:

(i) Dawson and Fabrycky (2010) outline a method for helping distinguish a physical signal from an alias which makes use of the generalized Lomb-Scargle (GLS) periodogram (Zechmeister and Kürster, 2009). The method involves comparing the

[21] Deeming (1975) shows that for a deterministic signal (as opposed to a stochastic signal)

$$\frac{1}{N^2}|F_N(f)|^2 = \frac{1}{N^2}F_N(f)F_N^*(f) = [F(f) * W(f)][F^*(f) * W^*(f)]$$
$$\neq [F(f)F^*(f)] * [W(f)W^*(f)]. \quad (3.22)$$

It is thus not meaningful to plot $|W(f)|^2$.

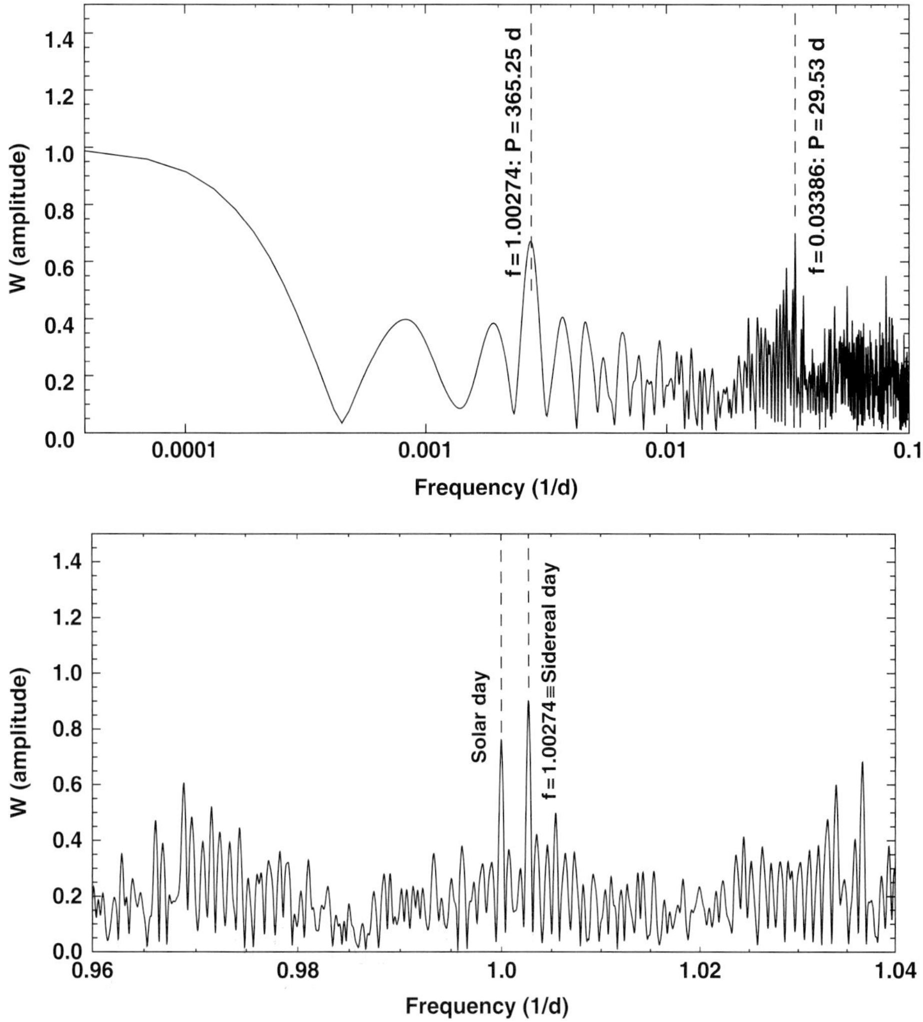

FIGURE 3.23 Amplitude of the spectral window function for the HD 208487 RV measurements.

periodogram of the true residuals of an n signal fit to periodograms of noise-free simulations of possible choices of the $n+1$ signal and includes a comparison of the GLS phase of each spectral peak. GLS improves on the Lomb-Scargle method by allowing for a floating offset and weights.

We illustrate the general idea of the Dawson-Fabrycky method in Figure 3.25 as it applies to our HD 208487 analysis for aliases arising from the strongest peaks in the window function. The top row shows three portions of GLS periodogram of the 1-planet MAP fit residuals with phase circle above several peaks of interest. In the first two columns, dashed lines show the locations of the 854 and 28 d candidate signals together with 1-year aliases, and the dotted lines show 1 synodic month aliases of the 28 d candidate signal (overlaps the 854 d signal) and 854 d signal. In the third column, the dashed lines show the aliases of the 1 solar day and 1 sidereal day window function peaks with the 28 d candidate signal, and the dotted lines are for the 854 d candidate signal. In the left two columns, the strongest peaks are for our two candidates for the physical signal, ~854 and 28.65 d, respectively.

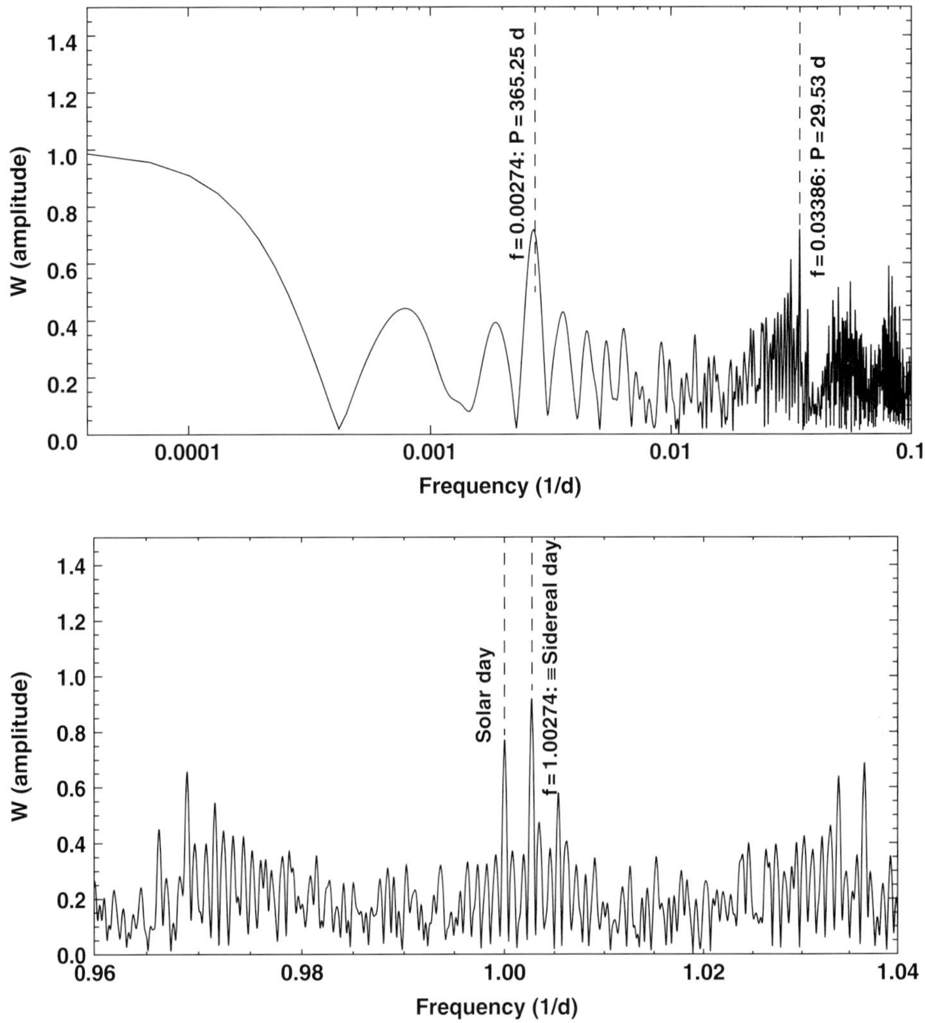

FIGURE 3.24 Amplitude of the weighted spectral window function for the HD 208487 RV measurements.

If the 28 d peak corresponds to the physical signal, then when convolved with the spectral window function[22] shown in Figures 3.23 and 3.24 additional aliases would be expected at $26.57 = 1/(1/28.65 + 1/365.25)$ and $31.09 = 1/(1/28.65 - 1/365.25)$. The alias at 26.57 d is clearly present but not the expected peak at 31.09. If the 854 d peak corresponds to the physical signal, then additional aliases would be expected near $256 = 1/(1/365.25 + 1/854)$, $638 = 1/(1/365.25 - 1/854)$ and $30.59 = 1/(1/29.53 - 1/854)$. The peak near to 256 d is at $267 = 1/(1/365.25 + 1/1{,}000)$ or just within the period uncertainty. There is no clear evidence for a peak near 638 d. In the third column the dashed lines show the four aliases of the 28 d candidate

[22] The dominant peaks in the spectral window function are at 1 sidereal day, one synodic month (29.53 d), and 1 year.

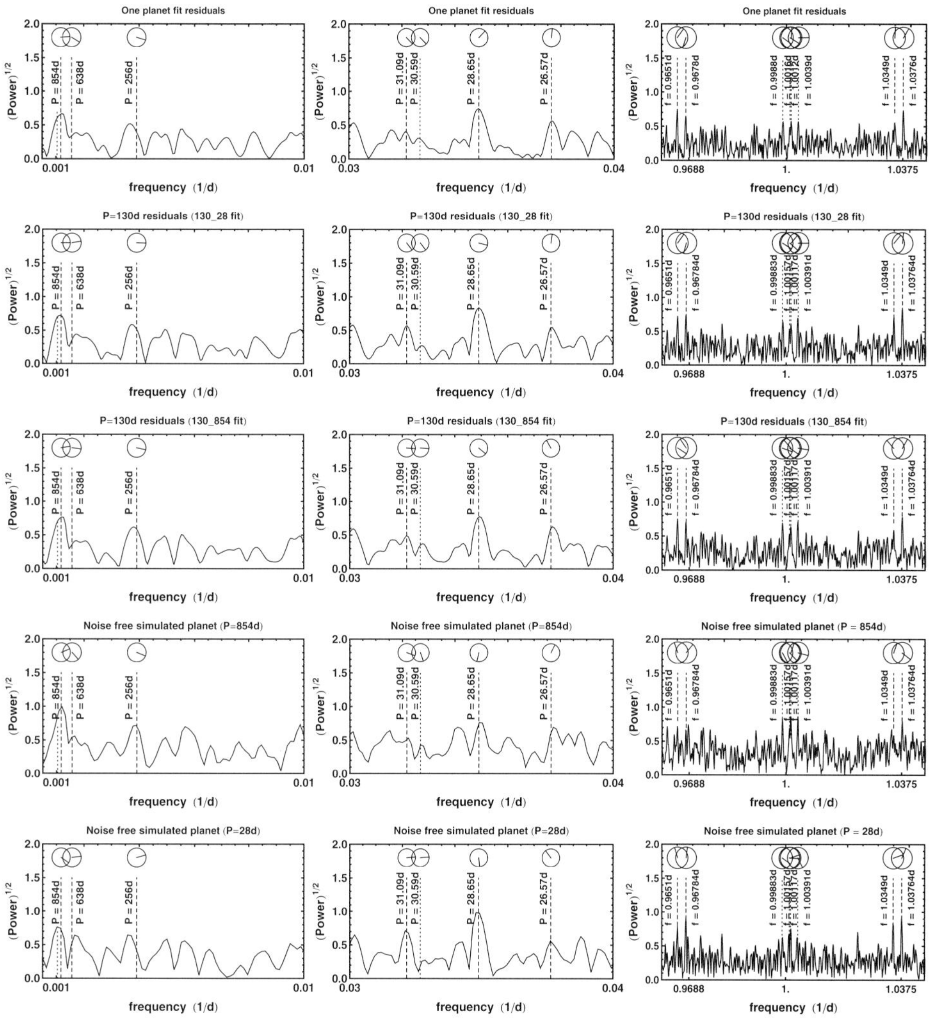

FIGURE 3.25 The Dawson-Fabrycky method for distinguishing an alias from a physical signal. Top row shows 3 portions of GLS periodogram of the 1-planet MAP fit residuals with a phase circle above the peaks of interest. In the first two columns, dashed lines show the locations of the 854 and 28 d candidate signals together with 1-year aliases, and the dotted lines show one synodic month aliases of the 28 d candidate signal (overlaps the 854 d signal) and 854 d signal. In the third column, the dashed lines show the aliases of the 1 solar day and one sidereal day window function peaks with the 28 d candidate signal, and the dotted lines are for the 854 d candidate signal. The second row shows 3 portions of GLS periodogram of the $P = 130$ d planet residuals obtained from a 2-planet ($P = 130$ d and $P = 28$ d) fit. The third row is a similar GLS periodogram of the $P = 130$ d planet residuals obtained from a 2-planet ($P = 130$ d and $P = 854$ d) fit. Row 4 is a GLS periodogram of a noise-free simulation of the 854 d signal. Row 5 is a GLS periodogram of a noise-free simulation of the 28 d signal.

signal, two for the 1 solar day and two for the 1 sidereal day peaks in the window function. Smaller peaks are just discernible for aliases of the 854 d candidate signal.

As shown in Figures 3.17 and 3.18, the typical K value of the dominant 130 d signal in the K versus eccentricity plot for the 1-planet fit is considerably higher than for the 2-planet fits regardless of whether the second real physical signal is 28 or 854 days. This suggests that it might be useful to construct two other typical 130 d planet residuals as show in rows 2 and 3. The second row shows the same 2 portions of GLS periodogram of the $P = 130$ d planet residuals obtained from a 2-planet ($P = 130$ d and $P = 28$ d) fit. The third row is for the $P = 130$ d planet residuals obtained from a 2-planet ($P = 28$ d and $P = 854$ d) fit. In the second row, a feature corresponding to the expected alias at 31.09 d increased in height, lending more support to 28 d as the physical signal. In the third row, the feature nearest the expected 256 d alias of a possible 854 d physical signal is stronger, peaking at $\sim 1/(1/365.25 + 1/940)$. However, there is still no clear feature near 638 d.

Row 4 is a GLS periodogram of a noise-free simulation of the 854 d signal and row 5 is a GLS periodogram of a noise-free simulation of the 28 d signal. An examination of all the rows in Figure 3.25 indicates that the 28-day aliases are slightly stronger. This analysis does not lead to a definite conclusion as to which is the true signal but slightly favours the 28 d candidate.

(ii) Both exhibit a preference for low eccentricities and the K value of the 900 d signal is slightly stronger. On this grounds it is more likely to be the real signal. However, noise may add coherently to the alias, causing the alias to be stronger.

(iii) In Section 3.5 we show how to do Bayesian model comparison using the Bayes factor. The Bayes factor favours a 2-planet Kepler model with periods of 130 and 900 d by a factor of 9.0 over the 28, 130 d combination. We also show that the Bayesian false-alarm probability for a 2-planet model (regardless of the true second period) is 4.4×10^{-3}, but that the false-alarm probability for the best candidate 2-planet model (130 and 900 d) is too high at 0.10 to conclude that the 900 d signal is the correct second signal.

(iv) Can we form a long-term, stable 2-planet system together with the 130 d Kepler orbit? Clearly if only 1 choice is viable this argues it is the real signal. An approximate Lagrange stability[23] analysis was carried out for both HD 208487 solutions, following the work of Tuomi (2012), which in turn is based on the work of Barnes and Grenberg (2006). This analysis indicates that both choices appear to be long-term stable, but full-scale numerical integrations are needed to confirm this.

3.4.5 Gliese 581 Example

In Section 3.5 we intercompare 3 different Bayesian methods for model comparison, making use of our FMCMC model fits to precision RV data for a range of models from 1 to 5 planets. In anticipation of this we re-analysed the latest HARPS data (Forveille et al., 2011) for Gliese 581 (Gl 581) for models spanning the range 3 to 6 planets using our latest version of FMCMC together with the priors discussed in Section 3.4.1. Gl 581 is an M dwarf 0.31 times the mass of the Sun at a distance of 20 light years, which received a lot of attention because of the possibility of 2 super-earths in the habitable zone where liquid water could exist (Vogt et al., 2010). Our earlier Bayesian analysis (Gregory, 2011b) of the HARPS (Mayor et al., 2009) and HIRES data (Vogt et al., 2010) did not support the detection of a second habitable-zone planet known at the time as Gl 581g. Subsequent

[23] Work in the 1970s and 1980s showed that the motions of a system of a star with 2 planets (not involved in a low-order mean motion resonance) would be bounded in some situations. Two dominant definitions of stability emerged: Hill stability and Lagrange stability. In Hill stability the ordering of the 2 planets in terms of distance from the central star is conserved. In addition, for Lagrange stability the planets remain bound to the star and the semimajor axis and eccentricity remain bounded.

analysis of a larger sample of HARPS data (Forveille et al., 2011) failed to detect more than 4 planets. Recent analyses of H_α stellar activity for Gliese 581 indicate a correlation between the RV and stellar activity, which leads to the conclusion that the 67 d signal (Gl 581d) is not planetary in origin (Robertson et al., 2014). In the context of comparing marginal likelihood estimators, we are not concerned about the origin of the signals but only how many signals are significant on the basis of the RV data and Keplerian models.

Our current analysis clearly detects the earlier periods of 3.15, 3.56, 12.9, and 67 days, and only hints at a fifth signal with a period of 192 d. Still it is an interesting model-comparison challenge to quantify the probability of this 5-signal model. With this in mind we show a variety of periodogram results for the 4- and 5-Keplerian-signal models.

Figure 3.26 shows a variety of 4-planet periodogram plots for the GL 581 data for a scale invariant orbital frequency prior $\propto f^{-1}$. The top left shows the Log_{10}[Prior × Likelihood] versus FMCMC iteration for every hundredth point. The bottom left shows the evolution of the 4 period parameters from their starting values, marked by the 4 dots that occur before the 0 on the iteration axis. It is clear that the FMCMC did not make transitions to any other peaks. The top-right panel shows a sample of the 4-period parameter values versus a normalised value of Log_{10}[Prior × Likelihood]. The bottom right shows a plot of eccentricity versus period.

Figures 3.27 and 3.28 show the 5-planet Kepler periodogram results using 2 different orbital frequency priors. The latter is scale invariant and the former employs a frequency prior $\propto 1/\sqrt{f}$, which helps with the detection of shorter period signals. The best set of parameters from the 4-planet fit were used as start parameters. The starting period for the fifth period was set = 30 d and the dominant fifth period found in both trials was ∼192 d, on the basis of the number of samples. As illustrated in these examples, the parallel tempering feature identifies not only the strongest peak but other potential interesting ones as well. In Figure 3.28 the MAP value of the fifth period is 192 d, which is >10 times larger than the next strongest with a period of 45 d. Two other peaks at 72 and 90 are consistent with 1-year aliases of each other. For the scale-invariant trial shown in Figure 3.28, 89% of the samples include the 192 d peak. The previous 4 periods in the 4-planet fit are clearly present in both trials.

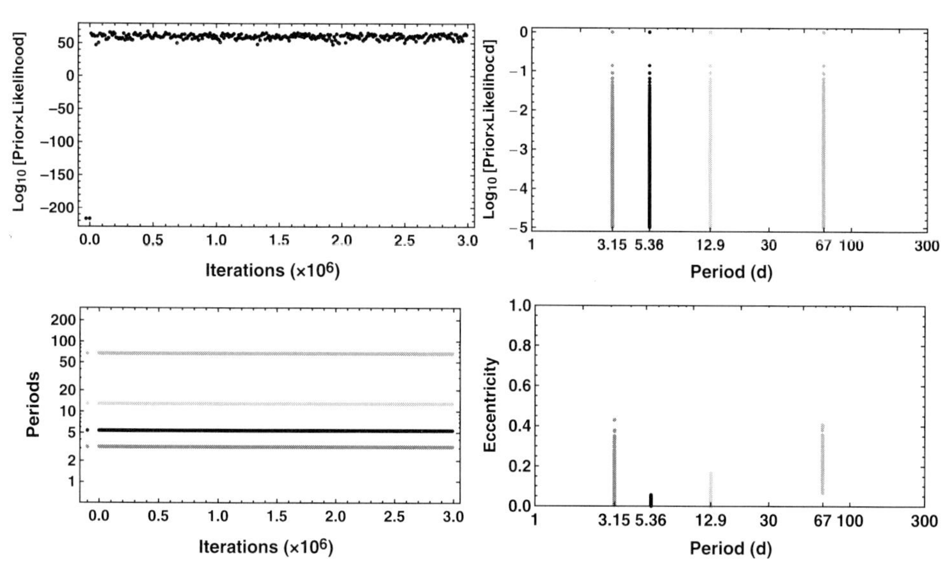

FIGURE 3.26 A variety of 4-planet periodogram plots for GL 581.

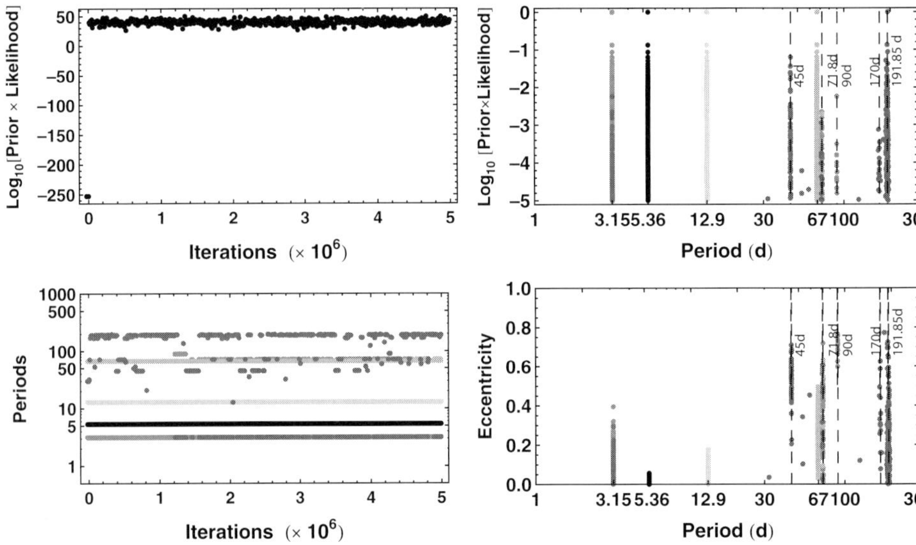

FIGURE 3.27 A variety of 5-planet periodogram plots for GL 581 for an orbital frequency prior $\propto 1/\sqrt{f}$.

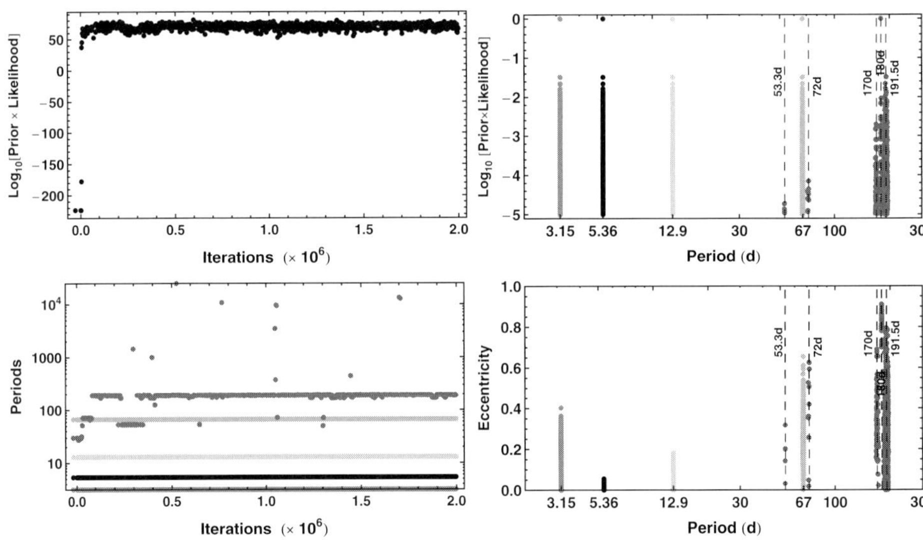

FIGURE 3.28 A variety of 5-planet periodogram plots for GL 581 for a scale-invariant orbital frequency prior $\propto f^{-1}$.

Looking at the eccentricity distributions of fifth-period candidate signals, it is clear that only the 192 d peak favours low eccentricities. Figure 3.29 shows a plot of a subset of the FMCMC-parameter marginal distributions for the 5-signal fit of the HARPS data after filtering out the post-burn-in FMCMC iterations that correspond to the 5 dominant period peaks at 3.15, 5.37, 12.9, 66.9, and 192 d. Still, on the basis of this data, the author is not inclined to claim this as a likely candidate planet. The main point of this exercise is taken up in Section 3.5, where we see what probability theory has to say about the relative probability of this particular 5-signal model to the 4-signal model for our choice of priors.

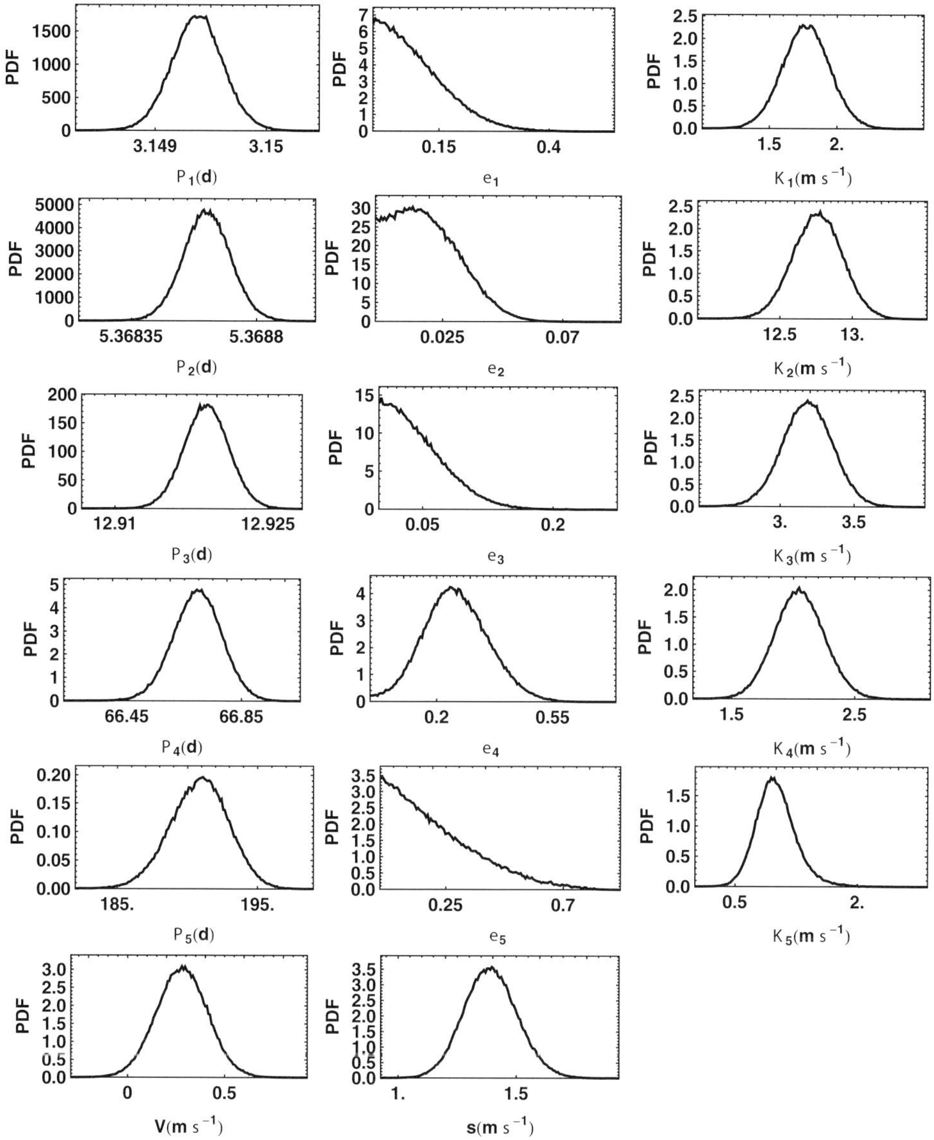

FIGURE 3.29 FMCMC parameter marginal distributions for the 5-planet fit of the HARPS data after filtering out the post-burn-in FMCMC iterations that correspond to the 5 dominant period peaks at 3.15, 5.37, 12.9, 66.9, and 192 d.

It is sometimes useful to explore the option for additional Keplerian-like signals beyond the point at which the false-alarm probability starts to increase. This is because the presence of other signals not accounted for in the model can give rise to an effective correlated noise that once removed can sometimes lead to significantly improved detections. Figure 3.30 shows the results for a 6-signal fit. As shown in Section 3.5.5, the false-alarm probability of the 6-signal case is lower than for the 5-signal case, but is still too high to be considered significant. Both 7-signal and 8-signal models were run but resulted in large false-alarm probabilities and are not shown.

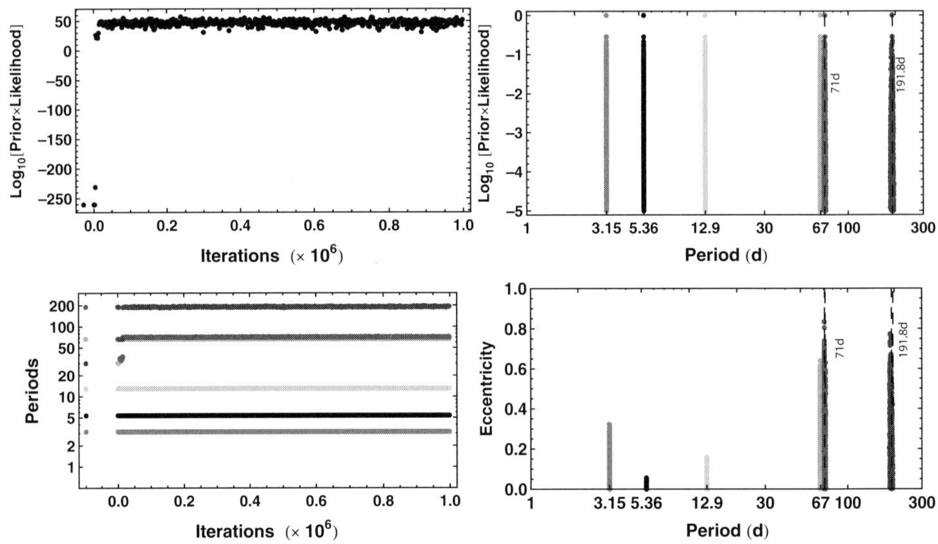

FIGURE 3.30 A variety of 6-planet periodogram plots for GL 581, with orbital-frequency prior $\propto \nu^{-1/2}$.

3.5 Model Comparison

One of the great strengths of Bayesian analysis is the built-in Occam's razor. More complicated models contain larger numbers of parameters and thus incur a larger Occam penalty, which is automatically incorporated into a Bayesian model-comparison analysis in a quantitative fashion (see, for example, Gregory 2005b, 45). The analysis yields the relative probability of each of the models explored.

To compare the posterior probability of the ith planet model[24] to the 1-planet model we need to evaluate the odds ratio $O_{i1} = p(M_i|D,I)/p(M_1|D,I)$, the ratio of the posterior probability of model M_i to model M_1. Application of Bayes' theorem leads to

$$O_{i1} = \frac{p(M_i|I)}{p(M_1|I)} \frac{p(D|M_i,I)}{p(D|M_1,I)} \equiv \frac{p(M_i|I)}{p(M_1|I)} B_{i1} \tag{3.24}$$

where the first factor is the prior odds ratio, and the second factor is called the *Bayes factor*, B_{i1}. The Bayes factor is the ratio of the marginal (global) likelihoods of the models. The marginal likelihood for model M_i is given by

$$p(D|M_i,I) = \int dX p(X|M_i,I) \times p(D|,M_i,I). \tag{3.25}$$

Thus Bayesian model comparison relies on the ratio of marginal likelihoods, not maximum likelihoods. The marginal likelihood is the weighted average of the conditional likelihood, weighted by the prior probability distribution of the model parameters and s. This procedure is referred to as marginalisation.

The marginal likelihood can be expressed as the product of the maximum likelihood and the Occam penalty (see e.g. Gregory, 2005b, 48). The Bayes factor will favour the more complicated model only if the maximum likelihood ratio is large enough to overcome

[24] More accurately these models assume different numbers of Kepler-like signals. As previously mentioned, stellar activity can also generate Kepler-like signals which need to be ruled out before the signal is ascribed to a planetary candidate.

this penalty. In the simple case of a single parameter with a uniform prior of width ΔX, and a centrally peaked likelihood function with characteristic width δX, the Occam factor is $\approx \delta X/\Delta X$. If the data is useful then generally $\delta X \ll \Delta X$. For a model with m parameters, each parameter contributes a term to the overall Occam penalty. The Occam penalty depends not only on the number of parameters but also on the prior range of each parameter (prior to the current data set D), as symbolized in this simplified discussion by ΔX. If two models have some parameters in common then the prior ranges for these parameters cancel in the calculation of the Bayes factor. To make good use of Bayesian model comparison, we fully specify priors that are independent of the current data D. The sensitivity of the marginal likelihood to the prior range depends on the shape of the prior and is much greater for a uniform prior than a scale-invariant prior (see e.g. Gregory, 2005b, 61). In most instances we are not particularly interested in the Occam factor itself, but only in the relative probabilities of the competing models as expressed by the Bayes factors. Because the Occam factor arises automatically in the marginalisation procedure, its effects will be present in any model-comparison calculation. Note: No Occam factors arise in parameter-estimation problems. Parameter estimation can be viewed as model comparison where the competing models have the same complexity so the Occam penalties are identical and cancel out.

The MCMC algorithm produces samples which are in proportion to the posterior probability distribution, which is fine for parameter estimation but one needs the proportionality constant for estimating the model marginal likelihood. Clyde et al. (2007) reviewed the state of techniques for model comparison from a statistical perspective and Ford and Gregory (2007) evaluated the performance of a variety of marginal likelihood estimators in the exoplanet context. Other techniques proposed include: nested restricted Monte Carlo (NRMC) (Gregory and Fischer, 2010), MultiNest (Feroz, Balan, and Hobson, 2011), annealing adaptive importance sampling (Loredo et al., 2011), and reversible jump Monte Carlo using a k-d tree (Farr et al., 2011). For 1-planet models with 7 parameters, a wide range of techniques perform satisfactorily. The challenge is to find techniques that handle high dimensions. A 6-planet model has 32 parameters and one needs to develop and test methods of handling at least 8 planets with 42 parameters. At present there is no widely accepted method to deal with this challenge.

In this work we compare the results from 3 marginal-likelihood estimators: (a) parallel tempering, (b) ratio estimator, and (c) restricted Monte Carlo. A brief outline of each method is presented in Sections 3.5.1, 3.5.2, and 3.5.3. A comparison of the 3 methods is given in Section 3.5.4.

3.5.1 Parallel Tempering Estimator

The MCMC samples from all (n_β) simulations can be used to calculate the marginal likelihood of a model according to Equation 3.26 (Gregory, 2005b). This method of estimating the marginal likelihood is commonly referred to as *thermodynamic integration*:

$$\ln[p(D|M_i, I)] = \int d\beta \langle \ln[p(D|M_i, X, I)] \rangle_\beta, \tag{3.26}$$

where $i = 0, 1, \ldots, m$ corresponds to the number of planets, and X represents the set of the model parameters, which includes the extra Gaussian noise parameter s. In other words, for each of the n_β parallel simulations, compute the expectation value (average) of the natural logarithm of the likelihood for post burn-in MCMC samples. It is necessary to use a sufficient number of tempering levels to estimate the above integral by interpolating values of

TABLE 3.2. Parallel tempering marginal likelihood estimate $p(D|M_2, I)_{PT}$ and fractional error versus β range for the 2-planet HD 208487 mode results.

β range	$p(D\|M_2, I)_{PT}$	Fractional Error
10^{-1}–1.0	3.290×10^{-52}	3×10^9
10^{-2}–1.0	4.779×10^{-60}	43
10^{-3}–1.0	2.817×10^{-61}	2.6
10^{-4}–1.0	1.635×10^{-61}	0.51
10^{-5}–1.0	1.306×10^{-61}	0.21
10^{-6}–1.0	1.148×10^{-61}	0.06
10^{-7}–1.0	1.091×10^{-61}	0.008
10^{-8}–1.0	1.083×10^{-61}	0.0008
10^{-9}–1.0	1.082×10^{-61}	0.0

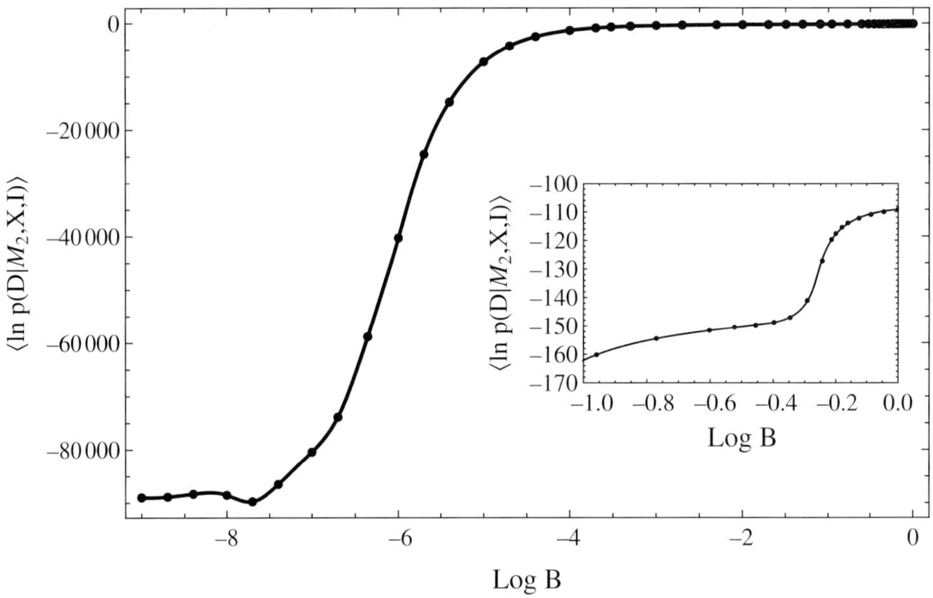

FIGURE 3.31 A plot of $\langle \ln[p(D|M_2, X, I)] \rangle_\beta$ versus β for the 2-planet HD 208487 model results. The inset shows a blow-up of the range $\beta = 0.01$–1.0.

$$\langle \ln[p(D|M_i, X, I)] \rangle_\beta = \frac{1}{n} \sum_t \ln[p(D|M_i, X_{t,\beta}, I)] \tag{3.27}$$

in the interval from $\beta = 0$–1, from the finite set, and where n is the number of post burn-in samples in each set. For this problem we use 44 tempering levels in the range $\beta = 10^{-9}$–1.0. Figure 3.31 shows a plot of $\langle \ln[p(D|M_2, X, I)] \rangle_\beta$ versus β for a 2-planet model fit to the HD 208487 RV data of Tinney et al. (2005). The inset shows a blow-up of the range $\beta = 0.1$–1.0.

The relative importance of different ranges of β can be judged from Table 3.2. The first column gives the range of β included. The second column gives the estimated marginal likelihood $p(D|M_2, I)$. The third column gives the fractional error relative to the $p(D|M_2, I)$ value derived from the largest β range extending from 10^{-9} to 1. Thus, neglecting the contribution from the β range extending from 10^{-6} to 10^{-9} would result

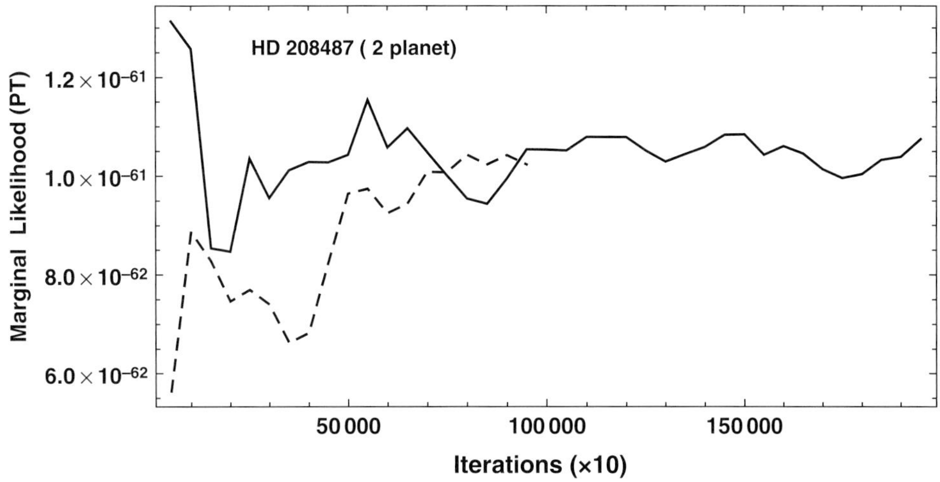

FIGURE 3.32 A plot of the marginal likelihood $p(D|M_2, X, I)_{PT}$ versus FMCMC iteration for the 2-planet HD 208487 model results for 2 trials.

in a fractional error of 0.06. The fractional error falls rapidly with each decade. If we want a good answer to ∼20% then a β range from 10^{-5} to 1 would suffice. Earlier 1-planet model results for HD 188133 yielded a fractional error of 0.26 for $\beta = 10^{-5}$–1. Later we see from a comparison of 3 different Bayesian marginal-likelihood methods that differences of order of a factor of 2 are not uncommon, so a β range of $10^{-5}-1.0$ will generally be sufficient.

Figure 3.32 show the dependence of the parallel tempering (PT) marginal-likelihood estimate versus FMCMC iteration number for the 2-planet HD 208487 model results for 2 trials. Only every tenth iteration is saved so the true number of iterations is a factor of 10 larger.

Figure 3.33 compares marginal-likelihood estimates for 1- to 5-planet RV fits. The 1- and 2-planet fits are to the HD 208487 data; the 3-, 4-, and 5-planet fits are for Gliese 581 data. Plots in the left-hand column show PT marginal-likelihood estimates versus iteration. For the 1- and 2-planet cases, where repeats were carried out, the agreement was good, to within 10%. For the 4-planet case, it is clear that the PT-derived marginal likelihood results did not reach an equilibrium value in 2.5×10^6 iterations. A PT-derived marginal likelihood for a 5-planet model was not attempted. Plots in the right-hand column are for the marginal likelihoods derived from the ratio estimator and NRMC methods, which are discussed in the next 2 sections.

3.5.2 Marginal Likelihood Ratio Estimator

Our second method was introduced by Ford and Gregory (2007).[25] It makes use of an additional sampling distribution $h(X)$. Our starting point is Bayes' theorem:

$$p(X|D, M_i, I) = \frac{p(X|M_i, I)p(D|M_i, X, I)}{p(D|M_i, I)}. \tag{3.28}$$

[25] Initially proposed by J. Berger, at an exoplanet workshop sponsored by the Statistical and Applied Mathematical Sciences Institute in January 2006.

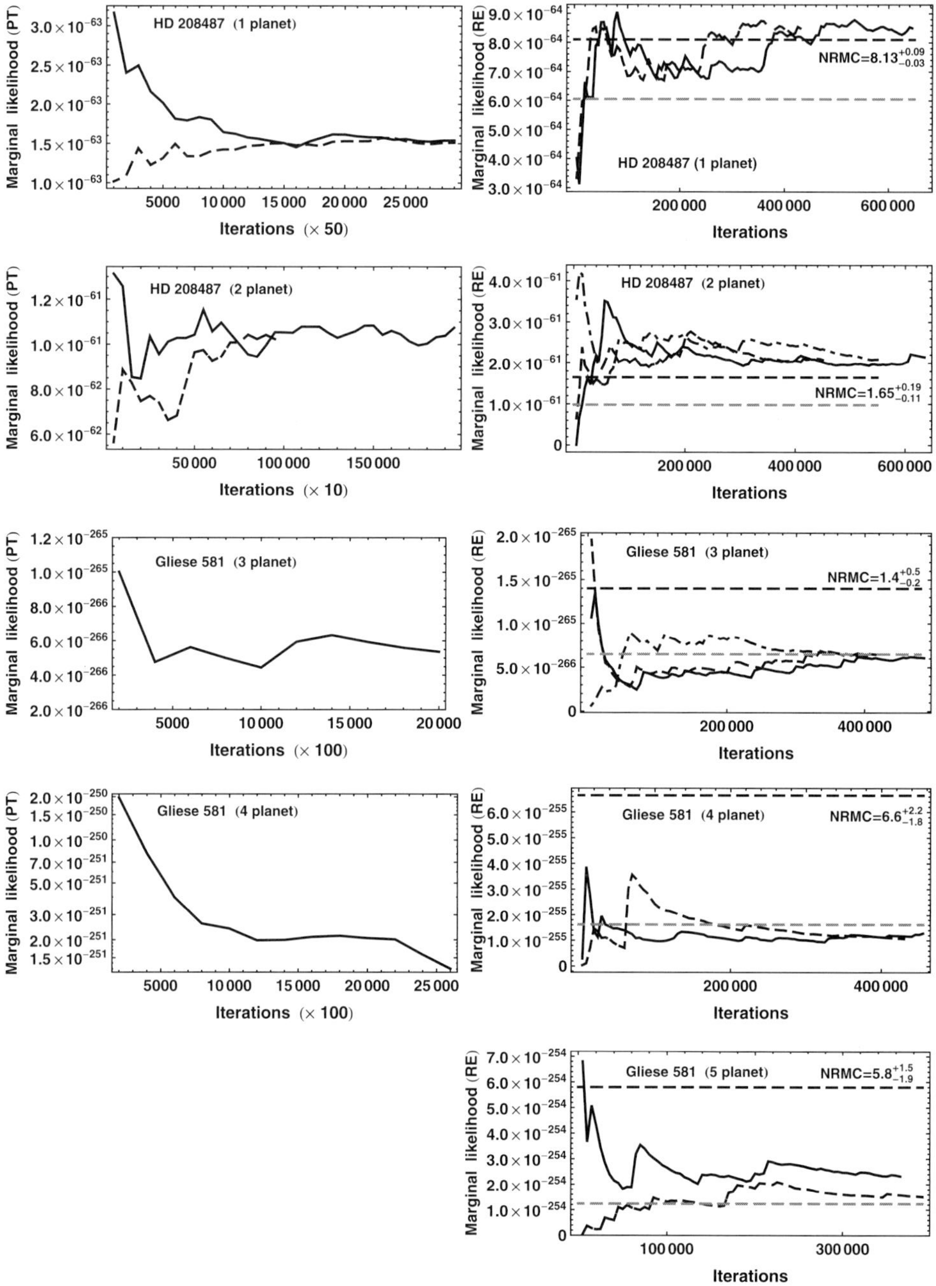

FIGURE 3.33 Comparisons of 3 different marginal-likelihood estimators versus iterations for 1- to 5-planet RV model fits. Plots in the left-hand column show parallel tempering marginal-likelihood values versus FMCMC iteration numbers. The curves in the right-hand column of panels show ratio estimator marginal-likelihood values versus iterations. The horizontal black dashed lines are the marginal likelihoods from the NRMC method together with the numerical value of the mean and range of 5 repeats. The horizontal grey dashed lines are the NRMC marginal-likelihood values within the 95% credible region of the model parameters.

Rearranging the terms and multiplying both sides by $h(X)$ we obtain

$$p(D|M_i, I)p(X|D, M_i, I)h(X) = \\ p(X|M_i, I)p(D|M_I, X, I)h(X). \tag{3.29}$$

Integrate both sides over the prior range for X:

$$p(D|M_i, I)_{RE} \int p(X|D, M_i, I)h(X)dX = \\ \int p(X|M_i, I)p(D|M_I, X, I)h(X)dX. \tag{3.30}$$

The ratio estimator of the marginal likelihood, $p(D|M_i, I)_{RE}$, is given by

$$p(D|M_i, I)_{RE} = \frac{\int p(X|M_i, I)p(D|M_i, X, I)h(X)dX}{\int p(X|D, M_i, I)h(X)dX}. \tag{3.31}$$

To obtain the marginal-likelihood ratio estimator, $p(D|M_i, I)_{RE}$, we approximate the numerator by drawing samples $\tilde{X}^1, \tilde{X}^2, \ldots, \tilde{X}^{n'_s}$ from $h(X)$ and approximate the denominator by drawing samples $X^1, X^2, \ldots, X^{n_s}$ from the $\beta = 1$ MCMC post burn-in iterations:

$$p(D|M_i, I)_{RE} = \frac{\frac{1}{n'_s}\sum_{i=1}^{n'_s} p(\tilde{X}^i|M_i, I)p(D|M_i, \tilde{X}^i, I)}{\frac{1}{n_s}\sum_{i=1}^{n_s} h(X^i)}. \tag{3.32}$$

The arbitrary function $h(X)$ was set equal to a multivariate normal distribution (multinormal) with a covariance matrix equal to twice the covariance matrix computed from a sample of the $\beta = 1$ MCMC output. We used $n'_s = 10^5$ and n_s from 10^4 to 2×10^5.[26] Some of the samples from a multinormal $h(X)$ can have non-physical parameter values (e.g. $K < 0$). Rejecting all non-physical samples corresponds to sampling from a truncated multinormal. The factor required to normalise the truncated multinormal is just the ratio of the total number of samples from the full multinormal to the number of physically valid samples. Of course we use the same truncated multinormal in the denominator of Equation 3.31 so the normalisation factor cancels.

Mixture Model

A single multinormal distribution cannot be expected to do a good job of representing the correlation between the parameters frequently evident. Following Ford and Gregory (2007), we improved over the single multinormal by using a mixture of multivariate normals by setting

$$h(X) = \frac{1}{n_c}\sum_{j=1}^{n_c} h_j(X). \tag{3.33}$$

We chose each mixture component to be a multivariate normal distribution, $h_j(X) = N(X|X_j, \Sigma_j)$, and determined a covariance matrix for each $h_j(X)$ using the posterior sample. As a first, compute $\vec{\rho}$, defined to be a vector of the sample standard deviations

[26] According to Ford and Gregory (2007), the numerator converges more rapidly than the denominator.

for each of the components of X, using the posterior sample.[27] Next, define the distance between the posterior sample X_i and the centre of $h_j(X)$, $d_{ij}^2 = \sum_k (X_{ki} - X_{kj})^2 / \rho_k^2$, where k indicates the element of X and $\vec{\rho}$. Now draw another random subset of $100n_c$ samples [28] from the original posterior sample (without replacement), select the 100 posterior samples closest to each mixture component, and use them to calculate the covariance matrix Σ_j for each mixture component. To compute the covariance matrix for each component we adopt the following approach. A random pair of the 100 posterior samples was selected with the intention of constructing a difference vector. For the angular parameters ψ and ϕ, we compute both the straight difference (d_1) and the difference of these components (d_2) after adding 2π to each. The smaller of these 2 differences avoids the wraparound problem mentioned in note. This process of selecting random pairs and computing their difference vector is repeated until we have 100 difference vectors. The covariance matrix for this mixture component is then computed from this set of difference vectors. Actually, it proves useful to employ this difference covariance matrix with components that are twice those of the true covariance matrix. Since the posterior sample is assumed to have fully explored the posterior, $h(X)$ should explore all regions of significant probability, provided that we use enough mixture components.

In the case of the 5-planet Kepler model fit to GL 581, the FMCMC analysis leads to multiple choices of 5 signal configurations. For both the RE and NRMC methods, it is possible to filter the FMCMC samples to select the individual signal configurations separately, allowing for a calculation of their relative probability. For a 5-planet fit to GL 581, the results reported here are only for the 3.15, 5.36, 12.9, 67, and 192 d period configuration. For the PT method this would not be possible; only the global marginal likelihood of the model can be evaluated.

Plots in the right-hand column of Figure 3.33 show RE marginal likelihoods versus FMCMC iterations for 1- to 5-planet model fits using a mixture model with 150 centres. The RE curve was computed twice (solid and dashed) to demonstrate the level of repeatability. In the worst case (5 planets) the agreement was within a factor of 2. Agreement with the other 2 marginal-likelihood estimators was best when the RE method was used with FMCMC data which was thinned sufficiently (by a factor of 50 to 100) so that the samples were essentially independent.

3.5.3 NRMC

Straight Monte Carlo (MC) integration can be inefficient because it involves random samples drawn from the prior distribution to sample the whole prior volume. The fraction of the prior volume of parameter space containing significant probability rapidly declines as the number of dimensions increases. For example, if the fractional volume with significant probability is 0.1 in 1 dimension then in 32 dimensions the fraction might be of order 10^{-32}. In restricted MC integration (RMC) this problem is reduced because the volume of parameter space sampled is greatly restricted to a region delineated by the outer borders of the marginal distributions of the parameters for the particular model. However, in high dimensions most of the MC samples fall near the outer boundaries of that volume,

[27] The angular parameters need to be treated in a special way because the PDF can pile up at both ends of the range with a big gap in the middle. The 2 ends of the PDF are actually close to one another in a wraparound sense because they are angular coordinates. Without allowing for this a simple variance calculation can lead to a misleadingly large value.

[28] This needs to be increased to $200n_c$ samples for a $\geq$5-planet model.

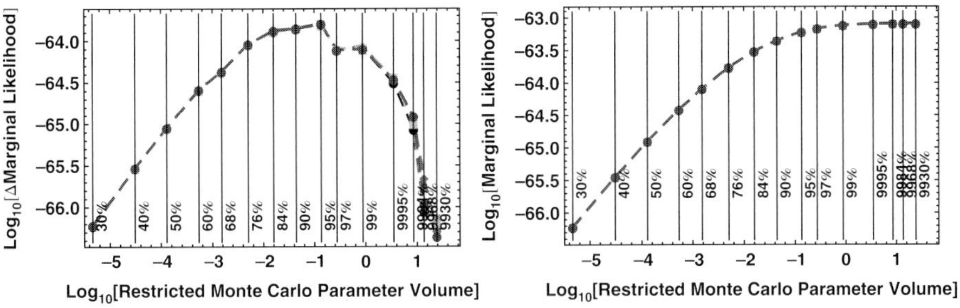

FIGURE 3.34 Left panel shows the contribution of the individual nested intervals to the NRMC marginal likelihood (for 5 repeats) based on a 1-planet model fit to the HD 208487 data. The right panel shows the sum of these contributions versus the parameter volume of the credible region.

so the sampling could easily under-sample interior regions of high probability leading to an underestimate of the marginal likelihood.

In NRMC integration (Gregory and Fischer, 2010; Gregory, 2013a), multiple boundaries of a restricted hypercube in parameter space are constructed based on credible regions ranging from 30% to ≥99%, as needed. To construct the $X\%$ hypercube we compute the $X\%$ credible region of the marginal distribution for each parameter of the particular model. The hypercube is delineated by the $X\%$ credible range of the marginal for each parameter. Note that the fraction of total probability of the joint posterior distribution contained within the hypercube will be greater than $X\%$, in part because the marginal distributions of the parameters will be broadened by any parameter correlations.

The next step is to compute the contribution to the total NRMC integral from each nested interval and sum these contributions. For example, for the interval between the 30% and 60% hypercubes, we generate random parameter samples within the 60% hypercube and reject any sample that falls within the 30% hypercube. Using the remaining samples we can compute the contribution to the NRMC integral from that interval. For NRMC, if there is more than 1 peak in the joint probability of the parameters that emerge from the FMCMC analysis, then NRMC must be performed for each peak separately.

The left panel of Figures 3.34 through 3.38 shows the NRMC contributions to the marginal likelihood from the individual intervals for 5 repeats of 1- and 2-planet fits to the HD 208487 data and 3-, 4-, and 5-planet fits to the GL 581 data. The right panel shows the summation of the individual contributions versus the volume of the credible region. The 5 repeats are shown by different greyscales which can go unnoticed as the agreement is very good in some cases. The credible region encoded as 9995% is defined as follows. Let X_{U99} and X_{L99} correspond to the upper and lower boundaries of the 99% credible region, respectively, for any of the parameters. Similarly, X_{U95} and X_{L95} are the upper and lower boundaries of the 95% credible region for the parameter. Then $X_{U9995} = X_{U99} + (X_{U99} - X_{U95})$ and $X_{L9995} = X_{L99} + (X_{L99} - X_{L95})$. Similarly,[29]

[29] Import details: Test that the extended, credible-region, outer boundary (encoded 9930) for each period parameter does not overlap the credible region of an adjacent period parameter in a multiple-planet fit. In the case of a probability distribution with multiple peaks it is advisable to define cutoffs in period parameter space around each peak to prevent this overlap. Even in the case of a single peak it is useful to define period cutoffs as follows. Note the combination of upper- and lower-period parameter values that contain all MCMC samples. Then define cutoff period intervals that are approximately 1% larger. In high dimensions this translates to a significant increase in parameter space volume.

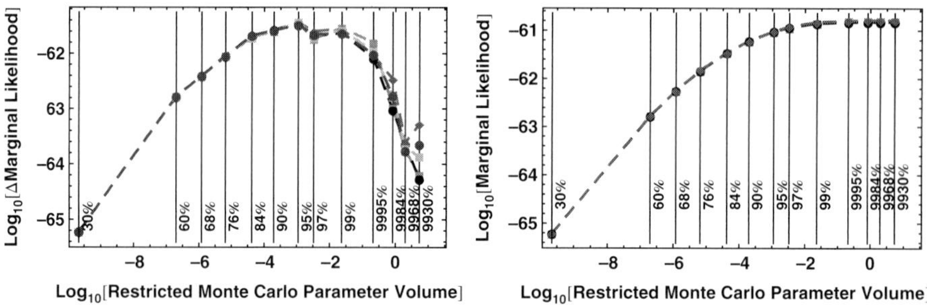

FIGURE 3.35 Left panel shows the contribution of the individual nested intervals to the NRMC marginal likelihood (for five repeats) based on a 2-planet model fit to the HD 208487 data. The right panel shows the sum of these contributions versus the parameter volume of the credible region.

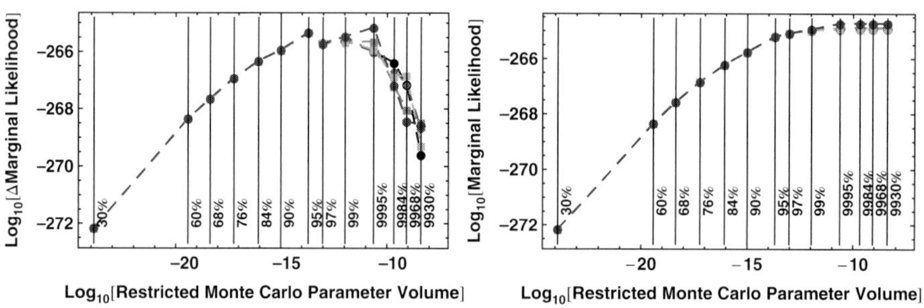

FIGURE 3.36 Left panel shows the contribution of the individual nested intervals to the NRMC marginal likelihood (for five repeats) based on a 3-planet model fit to the Gliese 581 data. The right panel shows the sum of these contributions versus the parameter volume of the credible region.

$X_{U9984} = X_{U99} + (X_{U99} - X_{U84})$. For the 3-planet fit the spread in results is within $\pm 23\%$ of the mean. For each credible region interval, 80,000 MC samples were used (4 repeats of 20,000 samples each). The Mathematica code parallelises the computation. The mean value of the prior $\times$ likelihood within the 30% credible region is a factor of 2×10^5 larger than the mean in the shell between the 97% and 99% credible regions. However, the volume of parameter space in the shell between the 97% and 99% credible regions is a factor of 8×10^{11} larger than the volume within the 30% credible region so the contribution from the latter to the marginal likelihood is negligible.

Figure 3.39 shows the fraction of the total NRMC marginal likelihood within the 95% and 99% credible regions versus the number of planets. The contribution to the marginal likelihood from a region bounded by the 95% credible region decreases systematically from 74% for a 1-planet fit to 22% for a 5-planet fit. The same trend is evident at a lower level for the region bounded by the 99% region with the exception of the last point.

What about the repeatability of the NRMC results? The 5 repeats span $\pm 1, 9, 23, 30,$ 30% for the 1-, 2-, 3-, 4-, and 5-planet fit, respectively. The biggest contribution to the spread in repeated NRMC marginal-likelihood estimates comes from the outer

Determining the marginal PDF boundaries of angular parameters needs to be treated in a special way because the PDF can pile up at both ends of the range with a big gap in the middle. The 2 ends of the PDF are actually close to one another in a wraparound sense because they are angular coordinates.

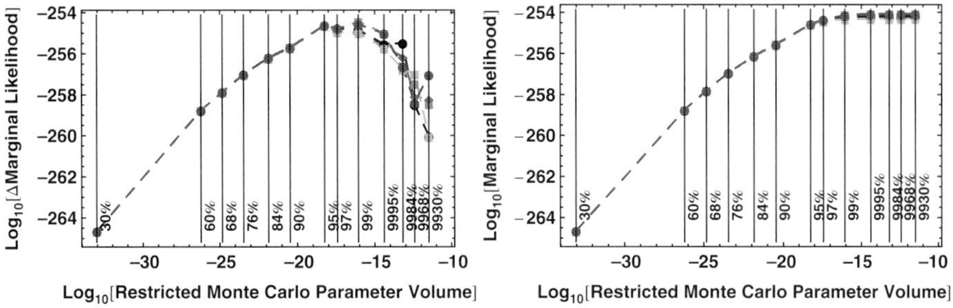

FIGURE 3.37 Left panel shows the contribution of the individual nested intervals to the NRMC marginal likelihood (for 5 repeats) based on a 4-planet model fit to the Gliese 581 data. The right panel shows the sum of these contributions versus the parameter volume of the credible region.

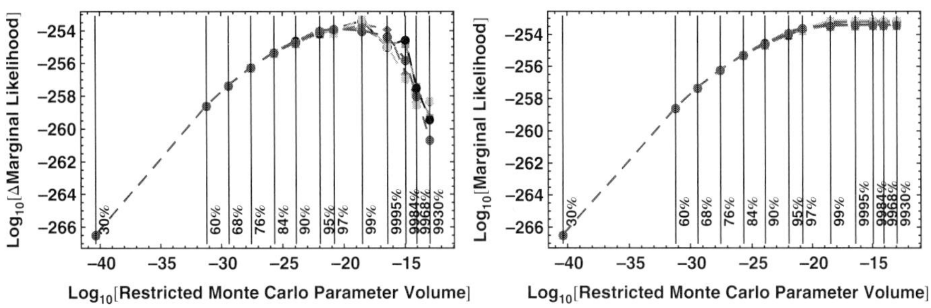

FIGURE 3.38 Left panel shows the contribution of the individual nested intervals to the NRMC marginal likelihood (for 5 repeats) based on a 5-planet model fit to the Gliese 581 data. The right panel shows the sum of these contributions versus the parameter volume of the credible region.

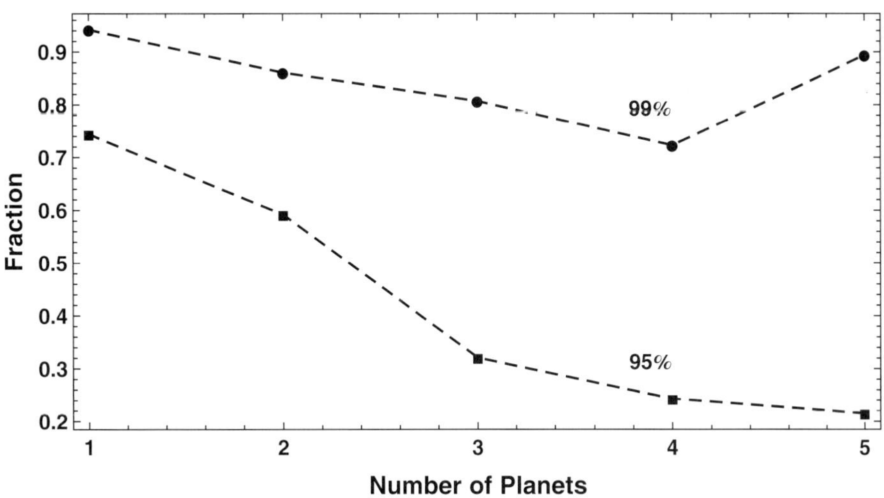

FIGURE 3.39 The fraction of the total NRMC marginal likelihood within the MCMC 95% and 99% credible regions versus the number of planets.

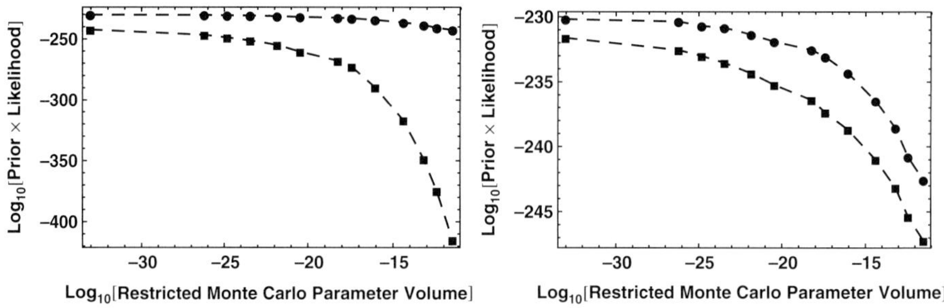

FIGURE 3.40 Left panel shows the maximum and minimum values of the Log_{10}[prior × likelihood] for each interval of credible region versus parameter volume for the NRMC 4-planet fit samples. The right panel shows the maximum and mean values of the Log_{10}[prior × likelihood] versus the parameter volume.

credible-region intervals starting around 99%. The reason for the increased scatter in the $\text{Log}_{10}[\Delta$ Marginal Likelihood] is apparent when we examine the NRMC samples. Figure 3.40 shows plots of the maximum and minimum values (left) and maximum and mean values (right) derived from the NRMC samples, for the 4-planet model, of the Log_{10}[prior × likelihood] for each interval of credible region, versus the parameter volume. The range of Log_{10}[prior × likelihood] values increases rapidly with increasing parameter volume starting around the 99% credible-region boundary. This makes the MC evaluation of the mean value more difficult in these outer intervals. Based on Figure 3.39 for the 4-planet model, the fraction of the total marginal-likelihood estimate that arises for the intervals beyond the 99% credible region is 28%. This fraction increases to 76% for all intervals beyond the 95% credible region.

What about the efficiency of the NRMC method? For example, when we sample the volume within the 95% credible region boundaries and reject all samples within the 90% we need a significant portion to fall in the region between the 2 boundaries. We also want some samples to be rejected to ensure that the sampling between the 2 boundaries extends throughout. This efficiency was examined as a function of the number of planets fit for the particular choice of boundaries used in the study. The average efficiency ranged from 60% for a 1- planet fit to 95% for 5-planet fit, which is quite reasonable. Extending this analysis to many more planets likely requires a finer-grained selection of boundaries to avoid 100% efficiency.

3.5.4 Comparison of Marginal Likelihood Methods

In my earlier discussion of the RMC method (Gregory, 2007c), I indicated that the method is expected to underestimate the marginal likelihood in higher dimensions and this underestimate is expected to become worse the larger the number of model parameters, i.e. increasing number of planets. This is because in high dimensions most of the random MC samples fall close the outer boundaries of that volume and so the sampling can easily under-sample interior regions of high probability leading to an underestimate of the marginal likelihood. NRMC overcomes this problem[30] and extends the outer credible-region boundary, allowing for a growing contribution (with increasing model complexity)

[30] In several subsequent publications (e.g. Gregory, 2013a) I mistakenly claimed that the NRMC method, like the RMC method, would be prone to underestimate the true marginal likelihood. The current analysis indicates that NRMC only underestimates because we neglect the

to the marginal likelihood from lower probability-density regions as demonstrated in Figure 3.39. Further confidence in this assertion comes from the comparison between the RE, NRMC, and PT marginal-likelihood estimators shown in Figure 3.33. For the 1-planet case, RE, NRMC, PT yielded values in the ratio 1.0:0.96:1.82, respectively. For the 2-planet case the values were in the ratio 1.0:0.75:0.52. For 3 planets the values were in the ratio 1.0:2.22:0.94. For 4 planets the values for the RE, NRMC methods were in the ratio 1.0:5.6, respectively. For 5 planets the values for the RE, NRMC methods were in the ratio 1.0:2.4, respectively. So for 3 planets and beyond the NRMC yields higher values by up to a factor of nearly 6 in the 4-planet case. What is going on here?

To help understand this, the plots in the right-hand column of Figure 3.33 show a dashed horizontal grey bar which corresponds to the NRMC marginal-likelihood contribution within the 95% credible region. For the 3- to 5-planet case this provides much better agreement with the RE estimate. The RE estimator depends on the MCMC samples. The NRMC sample density is proportional to the probability density of the target probability distribution, i.e. proportional to the Log_{10}[prior × likelihood]. As expected, we found that the range of MCMC Log_{10}[prior × likelihood] values was significantly reduced when a one-tenth subset of iterations was extracted. This suggests the possibility that dynamic range issues could limit marginal-likelihood estimators that depend on the MCMC samples, like the RE estimator. Further investigation of this issue is warranted. In NRMC we only use the MCMC samples as a guide for setting up nested hypercubes. The exact values of the credible region associated with any given hypercube is not important. We proceed to generate additional nested intervals until the contribution to the marginal likelihood is negligible. The contribution of lower probability-density regions to the NRMC estimator is responsible for the larger marginal-likelihood values.

It is clear from Figures 3.34 to 3.39, that the contribution from very low probability-density regions is much lower for the 1- and 2-planet cases so the agreement between the NRMC and RE methods is better.

What lessons can be learned from this?

- The NRMC method is not limited by possible dynamic range limitations of methods relying on the MCMC posterior samples (e.g. RE method), leading to a better estimate of the marginal likelihood enabling shorter MCMC simulations.
- Of the 3 methods the NRMC is not only conceptually simpler to appreciate but also much faster to compute, making it quick to run multiple trials to examine repeatability.
- In some situations the MCMC analysis leads to a multiple choice of signal configurations. For both the RE and NRMC methods, it is possible to filter the MCMC samples to select the individual signal configurations separately, allowing for a calculation of their relative probability. For the PT method this is not possible and only the global marginal likelihood of that model can be evaluated.
- The PT method is really only computationally feasible up to and including 3 planets. The PT method must also suffer from dynamic range limitations for a finite number of MCMC samples.
- The NRMC method works for even larger numbers of planets but care must be taken in the choice of credible region boundaries to ensure that some samples are always being rejected and ensure good sampling of each volume shell. To date, the author has successfully employed NRMC on up to 8-planet models involving 43 parameters.

possible contribution to the marginal likelihood from regions of parameter space outside those shown to be significant from the FMCMC exploration.

- As a general rule the repeatability spread increases with additional parameters, suggesting the need for more samples when dealing with larger numbers of parameters. However, keep in mind that in some cases we only need marginal likelihoods that are accurate to a factor of 2 because we usually require Bayes factors of >100 to achieve sufficiently low Bayesian false-alarm probabilities (see Section 3.5.5) to justify the more complicated model.

In the following sections our model comparison conclusions are based on Bayes factors computed from NRMC marginal-likelihood estimates.

3.5.5 Bayesian False-Alarm Probability

We can readily convert the Bayes factors, introduced in Equation 3.24, to a Bayesian false-alarm probability (FAP), which we define in Equation 3.34. For example, in the context of claiming the detection of m planets the FAP_m is the probability that there are actually fewer than m planets, i.e. $m-1$ or less:

$$\text{FAP}_m = \sum_{i=0}^{m-1} (\text{prob. of } i \text{ planets}). \tag{3.34}$$

If we assume *a priori* that all models under consideration are equally likely, then the probability of each model is related to the Bayes factors by

$$p(M_i \mid D, I) = \frac{B_{i1}}{\sum_{j=0}^{N} B_{j1}}, \tag{3.35}$$

where N is the maximum number of planets in the hypothesis space under consideration, and of course $B_{11} = 1$. For the purpose of computing FAP_m we set $N = m$. Suppose $m = 2$, then Equation 3.34 gives

$$\text{FAP}_2 = \frac{(B_{01} + B_{11})}{\sum_{j=0}^{2} B_{j1}}. \tag{3.36}$$

Let us now evaluate the Bayes factors and FAPs for our 2 example data sets, HD 208487 and Gliese 581.

HD 208487

Table 3.3 summarises the NRMC marginal-likelihood estimates for the models under consideration and the corresponding Bayes factors relative to model 1. Initially, we are interested in whether there is a single planet ($m = 1$) which yields a very small FAP of 1.4×10^{-4}. The question then shifts to an FAP of 2 planets ($m = 2$).

For HD 208487 there are 2 possible choices of 2-planet models which we label 2a for the 29, 130 d period combination, and 2b for the 130, 900 d combination. In the case of model 2b, Equation 3.36 becomes

$$\text{FAP}_2 b = \frac{(B_{01} + B_{11} + B_{2a1})}{(B_{01} + B_{11} + B_{2a1} + B_{2b1})} = 0.10. \tag{3.37}$$

We use the label 2 to represent a 2-planet model regardless of which of the 2-period configurations is true. In this case we can rewrite Equation 3.36 as

$$\text{FAP}_2 = \frac{(B_{01} + B_{11})}{(B_{01} + B_{11} + B_{2a1} + B_{2b1})} = 4.4 \times 10^{-3}. \tag{3.38}$$

TABLE 3.3. HD 208487 NRMC marginal-likelihood estimates, Bayes factors relative to model 1, and false-alarm probabilities. The quoted errors are the spread in the results for 5 repeats, not the standard deviation.

Model	Periods (d)	Marginal Likelihood	Bayes Factor Nominal	False Alarm Probability
M_0		1.44×10^{-68}	1.77×10^{-5}	
M_1	(130)	$(8.13^{+0.09}_{-0.03}) \times 10^{-64}$	1	1.4×10^{-4}
M_{2a}	(29, 130)	$(1.83^{+0.05}_{-0.03}) \times 10^{-62}$	22.5	0.90
M_{2b}	(130, 900)	$(1.65^{+0.19}_{-0.11}) \times 10^{-61}$	203	0.10
M_2	(29, 130) or (130, 900)	$(1.83^{+0.19}_{-0.11}) \times 10^{-61}$	225	4.4×10^{-3}

On this basis of the FAPs, there is significant evidence for a second planet-like signal but at present not enough data to determine which of the 2 period combinations is correct although the evidence currently favours the 130, 900 d combination.

Gliese 581

Table 3.4 summarizes the NRMC marginal-likelihood estimates for the models under consideration and the corresponding Bayes factors relative to model 4. The FAP makes a good case for up to 4 signals.[31] For a 5-signal case, we consider 2 alternatives. Case 5a corresponds to a fifth period of 192 days for which the contribution to the marginal likelihood for a 5-planet model was computed to be $(5.8^{+1.5}_{-1.9}) \times 10^{-254}$ using the NRMC method. Other choices for a fifth period were found (see Figure 3.27) and we represent them collectively as case 5b and designate the period as 'not 192 d'. We estimate their collective contribution to the marginal likelihood for a 5-planet model by multiplying the ratio of the post burn-in FMCMC samples that were not in the 192 d peak region compared to the number in the 192 d peak times the 192 d marginal-likelihood contribution for the scale-invariant prior case. The final, total 5-planet model marginal likelihood is then the sum of these 2 contributions. The FAP of our 'preferred' 192 d, fifth-period candidate is high at 0.19 so there is no reasonable case to be made for such a signal on the basis of our current state of knowledge. Also, the case for a 5-planet model of any period in our prior range does not pass our minimum threshold of an FAP of ≤ 0.01 to be considered significant.

It is sometimes useful to explore the option for additional Keplerian-like signals beyond the point at which the FAP starts to increase. Additional models ranging from 6 to 8 signals were also considered, e.g. see Figure 3.30. The FAP of the 6-signal case has decreased, compared to the 5-signal case, to 0.014. This is still too high to be considered significant. Both 7-signal and 8-signal models were run but resulted in large FAPs and are not listed.

[31] Recent analysis of H_α stellar activity for Gliese 581 indicate a correlation between the RV and stellar activity, which leads to the conclusion that the 67 d signal (Gl 581d) is not planetary in origin (Robertson et al., 2014, 4-planet candidates).

TABLE 3.4. Gliese 581 NRMC marginal-likelihood estimates, Bayes factors relative to model 4, and false-alarm probabilities. The quoted errors are the spread in the results for 5 repeats, not the standard deviation.

Model	Periods (d)	Marginal Likelihood	Bayes Factor Nominal	False Alarm Probability
M_0		5.32×10^{-393}	7.9×10^{-139}	
M_1	(5.37)	$(1.45 \pm 0.004) \times 10^{-295}$	2.2×10^{-41}	3.7×10^{-98}
M_2	$(5.37, 12.9)$	$(5.55^{+0.26}_{-0.09}) \times 10^{-273}$	2.6×10^{-19}	2.6×10^{-23}
M_3	$(5.37, 12.9, 66.9)$	$(1.40^{+0.5}_{-0.15}) \times 10^{-265}$	2.1×10^{-11}	3.9×10^{-8}
M_4	$(3.15, 5.37, 12.9, 66.9)$	$(6.7^{+2.2}_{-1.8}) \times 10^{-255}$	1.0	2.1×10^{-11}
M_{5a}	$(3.15, 5.37, 12.9, 66.9, 192)$	$(5.8^{+1.5}_{-1.9}) \times 10^{-254}$	8.7	0.19
M_{5b}	(3.15, 5.37, 12.9, 66.9,not 192)	$(0.7^{+0.18}_{-0.22}) \times 10^{-254}$	1.0	0.90
M_5	(3.15, 5.37, 12.9, 66.9,all)	$(6.5^{+1.7}_{-2.1}) \times 10^{-254}$	9.7	0.093
M_6	$(3.15, 5.37, 12.9, 66.9, 71, 190)$	$(5.21^{+1.5}_{-1.7}) \times 10^{-252}$	778	0.014

3.6 Impact of Stellar Activity on RV

In several earlier sections we referred to the challenges posed by stellar activity that can induce line-shape variations that mimic Keplerian RV signals on timescales similar to planetary signals. Many of the presentations at the Towards Other Earth II conference in Porto, Portugal (September 2014), concerned methods to deal with activity-induced RV variability. Currently a wide range of different approaches are being explored. They range from the simplest independent Gaussian noise stellar jitter approach to methods that allow for the natural signal correlation in time that results from stellar rotational modulation and the intrinsic evolution of magnetised regions. At the meeting, Xavier Dumusque (Harvard-Smithsonian Center for Astrophysics) proposed a blind competition using realistic fake RV data plus photometry and diagnostics to allow the community to determine the best strategy to distinguish real planetary signals from stellar activity-induced signals. The results of the competition were presented at the Extreme Precision Radial Velocity workshop at Yale University (6–8 July 2015).

3.7 Conclusion

The main focus of this chapter has been a new FMCMC approach to Bayesian nonlinear model fitting. In FMCMC the goal has been to develop an automated MCMC algorithm well suited to exploring multimodal probability distributions such as those that occur in the arena of exoplanet research. This has been accomplished by the fusion of a number of different statistical tools. At the heart of this development is a sophisticated control system that automates the selection of efficient MCMC proposal distributions (including for highly correlated parameters) in a parallel tempering environment. It also adapts to any new significant parameter set that is detected in any of the parallel chains or is bred by a genetic crossover operation. This controlled statistical-fusion approach has the potential to integrate other relevant statistical tools as required.

For some special applications it is possible to develop a faster, more specialized MCMC algorithm, perhaps for dealing with real-time analysis situations. In the current development of FMCMC, the primary focus has not been speed but rather to see how powerful a general-purpose MCMC algorithm we could develop and automate. That said, the Mathematica code does implement parallel processing on as many cores as are available. In real-life applications to challenging multimodal exoplanet data, FMCMC is proving to be a powerful tool. One can anticipate that this approach will also allow for the joint analysis of different types of data (e.g. RV astrometry, and transit information), giving rise to statistical fusion and data fusion algorithms.

In this chapter, considerable space has been devoted to Bayesian model comparison. In particular, significant new testing and comparison has been carried out for 3 Bayesian marginal-likelihood estimators: (1) parallel tempering (PT), (2) ratio estimator (RE), and nested restricted Monte Carlo (NRMC). All 3 are shown to be in good agreement for up to 17 parameters (3-planet model). PT ceased to be computationally practical for 4- or more planet models. Comparison between RE and NRMC was extended to the 5-planet case (27 parameters). On the basis of this comparison we recommend the NRMC method. Of the 4, NRMC is not only conceptually simpler to appreciate but also much faster to compute, making it quick to run multiple trials to examine repeatability. The NRMC method works for even larger number of planets but care must be taken in the choice of credible region boundaries to ensure that some samples are always being rejected and ensure good sampling of each volume shell.

Appendix 1: Exoplanet Priors

Frequency Search

For the Kepler model with sparse data, the target probability distribution can be very spiky. This is particularly a problem for the orbital period parameters which span roughly 6 decades.[32] The actual search in that domain is best implemented in frequency space for the following reasons. The width of a spectral peak, which reflects the accuracy of the frequency estimate, is determined by the duration of the data, the signal-to-noise (S/N) ratio, and the number of data points. More precisely (Bretthorst, 1988; Gregory, 2005b), for a sinusoidal signal model, the standard deviation of the spectral peak, δf, for an S/N >1, is given by

$$\delta f \approx \left(1.6 \frac{\mathrm{S}}{NT}\sqrt{N}\right)^{-1} \quad \mathrm{Hz}, \tag{3.39}$$

where T = the data duration in s, and N = the number of data points in T. Notice that the width of any peak is independent of the frequency of the peak. Thus the same frequency proposal distribution is efficient for all frequency peaks. This is not the case for a period search where the width of a spectral peak is $\propto P^2$. Not only is the width

[32] With the exception of a pulsar planet (PRS 1719-14 b) with a period of 0.09 d, the period range of interest is from ~0.2 d to ~1,000 yr. A period of 1,000 years corresponds roughly to a period where perturbations from passing stars and the galactic tide would disrupt the planet's orbit (Ford and Gregory, 2007). According to Exoplanet.eu (Schneider et al., 2011), the longest-period planets discovered to date are Fomalhaut b (320,000 d) and Oph 11 b (730,000 d).

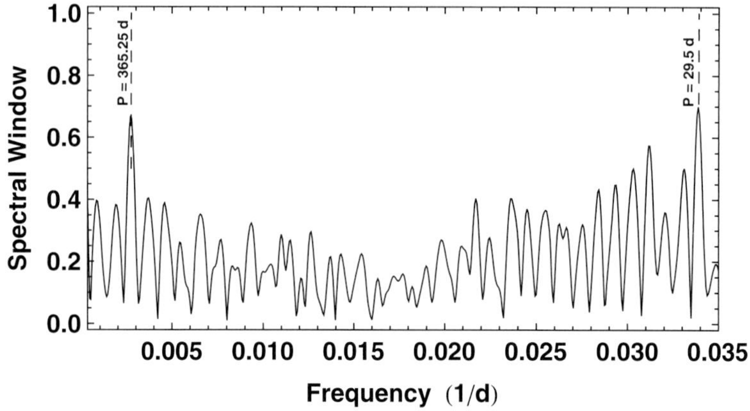

FIGURE 3.41 A portion of the spectral window function of the RV data for HD 208487 demonstrating the uniform spacing of peaks in frequency. The 29.5 d peak corresponds to the synodic month.

of the peak independent of f, but the spacing of peaks in the spectral window function is roughly constant in frequency, which is another motivation for searching in frequency space (Scargle, 1982; Cumming, 2004). Figure 3.41 shows a section of the spectral window function, described in Section 3.4.3, for the 35 samples of RV measurements for HD 208487. The true periodogram of the data is the convolution of the true spectrum with the spectral window function.

Choice of Frequency Prior for Multiplanet Models

In this section we address the question of what prior to use for frequency for multiplanet models. For a single-planet model we use a scale-invariant prior because the prior period (frequency) range spans almost 6 decades. A scale-invariant prior corresponds to a uniform probability density in $\ln f$. This says that the true frequency is just as likely to be in the bottom decade as the top. The scale-invariant prior can be written in 2 equivalent ways:

$$p(\ln f|M_1, I)\; d\ln f = \frac{d\ln f}{\ln(f_H/f_L)} \tag{3.40}$$

$$p(f|M, I)\; df = \frac{df}{f\;\ln(f_H/f_L)}. \tag{3.41}$$

What form of frequency prior should we use for a multiple-planet model? We first develop the prior to be used in a frequency search strategy where we constrain the frequencies in an n planet search such that $(f_L \leq f_1 \leq f_2 \cdots \leq f_n \leq f_H)$. From the product rule of probability theory and the above frequency constraints we can write

$$\begin{aligned} p(\ln f_1, \ln f_2, \ldots \ln f_n|M_n, I) &= p(\ln f_n|M_n, I) \\ &\times p(\ln f_{n-1}|\ln f_n, M_n, I) \cdots p(\ln f_2|\ln f_3, M_n, I) \\ &\times p(\ln f_1|\ln f_2, M_n, I). \end{aligned} \tag{3.42}$$

For model-comparison purposes we use a normalised prior which translates to the requirement that

$$\int_{\ln f_L}^{\ln f_H} p(\ln f_1, \ln f_2, \ldots \ln f_n | M_n, I) d\ln f_1 \ldots d\ln f_n = 1. \tag{3.43}$$

We assume that $p(\ln f_1, \ln f_2, \ldots \ln f_n | M_n, I)$ is equal to a constant k everywhere within the prior volume. We can solve for k from the integral equation

$$k \int_{\ln f_L}^{\ln f_H} d\ln f_n \int_{\ln f_L}^{\ln f_n} d\ln f_{n-1} \cdots \int_{\ln f_L}^{\ln f_2} d\ln f_1 = 1. \tag{3.44}$$

The solution to Equation 3.44 is

$$k = \frac{n!}{[\ln(f_H/f_L)]^n}. \tag{3.45}$$

The joint frequency prior is then

$$p(\ln f_1, \ln f_2, \ldots \ln f_n | M_n, I) = \frac{n!}{[\ln(f_H/f_L)]^n}. \tag{3.46}$$

Expressed as a prior on frequency, Equation 3.45 becomes

$$p(f_1, f_2, \ldots f_n | M_n, I) = \frac{n!}{f_1 f_2 \cdots f_n \, [\ln(f_H/f_L)]^n}. \tag{3.47}$$

Note that a similar result, involving the factor $n!$ in the numerator, was obtained by Bretthorst (2003) in connection with a uniform-frequency prior.

Two different approaches to searching in the frequency parameters were employed in this work. In the first approach (a) an upper bound on $f_1 \leq f_2$ ($P_2 \geq P_1$) was utilised to maintain the identity of the 2 frequencies. In the second, more successful approach (b) both f_1 and f_2 were allowed to roam over the entire frequency range and the parameters relabelled afterwards. In this second approach nothing constrains f_1 to always be below f_2 so that degenerate parameter peaks can occur. For a 2-planet model there are twice as many peaks in the probability distribution possible compared with (a). For a n-planet model, the number of possible peaks is $n!$ more than in (a). Provided the parameters are relabelled after the MCMC, such that parameters associated with the lower frequency are always identified with planet 1 and vice versa, the 2 cases are equivalent[33] and Equation 3.46 is the appropriate prior for both approaches.

Approach (b) was found to be more successful because in repeated blind-period searches it always converged on the highest posterior probability distribution peak, in spite of the huge period search range. Approach (a) proved to be unsuccessful in finding the highest peak in some trials and in those cases where it did find the peak it required many more iterations. Restricting $P_2 \geq P_1$ ($f_1 \leq f_2$) introduces an additional hurdle that appears to slow the MCMC period search.

K Prior

Table 3.1 shows the full set of priors used in our Bayesian calculations. The limits on K_i have evolved over time. Initially, the upper limit corresponded to the velocity of a planet with a mass = 0.01 M. in a circular orbit with a shortest period of 1 day or

[33] To date this claim has been tested for $n \leq 3$.

$K_{\max} = 2129\text{m s}^{-1}$. An upper bound of $K_{\max}\left(\frac{P_{\min}}{P_i}\right)^{1/3}$ was proposed at an exoplanet workshop at the Statistics and Applied Math Sciences Institute (spring 2006). An upper bound on P_i of 1,000 years was suggested based on galactic tidal disruption. Previously we used an upper limit of 3 times the duration of the data. Again, we set $K_{\max} = 2129$ m s^{-1}, which corresponds to a maximum planet-star mass ratio of 0.01.

Later, the upper limit on K_i was set equal to

$$K_{\max}\left(\frac{P_{\min}}{P_i}\right)^{1/3}\frac{1}{\sqrt{1-e_i^2}}, \tag{3.48}$$

based on Equation 3.49.

$$K = \frac{m \sin i}{M_*}\left(\frac{2\pi G M_*}{P}\right)^{1/3}\left(1+\frac{m}{M_*}\right)^{-2/3} \tag{3.49}$$

where m is the planet mass, M_* is the star's mass, and G is the gravitational constant. This is an improvement over $K_{\max}\left(\frac{P_{\min}}{P_i}\right)^{1/3}$ because it allows the upper limit on K to depend on the orbital eccentricity. Clearly, the only chance we have of detecting an orbital period of 1,000 years with current data sets is if the eccentricity is close to 1 and we are lucky enough to capture periastron passage. All the calculations in this supplement are based on Equation 3.48.

Eccentricity Prior

In the early years of the Bayesian analysis of exoplanets it was common to use a flat prior for eccentricity. It was soon realized that the effect of noise is to favour higher eccentricities. Gregory and Fischer (2010) provided the following explanation of this bias. To mimic a circular velocity orbit the noise points need to be correlated over a larger fraction of the orbit than they do to mimic a highly eccentric orbit. For this reason it is more likely that noise will give rise to spurious, highly eccentric orbits than low-eccentricity orbits. In a related study, Shen and Turner (2008) explored least-χ^2 Keplerian fits to synthetic RV data sets. They found that the best-fit eccentricities for low S/N ratio $K/\sigma \leq 3$ with a moderate number of observations $\text{N}_{\text{obs}} \leq 60$, were systematically biased to higher values, leading to a suppression of the number of nearly circular orbits. More recently, Zakamska, Pan, and Ford (2011) found that eccentricities of planets on nearly circular orbits are preferentially overestimated, with a typical bias of 1 to 2 times the median eccentricity uncertainty in a survey, e.g. 0.04 in the Butler et al. (2006) catalogue. When performing population analysis, they recommend using the mode of the marginalised posterior eccentricity distribution to minimise potential biases.

Kipping (2013) fit the eccentricity distribution of 396 exoplanets, detected through RV with high S/N, with a beta distribution. The beta distribution can reproduce a diverse range of PDFs using just 2 shape parameters (a and b). The large black dashed curve in Figure 3.42 is the best fit beta distribution (a = 0.867, b = 303) to all exoplanets in the sample. The dot-dashed black curve is the best-fit beta distribution (a = 1.12, b = 3.09) to exoplanets in the sample with $P < 382.3$ d. The dashed black curve is the best-fit beta distribution (a = 0.697, b = 3.27) to exoplanets in a sample with $P > 382.3$ d. One drawback with the first 2 of these beta distributions is that they are infinite at e = 0, so it is necessary to work with a prior in the form of a cumulative density distribution instead of a simpler PDF for MCMC work. The black curve is the eccentricity distribution utilised in this work, which is another beta distribution (a = 1, b = 3.1) intermediate

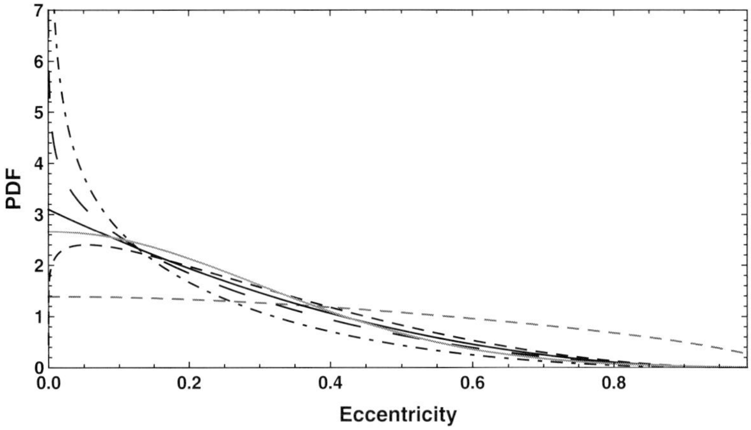

FIGURE 3.42 Exoplanet eccentricity priors. The large black dashed curve is the best fit beta distribution (Kipping, 2013) to the eccentricity data of 396 high S/N exoplanets. The dashed and dot-dashed black curves are Kipping's beta-distribution fits to the subsets with periods >382.3 d (median) and <382.3 d, respectively. The grey curve is the Gaussian eccentricity prior adopted by Tuomi (2012). The dashed grey curve is my earlier eccentricity prior which attempted a modest correction for noise-induced eccentricity bias. The black curve is the eccentricity prior employed in this work.

between Kipping's low- and high-period curves and is well behaved at e = 0. With these values the beta distribution prior simplifies to

$$p(e|I) = 3.1(1-e)^{2.1}. \tag{3.50}$$

The grey curve is the Gaussian prior adopted by Tuomi (2012). The dashed grey curve is my earlier eccentricity prior, which attempted a modest correction for noise-induced eccentricity bias.

BIBLIOGRAPHY

Aigrain, S., Pont, F., and Zucker, S. 2012. A simple method to estimate radial velocity variations due to stellar activity using photometry. *MNRAS* 419, 3147–3158.

Anglada-Escudé, G., Tuomi, M., Gerlach, E., et al. 2013. A dynamically-packed planetary system around GJ 667C with three super-Earths in its habitable zone. *Astron. Astrophys.* 556, A126.

Atchade, Y. F., Roberts, G. O., and Rosenthal, J. S. 2011. Towards optimal scaling of metropolis-coupled Markov chain Monte Carlo. *Stat. Comput.* 21(4), 555–568.

Barnes, R., and Grenberg, R. 2006. Stability limits in extrasolar planetary systems. *Astrophys. J.* 647, L-163–L166.

Baluev, R. V. 2013. The impact of red noise in radial velocity planet searches: only three planets orbiting GJ 581? *MNRAS* 429, 2052–2068.

Braak, C. J. F. ter 2006. A Markov chain Monte Carlo version of the genetic algorithm Differential Evolution: easy Bayesian computing for real parameter spaces. *Stat. Comput.* 16, 239–249.

Bretthorst, G. L. *Bayesian spectrum analysis and parameter estimation.* Springer-Verlag, New York, USA.

Butler, R. P., Wright, J. T., Marcy, G. W., et al. 2006. Catalog of nearby exoplanets. *Astrophys. J.* 646, 505–522.

Campbell, B., Walker, G. A. H., and Yang, S. 1988. A search for substellar companions to solar-type stars. *Astrophys. J.* 331, 902–921.

Clyde, M. A., Berger, J. O., Bullard, F., et al. 2007. Current challenges in Bayesian model choice. In *Statistical challenges in modern astronomy* IV' edited by Babu, G. J., and Feigelson, E. D., 224–240. ASP Conference Series 371.

Cumming, A. 2004. Detectability of extrasolar planets in radial velocity surveys. *MNRAS* 354, 1165–1176.
Cumming, A., and Dragomir, D. 2010. An integrated analysis of radial velocities in planet searches. *MNRAS* 401, 1029–1042.
Dawson, R. I., and Fabrycky, D. C. 2010. Radial velocity planets de-aliased: a new, short period for Super-Earth 55 Cnc e. *Astrophys. J.* 722, 937–953.
Deeming, T. J. 1975. Fourier analysis with unequally-spaced data. *Astrophys. Space Sci.* 36, 137–158.
Deeming, T. J. Fourier analysis with unequally-spaced data. *Astrophys. Space Sci.* 42, 257.
Farr, W. M., Sravan, N., Cantrell, A., et al. 2011. The mass distribution of stellar-mass black holes. *Astrophys. J.* 741, 103–122.
Feroz, F., Balan, S. T., Hobson, M. P. 2011. Detecting extrasolar planets from stellar radial velocities using Bayesian evidence. *MNRAS* 415, 3462–3472.
Fischer, D. A., Laughlin, G. L., Butler, R. P., et al. 2005. The N2K Consortium. I. A Hot Saturn Planet Orbiting HD 88133, *Astrophys. J.* 620, 481–486.
Ford, E. B. Quantifying the uncertainty in the orbits of extrasolar planets. *Astrophys. J.* 129, 1706–1717.
Ford, E. B. 2006. Improving the efficiency of Markov chain Monte Carlo for analzing the orbits of extrasolar planets. *Astrophys. J.* 642, 505–522.
Ford, E. B., and Gregory, P. C. 2007. Bayesian model selection and extrasolar planet detection. In *Statistical challenges in modern astronomy* IV, edited by Babu, G. J., and Feigelson, E. D., 189–204. ASP Conference Series 371.
Forveille, T., Bonfils, X., Delfosse, X., et al. 2011. The HARPS search for southern extra-solar planets? XXXII. Only 4 planets in the Gl 581 system, arXiv:1109.2505v1.
Geyer, C. J. 1991. Markov chain Monte Carlo. In *Computing science and statistics: 23rd symposium on the interface*, 156–163. Interface Foundation, Fairfax Station, Virginia, USA.
Gregory, P. C. 2005. A Bayesian analysis of extra-solar planet data for HD 73526. *Astrophys. J.* 631, 1198–1214.
Gregory, P. C. 2005. *Bayesian logical data analysis for the physical sciences: a comparative approach with Mathematica support.* Cambridge University Press, Cambridge, UK.
Gregory, P. C. 2007. A Bayesian Kepler periodogram detects a second planet in HD 208487. *MNRAS* 374, 1321–1333.
Gregory, P. C. 2007. A Bayesian periodogram finds evidence for three planets in HD 11964. *MNRAS* 381, 1607–1619.
Gregory, P. C. 2007. A Bayesian re-analysis of HD 11964: evidence for three planets. In *Bayesian inference and maximum entropy methods in science and engineering: 27th International Workshop*, edited by Knuth, K. H., Caticha, A., Center, J. L., et al., 307–314. AIP Conference Proceedings.
Gregory, P. C. Bayesian exoplanet tests of a new method for MCMC sampling in highly correlated parameter spaces. *MNRAS* 410, 94–110.
Gregory, P. C. 2011. Bayesian re-analysis of the Gliese 581 exoplanet system. *MNRAS* 415, 2523–2545.
Gregory, P. C. 2011. Commentary: Bayesian model selection and parameter estimation. In *Statistical challenges in modern astronomy V*, edited by Feigelson, E., and Babu, J., 117–125. Springer, New York, USA.
Gregory, P. C. 2013. Extrasolar planets via fusion MCMC. In *Astrostatistical challenges for hte new astronomy*, edited by Hilbe, J. M., 121–148. Springer Series in Astrostatistics 1. Springer, New York, USA.
Gregory, P. C., and Fischer, D. A. 2010. A Bayesian periodogram finds evidence for three planets in 47 Ursae Majoris. *MNRAS* 403, 731–747.
Gregory, P. C, Lawler, S. M., and Gladman, B. 2014. Additional Keplerian signals in the HARPS data for Gliese 667C: further analysis. In *Exploring the formation and evolution of planetary systems*, edited by Matthews, B., and Graham, J. Proceedings of the IAU Symposium 299. Cambridge University Press, Cambridge, UK.
Haywood, R. D., Collier Cameron, A., Queloz, D., et al. 2014. Planets and stellar activity: hide and seek in the CoRoT-7 system. *MNRAS* 443, 2517–2531.
Hukushima, K., and Nemoto, K. 1996. Exchange Monte Carlo method and application to spin glass simulations. *J. Phys. Soc. Jpn.* 65(4), 1604–1608.

Jaynes, E. T. 1957. How does the brain do plausible reasoning? Stanford University Microwave Laboratory Report 421. Reprinted in Maximum Entropy and Bayesian Methods in Science and Engineering, edited by Erickson, G. J., and Smith, C. R., 1–29, 1988. Kluwer Academic Press, Dordrecht, Netherlands.

Jaynes, E. T. 1987. Bayesian spectrum and chirp analysis. In *Maximum entropy and Bayesian spectral analysis and estimation problems*, edited by Smith, C. R., and Erickson, D., 1–37. Reidel, Dordrecht, Netherlands.

Jefferys, W. H. 2007. Discussion on Clyde et al. paper. In *Statistical challenges in modern astronomy IV*, edited by Babu, G. J., and FEigelson, E. D., 241–244. ASP Conference Series 371.

Kipping, D. M. 2013. Parameterizing the exoplanet eccentricity distribution with the Beta distribution. *MNRAS* 434, L51–55.

Lomb, N. R. 1976. Least squares frequency analysis of unevenly spaced data. *Astrophys. Space Sci.* 39, 447–462.

Loredo, T. 2004. Bayesian adaptive exploration. In *Bayesian inference and maximum entropy methods in science and engineering: 23rd International Workshop*, edited by Erickson, G. J., and Zhai, Y., 330–346. AIP Conference Proceedings 707.

Loredo, T. L., and Chernoff, D. 2003. Bayesian adaptive exploration. In *Statistical challenges in modern astronomy III*, edited by Feigelson, E. D., and Babu, G. J., 57–69. Springer-Verlag, New York, USA.

Loredo, T. L., Berger, J. O., Chernoff, D. F., et al. 2011. Bayesian methods for analysis and adaptive scheduling of exoplanet observations. *Stat. Methodol.* 9, 101–114.

Lovis, C. et al. 2011. The HARPS search for southern extra-solar planets. XXVIII. Up to seven planets orbiting HD 10180: probing the architecture of low-mass planetary systems. *Astron. Astrophys.* 528, 112L.

Marcy, G. W., and Butler R. P. 1996. A planetary companion to 70 Virginis. *Astrophys. J.* 464, L147–151.

Mayor M., and Queloz D. 1995. A Jupiter-mass companion to a solar-type star. *Nature* 378, 355.

Mayor, M., Bonfils, X., Forveille, T., et al. 2009. The HARPS search for southern extra-solar planets. XVIII. An Earth-mass planet in the GJ 581 planetary system. *Astron. Astrophys.* 507, 487.

Metropolis, N., Rosenbluth, A., Rosenbluth, M., et al. 1953. Equation of state calculation by fast computing machines. *J. Chem. Phys.* 21, 1087–1092.

Petigura, E. A., Marcy, G. W., and Howard, A. W. 2013. Prevalence of Earth-size planets orbiting Sun-like stars. *Proc. Natl. Acad. Sci. USA* 110(48), 19273–19278. www.pnas.org/cgi/doi/10.1073/pnas.1321363110.

Queloz, D., Henry, G. W., Sivan, J. P., et al. 2001. No planet for HD 166435. Astron. Astrophys. 379, 279–287.

Queloz, D., Bouchy, F., Moutou, C., et al. 2009. The CoRoT-7 planetary system: two orbiting super-Earths. *Astron. Astrophys.* 506, 303–319.

Roberts, G. O., Gelman, A., and Gilks, W. R. 1997. Weak convergence and optimal scaling of random walk Metropolis algorithms. *Ann. Appl. Probab.* 7, 110–120.

Robertson, P., Mahadevan, S., Endl, M., et al. 2014. Stellar activity masquerading as planets in the habitable zone of the M dwarf Gliese 581. *Science* 345, 440–444.

Saar, S. H., and Donahue, R. A. 1997. Activity-related radial velocity variation in cool stars. *Astrophys. J.* 485, 319–327.

Saar, S. H., Butler, R. P., and Marcy, G. W. 1998. Magnetic activity-related radial velocity variations in cool stars: first results from the Lick Extrasolar Planet Survey, *Astrophys. J.* 498, L153–157.

Scargle, J. D. 1982. Studies in astronomical time series analysis. II – Statistical aspects of spectral analysis of unevenly spaced data. *Astrophys. J.* 263, 835–853.

Schneider, J., Dedieu, C., Le Sidaner, P. et al. 2011. Defining and cataloging exoplanets: the exoplanet.eu database. *Astron. Astrophys.* 532, A79–89.

Shen, Y., and Turner, E. L. 2008. On the eccentricity distribution of exoplanets from radial velocity surveys. *Astrophys. J.* 685, 553.

Tinney, C. G., Butler, R. P., Marcy, G. W., et al. 2005. Three low-mass planets from the Anglo-Australian Planet Search. *Astrophys. J.* 623, 1171–1179.

Tuomi, M. 2012. Evidence for nine planets in the HD 10180 system. *Astron. Astrophys.* 543, 52–63.

Tuomi, M., Jones, H. R. A., Jenkins, J. S., et al. 2013. Signals embedded in the radial velocity noise. Periodic variations in the Ceti velocities. *Astron. Astrophys.* 551, A79.

Udry, S., Bonfils, X., Delfosse, X., et al. 2007. *Astron. Astrophys.* 469, L43.

Vogt, S. S., Butler, R. P., Rivera, E. J., et al. 2010. The Lick-Carnegie Exoplanet Survey: A 3.1 Earth mass planet in the habitable zone of the nearby M3V star Gliese 581 *Astrophys. J.* 723, 954–965.

Wittenmyer, R. A., Endl, M., Cochran, W. D., et al. 2009. A search for multi-planet systems using the Hobby-Eberly Telescope. *Astrophys. J. Suppl. S.* 182, 97–119.

Wolszczan, A., and Frail, D. 1992. A planetary system around the millisecond pulsar PSR1257 + 12. *Nature* 355, 145–147.

Wright, J. T. 2005. Radial velocity jitter in stars from the California and Carnegie Planet Search at Keck Observatory. *Publ. Astron. Soc. Pac.* 117, 657–664.

Wright, J. T. Marcy G. W., Fischer, D. A., et al. 2007. Four new exoplanets and hints of additional substellar companions to exoplanet host stars. *Astrophys. J.* 657, 533–545.

Zakamska, N. L., Pan, M., and Ford, E. B. 2011. Observational biases in determining extrasolar planet eccentricities in single-planet systems. *MNRAS* 410, 1895.

Zechmeister, M., and Kürster, M. The generalised Lomb-Scargle periodogram: a new formalism for the floating-mean and Keplerian periodograms. *Astron. Astrophys.* 496, 577–584.

4 Bayesian Cosmology

ROBERTO TROTTA

4.1 Introduction

4.1.1 Foundations of Cosmology

Cosmology studies the entire universe, its properties, its fundamental nature, constituents, origins, and fate. Distances, timescales, and energy scales span tens of orders of magnitude. Modern cosmology is entering the era when our knowledge about the universe is no longer limited by the amount of observational data we have, but by our ability to reduce them, process them, and understand their implications. Complex data require sophisticated statistical analysis tools in order to be compared with our theoretical models for the universe. This is where a more refined understanding of statistical inferential techniques can help: by allowing astrophysicists and theoretical physicists to explore more fully the parameter spaces of their models, and by accounting for subtle yet important aspects of the data-gathering process. This will help in reducing systematics and bringing cosmology into the epoch of accuracy, rather than merely 'precision'.

We begin with a brief overview of some foundational concepts that are needed to approach the problem of inference in cosmology. Observationally, most galaxy spectra are *redshifted*: emission lines from spectra of distant galaxies are moved towards the red end of the spectrum. The interpretation is that redshifted galaxies are receding (i.e. moving away) from us, in analogy with the Doppler effect (when a source is moving away, light loses energy). However, the physical origin of this *cosmological redshift* is not the speed of galaxies moving in space. Rather, it is a consequence of the expansion of the universe.

Edwin Hubble (1929) is credited with discovering the expansion of the universe, which is still called 'Hubble expansion'. Hubble's key observation was that the recession speed of galaxies, v, is proportional to their distance, d. The *Hubble constant* H_0 is the slope of that relationship:

$$v[\mathrm{km/s}] = H_0 \left[\frac{\mathrm{km/s}}{\mathrm{Mpc}}\right] d[\mathrm{Mpc}]. \tag{4.1}$$

H_0 expresses the current rate of expansion of the universe.

Hubble estimated the value of H_0 to be $\sim$500 km/s/Mpc, an order of magnitude off today's measurements. Today, we have measured H_0 locally with high precision, $H_0 = 73.8 \pm 2.4$ [km/s/Mpc] (Riess et al., 2011), by extending and improving Hubble's method with a so-called distance ladder.[1] A large H_0 implies a young universe, as the age of the universe $\sim 1/H_0$ and using the value adopted by Hubble one would get $1/H_0 \sim 2\,\mathrm{Gyr}$, much smaller than the age of the Earth, which is obviously absurd. Often, H_0 is written as $H_0 = 100h \left[\frac{\mathrm{km/s}}{\mathrm{Mpc}}\right]$, where h is a dimensionless quantity. It is useful to remember that $1/H_0 = 9.78/h\,\mathrm{Gyr}$.

[1] This value is in slight tension with the value inferred using the much higher redshift cosmic microwave background anisotropies; see Equation 4.50.

TABLE 4.1. Examples of relationship between redshift and lookback time

Redshift z	Object/Event	Approximate Lookback Time (Gyr)
0.1	Nearby galaxies	1.3
1.0	High-z supernovae	7.8
7.0	Highest-z quasar known	12.95
10	First stars	13.24
1,100	Cosmic microwave background emitted	13.72
∞	Big Bang	13.798 ± 0.037

A very important quantity is redshift, defined as

$$z \equiv \frac{\lambda_{\text{obs}} - \lambda_0}{\lambda_0}, \tag{4.2}$$

where λ_{obs} is the observed wavelength and λ_0 is the emitted wavelength (also called 'rest frame' wavelength). $z > 0$ for receding (i.e. redshifted) objects. In general, higher redshift means that the object is 'farther away' from us, but the relationship between redshift and distance is not trivial. The conversion between redshift and *lookback time* (i.e. the difference in the age of the universe between now and when the light was emitted) requires general relativity. It is highly nonlinear, with some examples shown in Table 4.1.

Penzias and Wilson (1965) accidentally discovered the existence of the *cosmic microwave background* (CMB): an almost perfectly isotropic (i.e. the same in all directions) blackbody radiation, with a temperature

$$T_{\text{CMB}} = 2.728 \pm 0.004 \text{ K} \tag{4.3}$$

(see Figure 4.1). The CMB is evidence that the universe expanded from a hot, highly energetic, causally connected state: the Big Bang model (as derogatorily described by Fred Hoyle on a BBC radio programme in 1949).

4.1.2 A Simple Model for the Universe

In order to satisfy the observational evidence described above, we want to build a simple model of the universe that is:

(i) isotropic (i.e. the same in all directions)
(ii) homogeneous (i.e. the same in all locations)
(iii) expanding

Notice that homogeneity follows from isotropy (Figure 4.2).

Here we consider a simple Newtonian approach. Consider a metric of space–time that is homogeneous and isotropic and allows for expansion: the Friedmann–Robertson–Walker (FRW) metric, which is given by

$$ds^2 = c^2 dt^2 - a^2(t) \left[dx^2 + dy^2 + dz^2\right], \tag{4.4}$$

where $a(t)$ is a dimensionless function of time that stretches or contracts the spatial part of the metric isotropically. Notice that the FRW metric is the same as Minkowski space–time, with the extra scale factor for the spatial part.

Consider an inflating 3-D sphere, as depicted in Figure 4.3. The coordinate values of A, B do not change as a function of time, but their physical separation increases:

$$\Delta\phi(t_0) = \Delta\phi(t_1) \quad (4.5)$$

$$R(t_1) = a(t_1)R_0 \quad (4.6)$$

$$\text{Distance A-B at } t_1\text{: } a(t_1)R_0\Delta\phi \text{ (rad)}. \quad (4.7)$$

It is important to distinguish between coordinate distance (*comoving distance*) and physical distance.

$$\text{Physical distance } = a \times \text{ comoving distance}. \quad (4.8)$$

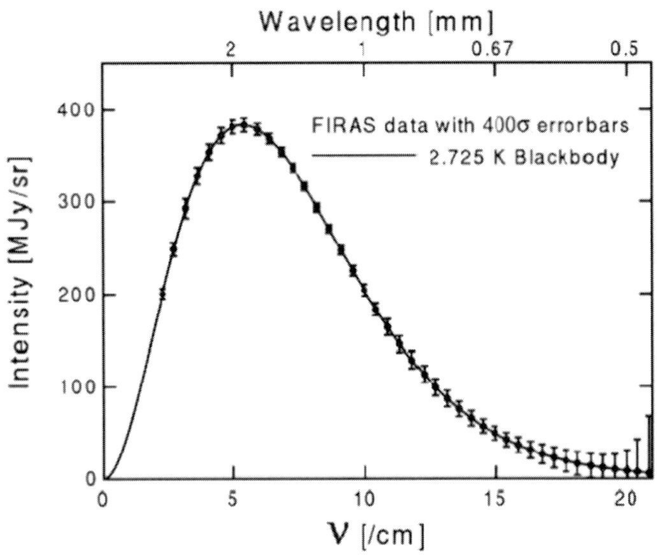

FIGURE 4.1 Blackbody spectrum of the cosmic microwave background, as measured by the Far Infrared Absolute Spectrophotometer (FIRAS) on-board the Cosmic Background Explorer satellite (Fixsen et al., 1996). Experimental error bars have been multiplied by 400 to make them visible.

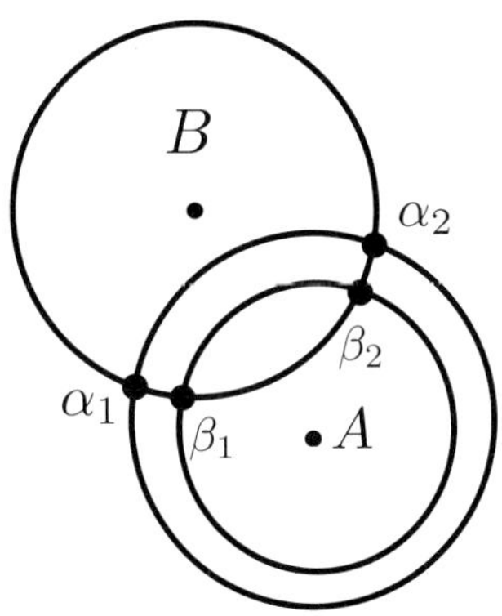

FIGURE 4.2 Homogeneity follows from isotropy. Because of isotropy around B, the density of the universe must be the same at $\alpha_1, \alpha_2, \beta_1, \beta_2$. But if we look from the point of view of observer A, it follows that the universe must be homogeneous, because β_1 and α_1, for example, are at different distances from A. But the density at α_1 is equal to the density at β_1. This holds true for spheres of any radius around A.

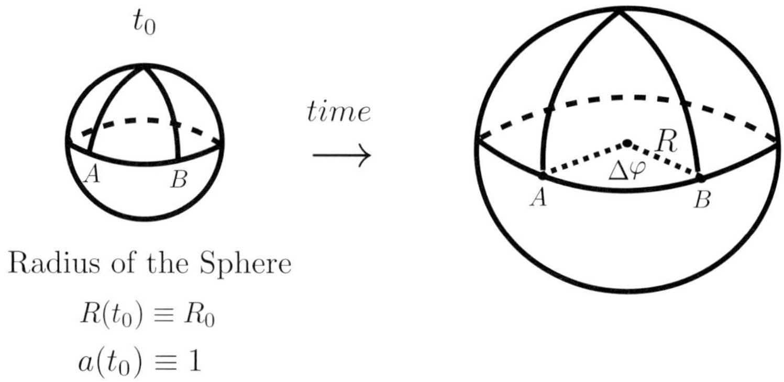

FIGURE 4.3 An inflating 3-D sphere.

We now show that Hubble's law follows from the FRW metric. Let us consider the recession velocity of a test particle, with $\dot{\phi} = \dot{\vartheta} = 0$ (no angular velocity). We typically work in spherical coordinates, with the origin of the coordinate system centred on us. An overdot denotes differentiation with respect to time.

The radial velocity of a particle on the surface of the expanding sphere is:

$$v_r = \frac{d}{dt} r(t) = \frac{d}{dt}\left(a(t)R\right), \tag{4.9}$$

where R is the comoving radial position. Carrying out the total derivative gives:

$$\begin{aligned} v_r &= \dot{a}(t)R + a(t)\dot{R} && (4.10)\\ &= \frac{\dot{a}}{a}(aR) + a(t)\dot{R} && (4.11)\\ &= H(t)r(t) + a(t)\dot{R}, && (4.12)\end{aligned}$$

where we have defined $H(t) = \dot{a}(t)/a(t)$. The first term is the velocity due to the expansion ($\dot{a} \neq 0$), while the second term is the velocity due to the movement in space (*peculiar velocity*, $\dot{R} \neq 0$).

If we consider a *comoving observer* (i.e. one that 'rides along' with the expansion, but with $\dot{R} = (0)$, we find:

$$v_r = H(t)r(t), \tag{4.13}$$

which is very similar to Hubble's law:

$$v = H_0 d. \tag{4.14}$$

Indeed, Hubble's law is recovered by setting $t = t_0$ (today) in Equation 4.13. Since H_0 is positive (positive slope in the Hubble diagram), it follows that $\dot{a} > 0$, hence the universe expands (the scale factor grows with time).

4.1.3 Dynamics of the Expansion

It is natural – and one of the central questions of modern cosmology – to ask whether the current expansion will continue forever. In other words: what is the fate of the universe?

The future behaviour of $a(t)$ is controlled by quantities called 'cosmological parameters', which we need to determine observationally. The Hubble constant H_0 is an example of a cosmological parameter. We want to determine what will happen for $t \gg t_0$. In particular, how do we distinguish between:

(i) $\dot{a} \to 0$ for $t \to \infty$: the expansion slows down and stops 'at infinity'.
(ii) $\dot{a} > 0$ for all t: the universe expands forever
(iii) $\dot{a} < 0$ for $t > t_{\max}$: the universe recollapses in a Big Crunch
(iv) is $\ddot{a} > 0$ possible? In which case, the universe accelerates in its expansion

Consider a sphere of comoving radius R, enclosing a mass M, with homogeneous density $\rho(t)$:

$$\rho(t) = \frac{M}{\frac{4}{3}\pi r^3(t)} = \frac{M}{\frac{4}{3}\pi R^3 a^3(t)} \tag{4.15}$$

so $\rho(t) = \rho_0/a(t)^3$, with $\rho_0 \equiv \frac{M}{\frac{4}{3}\pi R^3}$, since mass is conserved. Consider now a comoving test particle on the surface of the sphere. The acceleration due to gravity is:

$$\ddot{r}(t) = \frac{d^2 r}{dt^2} = -\frac{GM(R)}{r^2(t)}, \tag{4.16}$$

where the negative sign denotes acceleration towards the centre of the sphere, and $M(r) =$ const because of mass conservation. Then:

$$\ddot{r}(t) = -\frac{G\rho_0 \frac{4}{3}\pi R^3}{a^2 R^2} = \ddot{a} R \tag{4.17}$$

$$\Rightarrow \ddot{a} = \frac{\ddot{r}}{R} = -\frac{4\pi G}{3}\rho(t)a(t), \tag{4.18}$$

where we have used $\rho_0/a^2 = \rho(t)a(t)$. So we obtain the *acceleration equation*

$$\ddot{a} = -\frac{4\pi G}{3}\rho(t)a(t). \tag{4.19}$$

A number of important points are worth noticing:

- The right-hand side is negative $\Rightarrow \ddot{a} < 0$, hence the expansion decelerates.
- $\ddot{a}$ only depends on the density ρ and not on R.
- The above equation needs modifying in general relativity if the universe is filled with a fluid other than pressureless matter, by introducing a pressure term, p:

$$\frac{\ddot{a}}{a} = -\frac{4\pi G}{3}\left(\rho(t) + 3\frac{p}{c^2}\right). \tag{4.20}$$

By integrating the acceleration equation over time we obtain:

$$\frac{v^2}{2} - \frac{GM(R)}{r} = C\frac{R^2}{2}, \tag{4.21}$$

where $v = R\dot{a}$ is the recession velocity, the term $\frac{v^2}{2}$ can be interpreted as a kinetic energy per unit mass, the term $\frac{GM(R)}{r}$ is a potential energy per unit mass, and C is an integration constant. This equation expresses the conservation of energy. Let us rewrite the integration constant as $C = -\kappa c^2$, where κ has units $\left[\frac{1}{\text{length}^2}\right]$ or $1/\sqrt{\kappa}$ has units of length. Within general relativity, this is interpreted as the *curvature scale* of the universe. The (constant) value of κ determines the geometry of space (Figure 4.4):

(i) $\kappa = 0$: Euclidean, flat space (infinite)
(ii) $\kappa > 0$: closed universe (finite). $1/\sqrt{\kappa}$ is its curvature radius
(iii) $\kappa < 0$: hyperbolic (open, infinite) universe

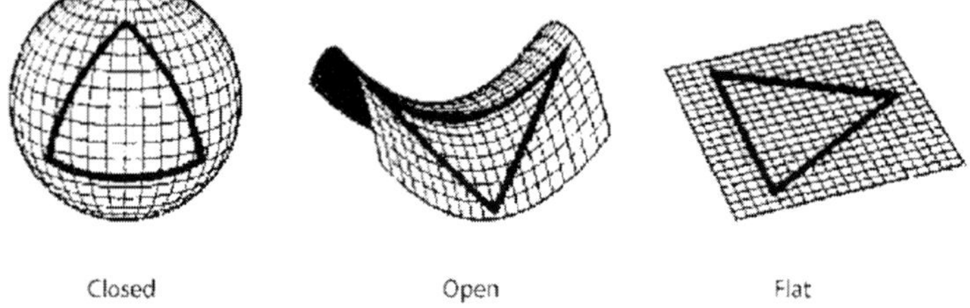

FIGURE 4.4 The 3 possible geometries for the spatial part of the universe (from Schneider, 2006).

For a given kinetic energy, changing the amount of mass (density) in the universe changes κ, and hence changes its geometry. This is a key aspect of general relativity: the mass density determines the curvature of space–time.

Rewrite Equation 4.21 as:

$$\frac{(R\dot{a})^2}{2} - G\frac{\rho_0 \frac{4}{3}\pi R^3}{Ra} = -\kappa c^2 \frac{R^2}{2} \tag{4.22}$$

$$\dot{a}^2 = \frac{8}{3}\pi G \frac{\rho_0}{a} - \kappa c^2 \tag{4.23}$$

$$\left(\frac{\dot{a}}{a}\right)^2 = \frac{8}{3}\pi G \frac{\rho_0}{a^3} - \frac{\kappa c^2}{a^2} \tag{4.24}$$

$$\left(\frac{\dot{a}}{a}\right)^2 = \frac{8}{3}\pi G \rho(t) - \frac{\kappa c^2}{a^2}. \tag{4.25}$$

This leads to the all-important *Friedmann equation*:

$$H^2(t) = \frac{8}{3}\pi G \rho(t) - \frac{\kappa c^2}{a^2}. \tag{4.26}$$

The solution to the Friedmann equation gives $a(t)$.

Now look at $\rho(t)$ in more detail: what if the universe does not contain just matter? Mass conservation for matter implies

$$\rho_m = \frac{\rho_{m,0}}{a^3}, \tag{4.27}$$

where a subscript 'm' denotes matter. But for radiation – subscript 'r' – (i.e. massless particles like photons and neutrinos[2]) the scale factor dependency of its energy density is modified:

$$\rho_r = \frac{\rho_{r,0}}{a^4}. \tag{4.28}$$

The extra $1/a$ factor comes from redshift, which reduces the energy of the photon. So in general

$$\rho = \frac{\rho_{m,0}}{a^3} + \frac{\rho_{r,0}}{a^4} + \cdots \tag{4.29}$$

[2] Neutrinos are actually massive (as indicated by oscillation experiments), but their mass can be neglected for the considerations here.

If the universe is spatially flat, then $\kappa = 0$ and the Friedmann equation, evaluated at $t = t_0$ (today), gives:

$$H_0^2 = \frac{8\pi G}{3}\rho(t_0). \tag{4.30}$$

We define the density on the right-hand side as the *critical energy density today*, denoted ρ_{crit}, i.e. the matter/energy density needed today so that the universe is precisely flat. It can be expressed as

$$\rho_{\text{crit}} \equiv \frac{3H_0^2}{8\pi G} = 1.88 \cdot 10^{-29} h^2 \text{ g/cm}^3 = 1.05 \cdot 10^4 h^2 \text{ MeV/m}^3. \tag{4.31}$$

Dividing Equation 4.26 by H_0^2 and using the above definition for ρ_{crit} we obtain:

$$\frac{H^2}{H_0^2} = \frac{8\pi G}{3}\frac{\rho}{\frac{8\pi G}{3}\rho_{\text{crit}}} - \frac{\kappa c^2}{a^2 H_0^2} \tag{4.32}$$

$$\Rightarrow H^2 = H_0^2\left[\frac{\rho_{m,0}}{\rho_{\text{crit}}}\frac{1}{a^3} + \frac{\rho_{r,0}}{\rho_{\text{crit}}}\frac{1}{a^4} - \frac{\kappa c^2}{H_0^2}\frac{1}{a^2} + \cdots\right] \tag{4.33}$$

$$\Rightarrow H^2 = H_0^2\left[\Omega_m\frac{1}{a^3} + \Omega_r\frac{1}{a^4} + \Omega_\kappa\frac{1}{a^2} + \cdots\right]. \tag{4.34}$$

In the last step we have defined the *cosmological parameters*

$$\Omega_x \equiv \frac{\rho_{x,0}}{\rho_{\text{crit}}}, \tag{4.35}$$

where the subscript x denotes the kind of matter/energy the quantity refers to. Also, the curvature parameter is

$$\Omega_\kappa \equiv -\frac{\kappa c^2}{H_0^2}. \tag{4.36}$$

The Ω_x are dimensionless quantities, expressing the energy density of component x in units of the critical energy density. By definition, $\Omega_x = 1$ corresponds to a flat universe. Note that if the universe is flat, $\Omega_\kappa = 0$ (i.e. no curvature) and the sum of the Ω_x must be 1:

$$\sum_x \Omega_x = 1 \quad \text{(for a flat universe)}. \tag{4.37}$$

Another term is allowed in Einstein's equations, which cannot be derived without general relativity. It leads to adding a new cosmological parameter, Ω_Λ, to the right-hand side of the Friedmann equation such that:

$$H^2 = H_0^2\left[\frac{\Omega_m}{a^3} + \frac{\Omega_r}{a^4} + \frac{\Omega_\kappa}{a^2} + \Omega_\Lambda\right]. \tag{4.38}$$

Here, Λ stands for the *cosmological constant*, and $\Omega_\Lambda \equiv \frac{\rho_\Lambda}{\rho_{\text{crit}}} = \frac{\Lambda}{3H_0^2}$. As we shall see, $\Lambda > 0$ leads to an exponential, accelerating expansion of the universe (repulsive effect).

4.2 Measuring Cosmological Parameters

4.2.1 The General Principle

Having introduced the most important cosmological parameters, we now turn to how to measure them observationally. For example, to measure H_0 from the Hubble law,

$$v = H_0 d, \tag{4.39}$$

we need to determine both v (the recession velocity) and d (the physical distance). The recession velocity can be measured easily. From the definition of redshift, Equation 4.2, we have that infinitesimally

$$dz = \frac{d\lambda}{\lambda} = \frac{dv}{c}. \tag{4.40}$$

So for small $z, v \Rightarrow z \approx v/c$ (if $v/c \ll 1$). Notice that this relationship is only valid for $z \ll 1$. However, it turns out that for $z \gtrsim 1.5$ the recession speed of galaxies is $v > c$, and always has been. This can be understood with general relativity (see Davis and Lineweaver, 2004).

Measuring distances in cosmology is the difficult part. The reason is that a given object (e.g. a galaxy) could be bright (or large) and far away, or dim (or small) and nearby, yet appear the same on the sky. So the apparent size or brightness is misleading, *unless* we know how big (or bright) the object is intrinsically. The solution is to use *standard candles* (objects having all the same luminosity) or *standard rulers* (objects having all the same physical size) to measure distances.

The Hubble law is an example of a local distance-redshift relation: the 2 are connected by the value of H_0, 1 of the cosmological parameters. general relativity tells us that

$$\text{Distance} = f(z, \Omega_m, \Omega_r, \Omega_\Lambda, \Omega_\kappa, H_0), \tag{4.41}$$

where 'distance' needs to be suitably defined (see below), and f is a known function of redshift z and of the 'cosmological parameters'

$$\Theta = \{\Omega_m, \Omega_r, \Omega_\Lambda, \Omega_\kappa, H_0\}. \tag{4.42}$$

So if we can measure distances to different z, since we know f we can reconstruct the cosmological parameters.

4.3 Distances in Cosmology

There are 2 definitions of distance in cosmology: the luminosity distance, d_L, and the angular diameter distance, d_A. The 2 are the same at low redshift, $z \ll 1$, but are different at large redshift, where the curvature of space-time cannot be neglected. It is important to realize that what 'distance' means depends on its observational definition: there is no unique, 'correct' definition of distance!

4.3.1 The Luminosity Distance

As the name implies, the luminosity distance d_L is defined in terms of how bright an object appears.

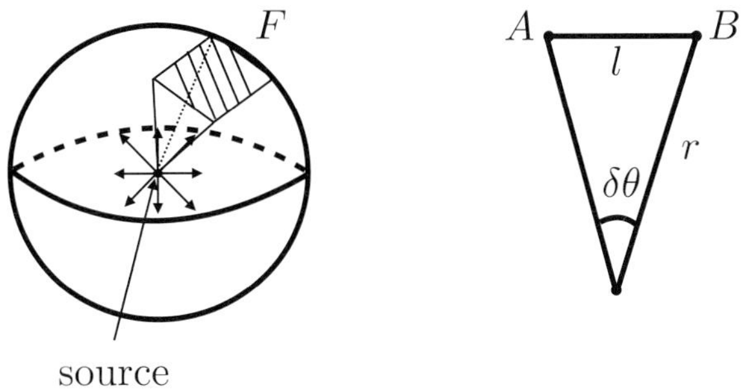

FIGURE 4.5 Left: Definition of luminosity distance. Right: Definition of angular diameter distance.

Definition: Consider a source of luminosity $L\left[\frac{\text{energy}}{\text{time}}\right]$ at redshift z. Its observed flux at our location is $F\left[\frac{\text{energy}}{\text{time}\times\text{length}^2}\right]$. In analogy with the usual inverse-square law, we define d_L as

$$F = \frac{L}{4\pi d_L^2(z)}, \tag{4.43}$$

where $4\pi d_L^2(z)$ would be the surface area of the 3-sphere (Figure 4.5, left).

4.3.2 The Angular Diameter Distance

The angular diameter distance, d_A, is defined from the angle subtended by an object on the sky.

Definition: Consider an object of physical length ℓ, subtending an angle $\delta\vartheta$ on the sky. The separation between A and B (Figure 4.5, right) is given by

$$\ell = r \cdot \delta\vartheta \Rightarrow \delta\vartheta = \frac{\ell}{r}, \tag{4.44}$$

where r is the physical distance to A and B. We thus define the angular diameter distance as:

$$d_A(z) = \frac{\ell}{\delta\vartheta}, \tag{4.45}$$

where ℓ is the physical size of an object at redshift z.

In Euclidean space, d_A and d_L are the same. In cosmology, they match $z \ll 1$ but are different at large redshift. It can be shown that in any metric theory of gravity (including general relativity) the 2 distances are related by:

$$d_L(z) = (1+z)^2 d_A(z). \tag{4.46}$$

From general relativity, one can compute theoretically $d_A(z)$ and $d_L(z)$. Those functions depend on the value of the cosmological parameters. Dimensional analysis tells us that the 'Hubble time' $1/H_0$ gives the scale for the age of the universe, while the typical size of the universe is set by c/H_0, the distance covered by light in a Hubble time. This is called the 'Hubble radius', and denoted by $R_H = c/H_0 = 2997/h$ Mpc. We can write the distances as

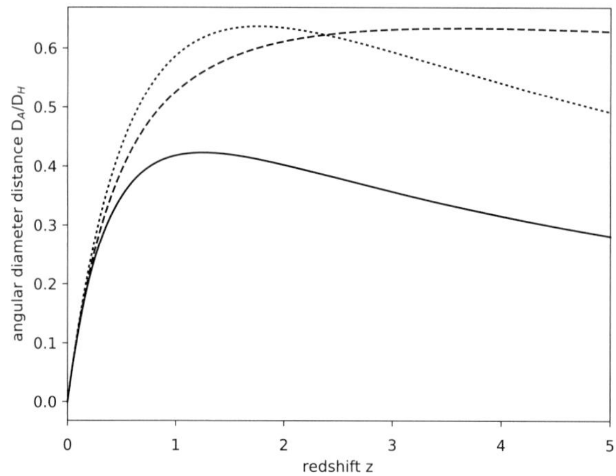

FIGURE 4.6 Examples of the angular diameter distance $d_A(z)$ for 3 choices of cosmological parameters (credit: Jonathan Pritchard).

$$d_A(z, \Theta) = \frac{c}{H_0} f_A(z, \Theta) \tag{4.47}$$

$$d_L(z, \Theta) = \frac{c}{H_0} (1+z)^2 f_A(z, \Theta), \tag{4.48}$$

where Θ is the cosmological parameters and $f_A(z, \Theta)$ is a known function found in any cosmology textbook (see Figure 4.6 for a few examples).

Notice that f_A is not necessarily a monotonically increasing function of z, as one might imagine: for values $z > z_{\max}$, it starts decreasing again (for most choices of cosmological parameters). This means that there is a minimum angle subtended on the sky by an object of a given size as a function of redshift. On the other hand, the apparent brightness of objects continues to decrease $\propto (1+z)^{-2}$ and the photons' energy is redshifted, hence objects become more and more difficult to see at large z.

The aim of observational cosmology is to measure 'standard rulers' (CMB and baryonic acoustic oscillations) or 'standard candles' (supernovae type Ia) as a function of z. This gives a measurement of $d_L(z)$ and $d_A(z)$, which in turn can be converted into a measurement of the cosmological parameters. Other techniques (such as standard clocks, standard sirens, and Alcock-Packinski test) are not covered here. After decades of effort, the cosmological parameters have been measured with great accuracy: we are in the era of 'precision cosmology', where most cosmological parameters are measured with better than 1% error. The latest data from the Planck CMB satellite (Planck Collaboration et al., 2015) give:

$$\Omega_m = 0.316 \pm 0.009 \tag{4.49}$$

$$H_0 = 67.8 \pm 0.9\,\text{km/s/Mpc} \tag{4.50}$$

$$\Omega_\Lambda = 0.684 \pm 0.009 \tag{4.51}$$

$$|\Omega_\kappa| < 0.005 \tag{4.52}$$

$$\text{Age} = 13.796 \pm 0.029\,\text{Gyr} \tag{4.53}$$

Notice that the value quoted above for H_0 comes from CMB measurements and is somewhat in tension with the slightly higher value found using local methods.

4.4 Supernovæ Type Ia Cosmology

4.4.1 Observational Definition and Formation Mechanism

Supernovæ (SN) are explosions of stars that reach a very high luminosity, $L \sim 10^9 L_\odot$ (where $\odot$ denotes the Sun), comparable to that of their host galaxy. They then fade away over a timescale of a few weeks. SN are traditionally classified according to their spectra:

- SN type I: no H lines in the spectrum, further subdivided in:
  - type Ia (SNIa): presence of strong Si II line at $\lambda = 615\,\text{nm}$
  - type Ib,c: (SNIb,c) no Si II lines
- SN type II (SNII): contain H lines

SNIb,c and SNII are 'core collapse' SN: they are the final stage of life of massive stars ($M \gtrsim 8M_\odot$), when iron fusion is reached in their core an no further energy can be gained from fusion. This means that there is no longer a source of pressure to counterbalance gravitational infall. The star implodes onto itself, then bounces back because of neutron degeneracy, leaving a neutron star as a remnant (or a black hole).

SNIa's are the explosion of white dwarfs (WD), compact stars made of C and O, the end of the evolution of less massive stars ($M \lesssim 2.5M_\odot$). Schematically, the astrophysics of WD formation goes as follows:

- H burning → explosive He burning
- mass ejection → He shell burning
- formation of a red giant/supergiant
- ejection of outer envelope
- central star contracts; core made of CO. Temperature is too low to ignite C fusion, but electron degeneracy pressure (from the Pauli exclusion principle) stops infall → high temperature, compact WD

So WDs are stars with typical size similar to the Earth's, mass $M \sim 0.6M_\odot$, density $\sim$ 1 tonne/cm^3.

If a WD accretes mass from a companion star, and if its mass approaches the 'Chandrasekhar limit' of $1.4M_\odot$ (past which electron degeneracy pressure cannot sustain the star any longer), C fusion is ignited in the core. The temperature of the star rises quickly, but the star cannot expand as its pressure does not change. This increases the temperature even faster and a runaway nuclear explosion destroys the star almost instantaneously. This process emits a large amount of energy, $E \sim 10^{44}$ J (by comparison, the yearly energy output of the Sun is $\sim 10^{34}$ J).

SNIa's are thus bright explosions, powered by a physical mechanism with a common underlying critical threshold, namely the Chandrasekhar mass. This makes them good candidates to act as standard candles.

The mechanism by which mass accretion occurs remains uncertain. There are 2 possibilities: the single-degenerate scenario (where mass accretion occurs from a massive, non-degenerate companion star), and the double-degenerate scenario (collision of 2 WDs in a close binary system).

The important fact is that by themselves SNIa's are *not* sufficiently accurate standard candles: their peak magnitude (in the B band) shows a variability of $\sim 0.4\,\text{mag}$, which is too large to use without further analysis.

We digress briefly to recall the concept of 'magnitude', used by astronomers to measure flux/luminosity.

Definition: Apparent magnitude (in a given passband), m, is a logarithmic measure of the flux observed from an object from the Earth, F:

$$m \equiv -2.5 \log_{10} \frac{F}{F_{\text{ref}}}, \tag{4.54}$$

where F_{ref} is a reference flux.

Definition: The absolute magnitude, M, is a logarithmic measure of the object's intrinsic brightness. It is the apparent magnitude of the object as seen from 10 pc away:

$$M = -2.5 \log_{10} \frac{L}{L_{\mathrm{ref}}}, \tag{4.55}$$

where L is the luminosity of the object and L_{ref} is a reference luminosity. Remember that, because of the minus sign in the definition, large brightness corresponds to small magnitude (which can be negative!). Mnemonic: 'large is small'. Also, in typical astronomy plots showing e.g. apparent magnitude over time, the vertical axis scale is reversed, so that small magnitudes (i.e. bright) are at the top!

The distance modulus μ is the difference between apparent and absolute magnitudes:

$$\mu \equiv m - M. \tag{4.56}$$

Starting from the definition of d_L, Equation 4.43, we divide both sides by $F_{\mathrm{ref}} = \frac{L_{\mathrm{ref}}}{4\pi(d_L = 10\ \mathrm{pc})^2}$ and take $-2.5\log_{10}$ of both sides, obtaining:

$$m - M = \mu = 5 \log_{10} \frac{d_L}{\mathrm{Mpc}} + 25. \tag{4.57}$$

This very important relation tells us that we can measure $d_L(z)$ from m (observed) if we know M. We do not actually need to know M, provided it is the same for all objects, in which case we measure d_L up to a multiplicative constant (H_0).

4.4.2 Standardization of SNIa

As mentioned above, M varies too much (at peak) between different SNIa's, as established by looking at a low-z calibration sample, for which the distance is known via independent means and hence observations of m can be converted into values of M.

However, Phillips (1993) discovered an empirical relation between M and the time it takes for an SNIa to fade. SNIa's that are intrinsically brighter (smaller M at their peak) take longer to fade (see Figure 4.7, upper plot). The drop in magnitude after 15 days, measured by the quantity Δm_{15}, is larger for dimmer SNIa's, i.e. $\Delta m_{15}^{\mathrm{dim}} > \Delta m_{15}^{\mathrm{bright}}$. Plotting the value of M at peak brightness versus Δm_{15} for many SNIa's, one obtains a linear relationship (as shown in Figure 4.7, lower plot).

The key idea behind 'standardising' SNIa's is to obtain observations of their light curve shape, determine from them Δm_{15} (or equivalent parameters), and use the empirical linear relation with M to 'correct' M, to achieve a much-reduced scatter of about ~ 0.12 mag (see Figures 4.8 and 4.9). The cosmological parameters entering into the theoretical expression for the luminosity distance can thus be determined. This led in the late 1990s to the discovery of the accelerated expansion of the universe, i.e. that $\Omega_\Lambda > 0$ (see Figure 4.10).

4.5 The CMB

4.5.1 The Origin of the CMB

The CMB is one of the most important tools for our detailed understanding of the universe for 2 main reasons: we have very accurate theoretical models that are based on general relativity and low-energy, well-understood (mostly) linear physics, and we have exquisite observations, with very small uncertainties.

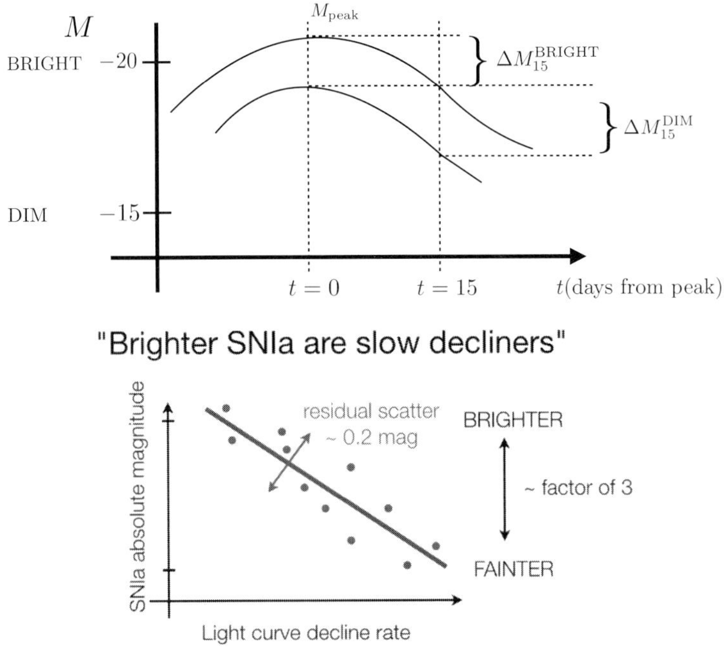

FIGURE 4.7 Top: SNIa's that are intrinsically brighter (smaller M at their peak) take longer to fade. Bottom: Linear relationship between peak intrinsic magnitude and drop in magnitude 15 days after peak.

The CMB originates from the early, hot universe and its temperature (photon wavelength) has been reduced (stretched) by the cosmological expansion. It can be shown that the temperature of a blackbody scales as a function of redshift as:

$$T(z) = T_0(1+z) = \frac{T_0}{a}, \tag{4.58}$$

where T_0 is its temperature today. Hence for large z (i.e. in the early universe) the CMB temperature was higher.

In the early universe, the baryonic matter was made mostly of H. hydrogen atoms were ionised because of the interaction with high-energy photons from the CMB:

$$e^- + \gamma \longrightarrow e^- + \gamma \tag{4.59}$$

This process proceeded via Thomson scattering, in which the energy of the photon is unchanged as it scatters off an electron virtually at rest. Only the photon's direction changes. This is valid as long as the photon energy, E_γ, obeys $E_\gamma \ll m_e c^2 \sim 0.5$ MeV.

As the universe expands, $a(t)$ grows and hence T_{CMB} (and therefore $\langle E_\gamma \rangle$) decreases. When the temperature is low enough, H^+ becomes neutral, by capturing a free electron:

$$p + e^- \longrightarrow \text{H} + \gamma. \tag{4.60}$$

Naively, one could imagine that the energy scale for this process, called 'recombination', would be set by the ionisation energy of H, i.e.

$$E_{\text{H}^+} = (m_e + m_p - m_H)c^2 = 13.6 \text{ eV} \approx 10^5 \text{ K}. \tag{4.61}$$

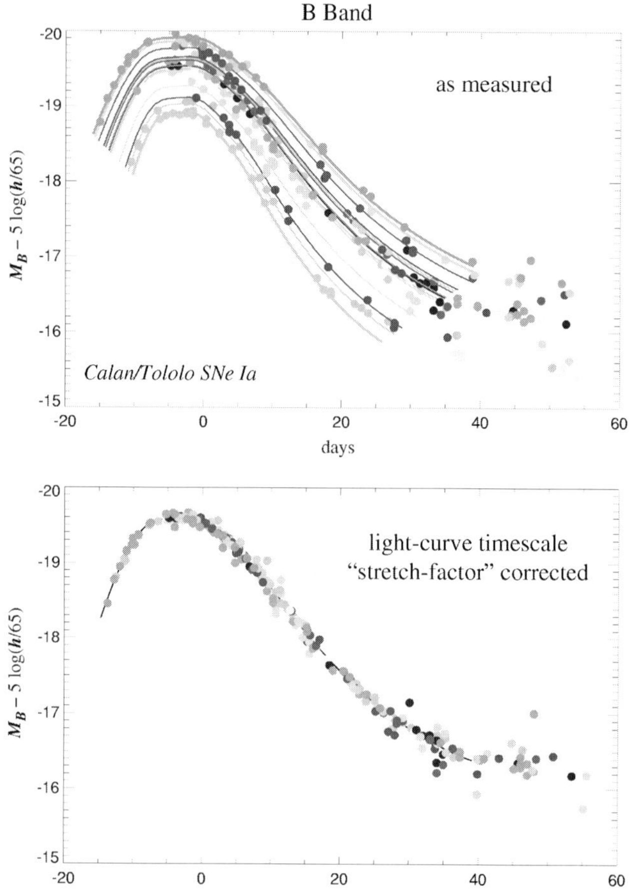

FIGURE 4.8 Up: Observed light curves of SN1a whose distance is known: the vertical axis shows absolute magnitude as a function of time. Down: After 'stretch correction', the light curves (and their magnitude) are standardized (credit: Saul Perlmutter).

However, we must keep in mind the high-energy Wien tail of the Planck distribution of photons, and the fact that there are many more photons than baryons (and hence, electrons, for the universe as a whole is neutral). The ratio of baryon to photon number density is denoted by

$$\eta_{10} \equiv \frac{n_b}{n_\gamma} \sim 10^{-9}. \tag{4.62}$$

As a consequence, recombination is delayed until the temperature has dropped much farther to

$$T_{\rm rec} \approx 3{,}000\,{\rm K} \approx 0.26\,{\rm eV}. \tag{4.63}$$

Since $\frac{T(z)}{T_0} = 1 + z$ and $T_0 = 2.72\,{\rm K}$, it follows that $z_{\rm rec} \approx 1{,}100$. Therefore the universe became neutral and transparent to photons for the first time at a redshift $z_{\rm rec} \approx 1{,}100$, corresponding to an age of the universe $t_{\rm rec} \approx 380{,}000$ years.

The ionisation history of the universe is summarised in Figure 4.11, where the ionisation fraction

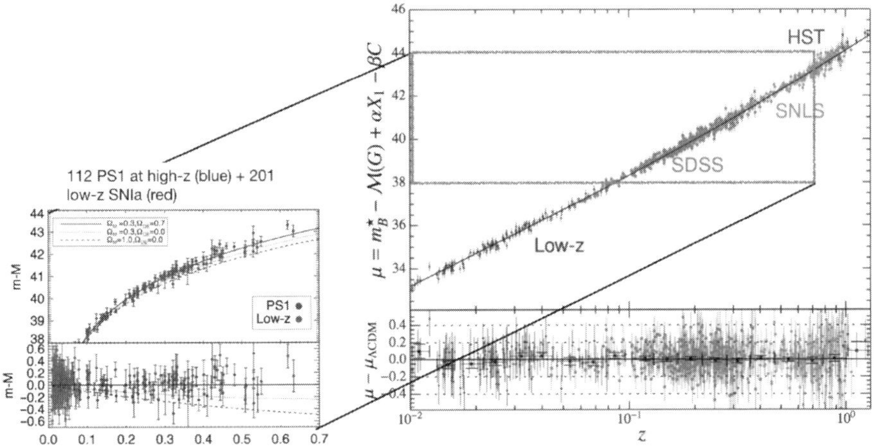

FIGURE 4.9 A compilation of some modern SNIa observations. Left: Rest et al. (2014). Right: Betoule et al. (2014).

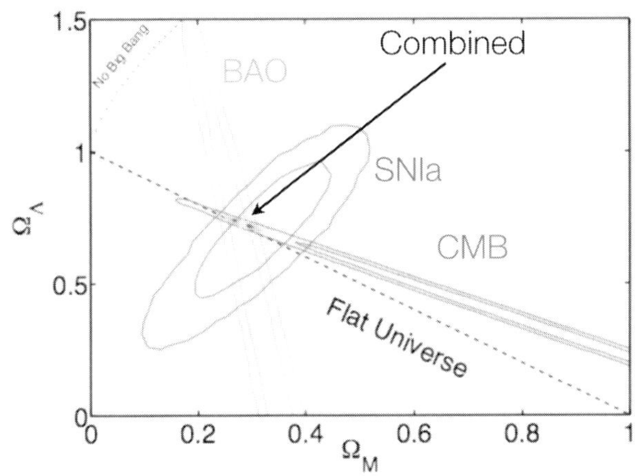

FIGURE 4.10 Combined constraints from SNIa, CMB, and baryon acoustic oscillations data (March et al., 2011).

$$x_e = \frac{n_e}{n_p + n_H} \tag{4.64}$$

is plotted as a function of z. Here, n_e is the comoving number density of free electrons, n_p the comoving number density of protons, and n_H the comoving number density of neutral hydrogen atoms. In the early universe, there is about 1 He nucleus for each 7 H nuclei, and so when He is fully ionised this frees up 2 electrons per He nucleus. This means that $x_e > 1$ at that epoch.

Starting from high redshift (early universe, right side of the plot), He recombines first, because its ionisation energy is larger than H. Hydrogen recombination happens at $z \approx 1{,}100$. When cosmologists talk about 'recombination', they always mean H recombination. Recombination is relatively quick, but not instantaneous, taking about $\Delta z \approx 80$ to complete. The universe remains almost completely neutral until $z \approx 8$, when UV radiation from the first stars photo-dissociates H atoms again and the universe is reionized.

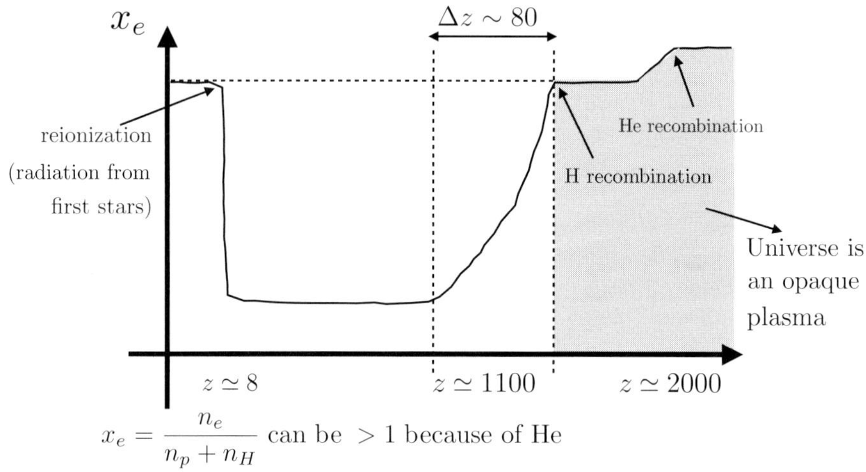

FIGURE 4.11 The ionisation history of the universe summarised. The ionization fraction x_e is plotted as a function of redshift, z.

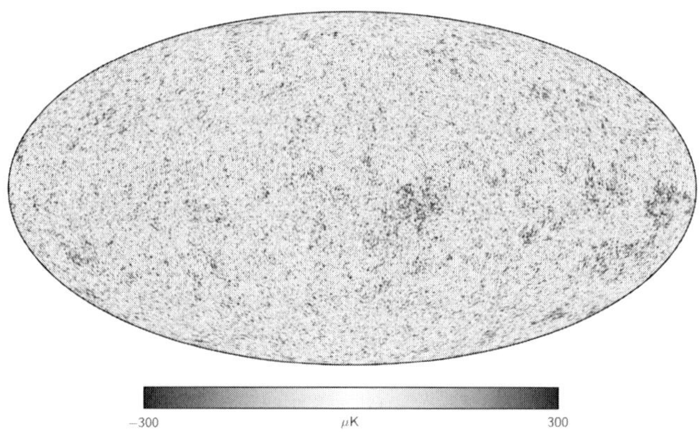

FIGURE 4.12 Full-sky CMB temperature anisotropy map measured by the Planck satellite (source: `www.cosmos.esa.int/web/planck`).

4.5.2 CMB Anisotropies

Very importantly, the energy density of the CMB is not exactly isotropic. There are small temperature fluctuations of the order

$$\frac{\delta T(\mathbf{n})}{T_0} \sim 10^{-5}, \tag{4.65}$$

where $\mathbf{n}$ is a unit vector giving the direction on the sky. These fluctuations were discovered in 1994 by the Cosmic Background Explorer satellite (see Smoot et al., 1992) – a discovery that was awarded the Nobel Prize in Physics in 2006. The fluctuations have since been mapped out in great detail, most notably by the Wilkinson Microwave Anisotropy Probe satellite, and recently by Planck (see Figure 4.12).

Observationally, these 'temperature anisotropies' can be distinguished from foregrounds (e.g. galactic emission) by exploiting the blackbody spectrum and by using multifrequency observations: while the CMB is a blackbody, foregrounds are not.

Theoretically, the statistical properties of the anisotropies can be predicted with very high accuracy because they are the result of low-energy, well-understood physics in an expanding space–time, and they are very small. This means that first-order perturbation theory is mostly adequate to describe them.

The physical origin of the temperature anisotropies is thought to be quantum fluctuations right after the Big Bang which have been stretched to cosmological scales by inflation. The temperature anisotropies we see today are the results of 4 main physical effects:

$$\left.\frac{\delta T(\mathbf{n})}{T_0}\right|_{\text{today}} = \left.\frac{\delta T}{T_0}\right|_{\text{rec}} + \Psi\Big|_{\text{rec}} + \mathbf{n}\cdot\mathbf{v}\Big|_{\text{rec}} + \int_{t_{\text{rec}}}^{t_0} 2\dot{\Psi}dt, \tag{4.66}$$

where Ψ is the gravitational potential, $\mathbf{v}$ is the photon-baryon velocity, and $\mathbf{n}$ is the direction in the sky. We briefly discuss each of the terms in turn.

(i) $\left.\frac{\delta T}{T_0}\right|_{\text{rec}}$: This represents temperature fluctuations in the CMB photons at the moment when they were emitted. A higher temperature is a reflection of a higher density in the photon-baryons fluid.

(ii) $\Psi\big|_{\text{rec}}$: The presence of fluctuations in the gravitational potential (Ψ) at the moment when the photons are emitted induces a redshift (when the potential is deeper than average) or blueshift (when it is shallower than average) in the photons.

(iii) $\mathbf{n}\cdot\mathbf{v}\big|_{\text{rec}}$: This is a Doppler effect from the velocity of the photon-baryon fluid when the photons were emitted. Only the component parallel to the line of sight ($\mathbf{n}$) contributes.

(iv) $\int_{t_{\text{rec}}}^{t_0} 2\dot{\Psi}dt$: This 'integrated Sachs-Wolfe effect' represents a change in the energy of the photon brought about by the time variation of the gravitational potential along the line of sight. If the gravitational potential changes with time ($\dot{\Psi} \neq 0$), the gravitational blueshift of the photon falling into the potential is no longer exactly compensated by the gravitational redshift when it climbs out of it.

All of the above contribute to what we see today, but the single most important effect is (i): the *acoustic oscillations* phenomenon. The underlying physics can be understood very simply: (slightly) overdense regions attract matter (both baryonic and dark) through gravitational collapse. However, baryons are coupled to photons via Thomson scattering. The photons' pressure hence becomes a restoring force which counteracts gravity. Dark matter is unaffected by photon pressure, as it does not interact electromagnetically, and collapses further. However, the photon-baryon fluid undergoes periodic oscillations: acoustic waves (i.e. sound). The sound horizon λ_S is the distance that a sound wave in the early universe can cover from $t = 0$ until the time of recombination, t_{rec}. The comoving sound horizon is given by the expression

$$\lambda_S^{\text{com}} = \int_0^{t_{\text{rec}}} \frac{c_s}{a}dt = \frac{c}{\sqrt{3}} \int_0^{(1+z_{\text{rec}})^{-1}} \frac{da}{Ha^2} \tag{4.67}$$

where the sound speed c_s is approximately given by $c_s = c/\sqrt{3}$. By using the expression for the Hubble function in a matter-dominated universe,

$$H^2 = H_0^2\frac{\Omega_m}{a^3}, \tag{4.68}$$

we obtain

$$\lambda_s^{\text{com}} \approx \frac{2c}{\sqrt{3}H_0\sqrt{\Omega_m}}(1+z_{\text{rec}})^{-1/2} \approx 150 \text{ Mpc}. \tag{4.69}$$

The sound horizon at recombination gives us a standard ruler (of huge size!) that we can use to measure $d_A(z_{\rm rec})$.

After recombination, electrons get bound to protons and photons can propagate freely. This is called 'decoupling'. The photon pressure in the plasma disappears, and gravity makes overdense regions collapse further. Regions of higher density (corresponding to crests in the sound waves) also exhibit higher $\delta T/T$ in the CMB photons. We say that the acoustic oscillations are frozen in the CMB at the moment of decoupling.

The fluctuations we see in the sky are the superposition of many waves. Therefore, the acoustic scale can only be extracted statistically, by computing the 'power spectrum', representing the Fourier transform of the 2-point correlation function of the temperature anisotropies on the sky. The power spectrum is usually plotted as a function of 'multiple number' ℓ, which is inversely proportional to the angular separation on the sky, ϑ.

The first acoustic peak in the CMB power spectrum corresponds to the 'first harmonic' of the sound waves – higher peaks correspond to higher harmonics. The first peak is measured at $\ell \approx 200 \approx 1\,\text{deg}$, and this allows us to measure the angular diameter distance to the epoch of recombination:

$$d_A(z_{\rm rec}) = \frac{\lambda_s^{\rm phys}}{\vartheta_{\rm peak}}, \tag{4.70}$$

where $\lambda_s^{\rm phys}$ is the known physical size of the acoustic oscillations at recombination, $\lambda_s^{\rm phys} = \lambda_s^{\rm com}/(1+z_{\rm rec})$ and $\vartheta_{\rm peak} \approx 1\,\text{deg}$ is the measured angle of the first acoustic peak in the power spectrum. We can also compute $d_A(z_{\rm rec})$ from out theoretical model. For a $\Omega_\Lambda = 0$ universe, one finds that:

$$d_A(z_{\rm rec}) \approx \frac{c}{H_0}\frac{2}{\Omega_m z_{\rm rec}}, \tag{4.71}$$

showing explicitly that, as expected, $d_A(z_{\rm rec})$ depends on the cosmological parameters H_0, Ω_m. The physical size of the acoustic scale is given by

$$\lambda_s^{\rm phys} = \frac{\lambda_s^{\rm com}}{(1+z)} \approx \frac{2}{\sqrt{3}}\frac{c}{H_0}\frac{1}{\sqrt{\Omega_m}}\frac{1}{(1+z_{\rm rec})^{3/2}}. \tag{4.72}$$

Putting the 2 expressions together we find:

$$\vartheta_{\rm peak} \approx \sqrt{\frac{\Omega_m}{z_{\rm rec}}\frac{1}{\sqrt{3}}} \approx \sqrt{\Omega_m} \times 1.1\,\text{deg}. \tag{4.73}$$

Since $\vartheta_{\rm peak} \approx 1\,\text{deg}$, it follows that $\Omega_m \approx 1$ (in a $\Omega_\Lambda = 0$ universe). In conclusion, the CMB acoustic scale subtends an angular scale of approximately 1 deg on the sky today. This allows us to measure the angular diameter distance to recombination, and it implies that $\Omega_m \approx 1$, i.e. that the universe is flat. A much more refined analysis (Planck Collaboration et al., 2015) yields for the curvature parameter (remember, $\Omega_\kappa = 0$ means a flat universe):

$$\Omega_\kappa = 0.0008 \pm 0.0040. \tag{4.74}$$

The above, simplified calculation is an example of how we can extract cosmological parameters from present-day CMB measurements. Other cosmological parameters have distinct effects on the shape of the CMB power spectrum (see Figure 4.13). By measuring the power spectrum to high precision (Figure 4.14), the value of the cosmological parameters can be inferred.

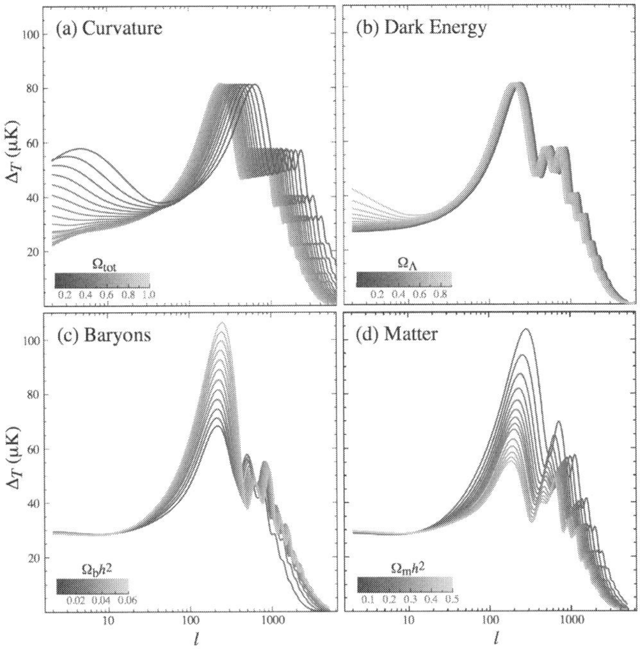

FIGURE 4.13 Impact of some cosmological parameters on the shape of the CMB power spectrum (Hu and Dodelson, 2002).

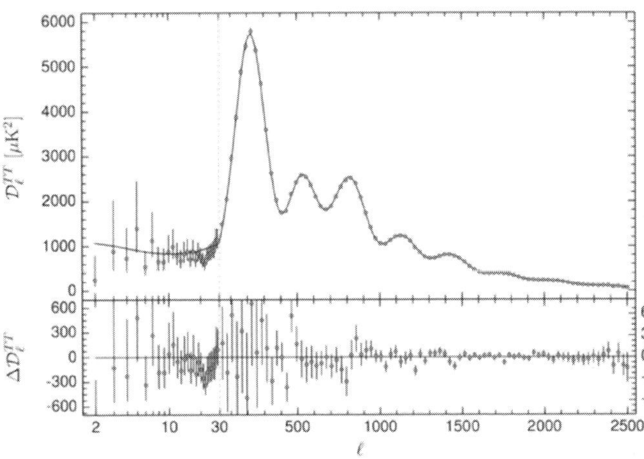

FIGURE 4.14 The CMB temperature power spectrum exhibiting acoustic oscillations, as measured by Planck. Notice the position of the first peak, at $\ell \approx 200$. The red curve is a 6-parameters theoretical model fit, the ΛCDM model – where Λ stands for a cosmological constant and CDM for 'cold dark matter' (Planck Collaboration et al., 2015).

Further information is encoded in the polarisation of CMB photons. This is much harder to measure, as only ≈ 1% of the photons are polarised, and it is subject to the confounding effects of foregrounds. The important point to remember is that much of our successes with modern-day precision cosmology relies on our detailed understanding of CMB physics.

4.6 Structure Formation

4.6.1 The Large Scale Structure of the Universe

Thus far, we have considered a cosmological model that is isotropic and homogeneous, with small departures from a perfectly smooth distribution of matter-energy density in the CMB anisotropies.

Clearly, the baryonic matter distribution around us is far from homogeneous: galaxies, clusters, and superclusters form structures with density far greater than average. We thus need to understand the structure of the matter distribution in the universe on large scales. This is called 'large-scale structure' (LSS[3]) of the universe. The observational study of the LSS of the universe is done via 'galaxy redshift surveys': the idea is to measure the position on the sky and the redshift of as many galaxies as possible, leading to a 3-D map of the universe.

Measuring the angular position on the sky can be done via imaging, i.e. by taking a picture of a patch of the sky – usually with several colour filters – and determining the position of the galaxies there. Determining redshift with high accuracy ($\delta z/z \sim 10^{-3}$) requires high-resolution spectra of the galaxies, which is time consuming. In the last 20 years, dedicated multiobject spectrographs have been developed that can take the spectrum of many galaxies at once. Examples of important galaxy redshift surveys include:

- The first CfA (Harvard) Galaxy Redshift Survey (1977–1982) measured galaxies in the nearby universe, where Hubble's law applies. It obtained spectra for $\sim 1,100$ galaxies, out to a maximum distance $d \sim 200\,\text{Mpc}$, with a limiting magnitude $m \approx 14.5$.
- The 2dF Galaxy Redshift Survey (1997–2002) measured 250,000 galaxies out to $z \sim 0.3$, to a maximum distance $d_A \sim 600/h\,\text{Mpc}$, with a limiting magnitude $m \approx 19.5$ (Figure 4.15).
- The Sloan Digital Sky Survey (SDSS) (2000–ongoing) started in 2000, and this highly successful programme continues to this day. It has measured the position and redshift of $\sim$ 1.5M galaxies, out to a redshift $z \sim 0.7$ with a limiting magnitude $m \approx 22.0$. All SDSS data are publicly available from `sdss.org`.

The distribution of galaxies in the sky is clearly not random, (Figure 4.15). The galaxy shows structures of size up to $\sim 100/h\,\text{Mpc}$, but there seems not to be any structures on scales above $R_{\text{hom}} \approx 200/h\,\text{Mpc}$, the 'homogeneity scale'. Comparing this scale with the Hubble radius R_H, we find that

$$R_{\text{hom}} \ll R_H, \tag{4.75}$$

and inside a Hubble volume we can fit:

$$\left(\frac{R_H}{R_{\text{hom}}}\right)^3 \sim 15^3 \approx 3{,}000 \tag{4.76}$$

'independent' volume elements. Averaged over scales $\gg R_{\text{hom}}$, the universe becomes homogeneous.

[3] The acronym LSS is routinely used to denote both the large-scale structure and the 'last scattering surface', i.e. the moment of CMB decoupling. Context should usually be sufficient to understand which one is meant.

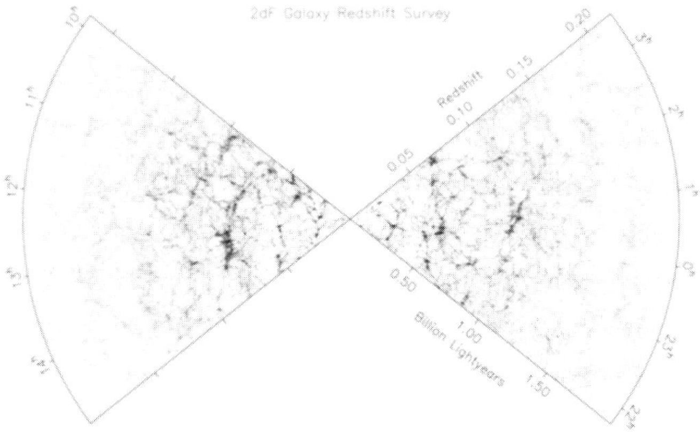

FIGURE 4.15 Completed in 2002, the 2dF redshift survey obtained spectra of 250k galaxies, revealing the large-scale structure of the universe. Each of the dots indicates the position of a galaxy (credit: 2dF Galaxy Redshift Survey team).

4.6.2 Gravitational Instability

We know from the CMB that at $z_{\rm rec} \approx 1,100$ the distribution of matter was not uniform, with fluctuations of order

$$\frac{\delta\rho}{\rho} \sim \frac{\delta T}{T} \sim 10^{-5}. \tag{4.77}$$

Today, density fluctuations are clearly much larger: a massive galaxy cluster with a typical radius $r \sim 1.5/h\,\text{Mpc}$ contains ~ 200 times the average density of the universe. The natural question to ask is: how did these structures form? The answer is gravitational instability, i.e. the process by which slightly overdense regions attract surrounding matter slightly more, thus growing denser, hence increasing their gravitational attraction, and so forth. This is the phenomenon of *gravitational collapse.*

To understand the basics of gravitational collapse, we start by defining the relative matter density contrast, $\delta(\mathbf{x}, t)$:

$$\delta(\mathbf{x}, t) \equiv \frac{\rho(\mathbf{x}, t) - \bar{\rho}(t)}{\bar{\rho}(t)}, \tag{4.78}$$

where $\bar{\rho}(t)$ is the spatially averaged, time-dependent, homogeneous density (see Figure 4.16). By definition, $\delta \geq -1$, because obviously $\rho \geq 0$, as you cannot get more empty than empty. At recombination, $|\delta| \ll 1$, but today it can be arbitrarily large: for example, the overdensity on Earth is of the order $\delta \sim 2 \cdot 10^{30}$.

Consider a spherical region of comoving radius R in the early universe, around $t_{\rm rec}$, in a homogeneous, matter-only universe. We add or subtract some matter to the sphere (see Figure 4.17), and compute the acceleration due to gravity of a test particle on the surface of the sphere (remember that the particle only feels the gravity of the mass *inside* the sphere):

$$\ddot{r} = \ddot{a}R = -\frac{GM(R)}{r^2(t)}, \tag{4.79}$$

where we have that $r = aR$. With $M(R) = \frac{4\pi}{3}(\bar{\rho}(1+\delta))r^3$ we obtain

$$\frac{\ddot{r}}{r} = \frac{\ddot{a}}{a} = -\frac{4\pi G}{r}\bar{\rho} - \frac{4\pi}{3}G\bar{\rho}\delta. \tag{4.80}$$

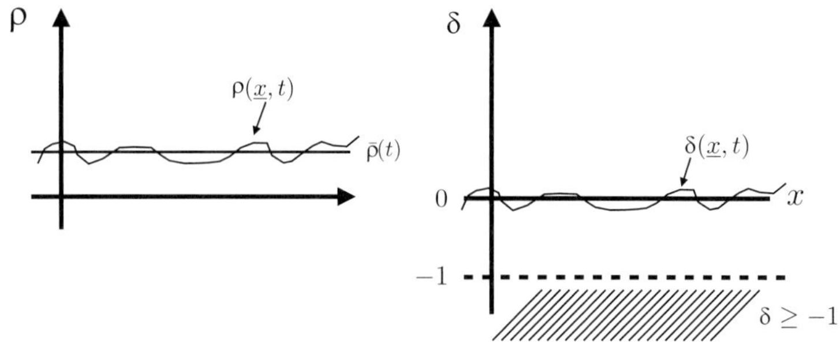

FIGURE 4.16 Definition of the density contrast.

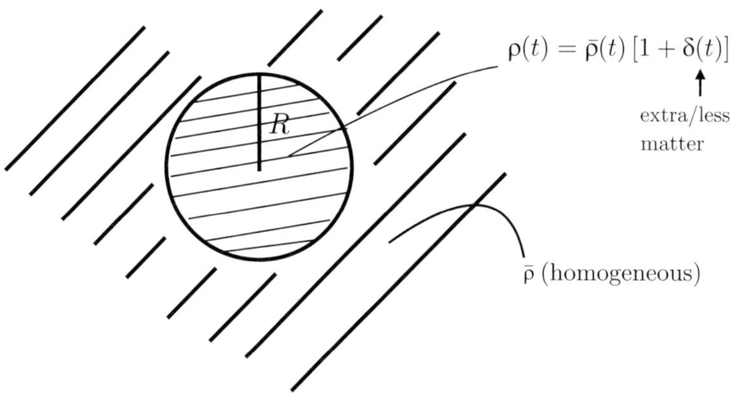

FIGURE 4.17 Analysis of spherical density contrast evolution.

Mass conservation dictates that the mass inside the sphere is constant:

$$M = \frac{4\pi}{3}\left(\frac{\rho_0}{a^3}(1+\delta)\right) r^3 = \text{ constant}, \tag{4.81}$$

where we have used $\bar{\rho}(t) = \rho_0/a(t)^3$. So the physical radius of the sphere scales as

$$r \propto a \left(\frac{1}{1+\delta}\right)^{1/3}. \tag{4.82}$$

If $\delta > 0$ (overdensity), the radius of the sphere grows slightly less rapidly than a, and, vice versa, if $\delta < 0$ (underdensity), the radius of the sphere grows more rapidly than a.

We now take the second derivative of Equation 4.82 and divide by r:

$$\frac{\ddot{r}}{r} = \frac{1}{r}\frac{d^2}{dt^2}(a \cdot b), \tag{4.83}$$

where $b \equiv (\delta + 1)^{-1/3}$. It follows:

$$\frac{\ddot{r}}{r} = \frac{\frac{d^2}{dt^2}(a \cdot b)}{a \cdot b} = \frac{\ddot{a}}{a} + \frac{2\dot{a}\dot{b}}{a \cdot b} + \frac{\ddot{b}}{b} \tag{4.84}$$

and, since $|\delta| \ll 1$, we only keep terms that are linear in the perturbation δ or its time derivative. Thus:

$$\dot{b} = -\frac{1}{3}\frac{\dot{\delta}}{(\delta+1)^{4/3}} \approx -\frac{1}{3}\dot{\delta} \text{ (to linear order)} \tag{4.85}$$

$$\ddot{b} \approx -\frac{1}{3}\ddot{\delta} \text{ (to linear order).} \tag{4.86}$$

It follows that:

$$\begin{aligned} \frac{\ddot{r}}{r} &= \frac{\ddot{a}}{a} - \frac{2}{3}\frac{\dot{a}}{a}\dot{\delta} - \frac{1}{3}\ddot{\delta} \\ &= -\frac{4\pi G}{3}\bar{\rho} - \frac{4\pi G}{3}\bar{\rho}\delta. \end{aligned} \tag{4.87}$$

Because of the homogeneous equation solution (for $\delta = 0$) we have that

$$\frac{\ddot{a}}{a} = -\frac{4\pi G}{3}\bar{\rho} \tag{4.88}$$

and so we obtain

$$-\frac{1}{3}\ddot{\delta} - \frac{2}{3}H\dot{\delta} = -\frac{4\pi G}{3}\bar{\rho}\delta. \tag{4.89}$$

This leads to the density contrast evolution equation:

$$\ddot{\delta} + 2H\dot{\delta} = 4\pi G\bar{\rho}\delta. \tag{4.90}$$

This linearised differential equation for the density contrast $\delta(\mathbf{x}, t)$ does not involve any derivative with respect to $\mathbf{x}$. This means that we can separate $\delta(\mathbf{x}, t)$ with the following Ansatz:

$$\delta(\mathbf{x}, t) = D(t)\tilde{\delta}(\mathbf{x}), \tag{4.91}$$

where $D(t)$ is a function of time only, and $\tilde{\delta}(\mathbf{x})$ is the spatial-only perturbation.

If we start at time t_1 with an arbitrary spatial distribution of inhomogeneities, $\tilde{\delta}(\mathbf{x})$, gravitational collapse will make $\delta(\mathbf{x}, t)$ grow with time without changing the spatial shape (in comoving coordinates). When $|\delta|$ becomes of order ~ 1, linear perturbation theory of the kind we used breaks down. Numerical simulations are needed to follow non-linear gravitational collapse. In the density contrast evolution equation, Equation 4.90, the term $\propto H\dot{\delta}$ acts as a friction term: the expansion of the universe ($H > 0$) slows down gravitational collapse.

We can solve Equation 4.90 easily in a flat, matter-only universe, where

$$\bar{\rho}_m = \rho_{\text{crit}}. \tag{4.92}$$

In such a universe, we have that

$$H(t) = \frac{2}{3}\frac{1}{t}. \tag{4.93}$$

Also, $\bar{\rho}_m = \bar{\rho}$ (since there is only matter) and $\bar{\rho} = \rho_{\text{crit}}$ (since it is a flat universe). We also generalise the definition of critical energy density to a time-dependent case:

$$\frac{8\pi G}{3}\rho_{\text{crit}}(t) = H^2(t). \tag{4.94}$$

With this we can write

$$4\pi G\bar{\rho} = 4\pi G\rho_{\text{crit}}(t) = \frac{3}{2}H^2(t) = \frac{2}{3}\frac{1}{t^2}. \tag{4.95}$$

So the density contrast equation becomes:

$$\ddot{\delta} + \frac{4}{3t}\dot{\delta} - \frac{2}{3t^2}\delta = 0. \tag{4.96}$$

We make use of the Ansatz, with $D(t) = t^n$:

$$\delta(\mathbf{x}, t) = \tilde{\delta}(\mathbf{x})t^n, \tag{4.97}$$

which leads to the 2 independent solutions for the time-dependent part:

$$D(t) = D_1 t^{2/3} + D_2 t^{-1}. \tag{4.98}$$

The solution is a superposition of a growing mode ($\propto t^{2/3}$) and a decaying mode ($\propto 1/t$), which can be ignored as it will disappear over time, since $D_2 t^{-1} \to 0$ for $t \to \infty$.

Therefore we conclude that in a matter-only flat universe density perturbations grow with time as

$$\delta \propto t^{2/3} \propto a \propto \frac{1}{1+z}. \tag{4.99}$$

At the time of recombination, $|\delta| \sim 10^{-5}$, at a redshift $z_{\text{rec}} \sim 10^3$. This means that from the above calculation we would expect that density perturbations today should be of order:

$$|\delta(t_0)| = |\delta(t_{\text{rec}})|(1 + z_{\text{rec}}) \sim 10^{-5} \times 10^3 \sim 10^{-2}. \tag{4.100}$$

This is obviously not what we observe today, when $\delta \gg 1$. Notice that this discrepancy is even stronger for a universe that – like ours – contains a non-0 cosmological constant, $\Omega_\Lambda > 0$. When the cosmological constant starts to dominate the energy density, the growth of perturbations almost stops. With the definition of the time-dependent matter density parameter $\Omega_m(t)$

$$\Omega_m(t) = \frac{\bar{\rho}_m(t)}{\rho_{\text{crit}}(t)} = \frac{8\pi G\bar{\rho}_m}{3H^2}. \tag{4.101}$$

We can rewrite the right-hand side of Equation 4.90 as follows:

$$4\pi G\bar{\rho}_m\delta = \frac{3}{2}\Omega_m(t)H^2\delta. \tag{4.102}$$

This leads to reformulating the density contrast evolution equation as:

$$\ddot{\delta} + 2H\dot{\delta} - \frac{3}{2}\Omega_m(t)H^2\delta = 0. \tag{4.103}$$

When the universe is cosmological constant dominated, $\Omega_m(t) \ll 1$ and the term $\propto \delta$ can be neglected. The Hubble function during a cosmological constant (Λ) phase is given by the constant value

$$H(t) = H_\Lambda \equiv (\Lambda/3)^{1/2} \tag{4.104}$$

and therefore the density contrast evolution equation becomes:

$$\ddot{\delta} + 2H_\Lambda\dot{\delta} = 0, \tag{4.105}$$

with solution

$$\delta(t) \approx C_1 + C_2 e^{-2H_\Lambda t}. \tag{4.106}$$

This shows that, during a cosmological constant–dominated phase, fluctuations in the matter density stop growing (the non-decaying mode, C_1, is constant).

The solution to the problem of the growth of perturbations is dark matter: since dark matter is neutral, it is decoupled from photons in the early universe, when perturbations in baryons cannot grow due to the photons restoring pressure (which leads instead to acoustic oscillations). It turns out that perturbations in the dark-matter distribution only grow logarithmically (hence, very slowly) with time while the universe is dominated by radiation, i.e. for the first $\sim$ 47,000 years after the Big Bang.

After the epoch of matter-radiation equality, $z_{\rm eq}$, the universe becomes matter dominated and dark matter perturbations start to grow (and our analysis above applies). Dark matter perturbations have a 'head start' on baryonic perturbations. Once the universe becomes neutral at recombination, baryons can fall into the gravitational wells created by dark matter, thus 'catching up' much faster than gravitational collapse under their own mass would imply.

4.7 Baryonic Acoustic Oscillations

After recombination, baryons 'catch up' with the dark matter by falling into the gravitational potential dominated by dark matter – remember that $\Omega_b/\Omega_{\rm DM} \sim 1/5$, i.e. dark matter (DM) is about 5 times more abundant than baryons. This is why the gravitational perturbations are dominated by DM.

However, we expect overdensities in baryons to occur at a preferential separation of $\lambda_s^{\rm com} \approx 150\,{\rm Mpc}$ (comoving), corresponding to the scale of the sound horizon. This is because an overdensity of baryons in correspondence with the crests of the acoustic sound waves occurs (see Figure 4.18 for a sketch).

The acoustic scale thus marks a preferential scale in the separation of galaxies. This translates into a slightly larger probability of finding 2 galaxies if they are separated by $\lambda_s^{\rm com} \approx 150\,{\rm Mpc}$. This can be quantified statistically with the 2-point correlation function.

Let us define P_1 as the probability of finding a galaxy in volume element dV at a location $\mathbf{x}$. If $\bar{n}$ is the average number density of galaxies, then $P_1 = \bar{n}dV$.

We define P_2 as the probability of finding a second galaxy in the volume element dV at location $\mathbf{y}$, given the first galaxy at $\mathbf{x}$. We have that

$$P_2 = (\bar{n}dV)^2(1 + \xi_g(\mathbf{x}, \mathbf{y})), \tag{4.107}$$

where ξ_g is the galaxy–galaxy correlation function. If galaxies were distributed at random, then

$$P_2 = (P_1)^2 \text{ and } \xi_g = 0. \tag{4.108}$$

But galaxies are not distributed at random: given a galaxy at location $\mathbf{x}$, it is more probable to find another galaxy nearby. Also, there is an increase in the probability in correspondence with a separation λ_s, which appears as a 'bump' in the 2-point correlation function (see Figure 4.19). Because of homogeneity, ξ_g can only depend on the difference $\mathbf{x} - \mathbf{y}$. Because of isotropy, it can only depend on $r = |\mathbf{x} - \mathbf{y}|$.

The baryon acoustic oscillations (BAO) can be used as a standard ruler by measuring the galaxy–galaxy correlation function at $z_{\rm BAO}$ (the average redshift of the galaxy survey – although in the future several 'redshift slices' will be used). The comoving size $\lambda_s^{\rm com}$

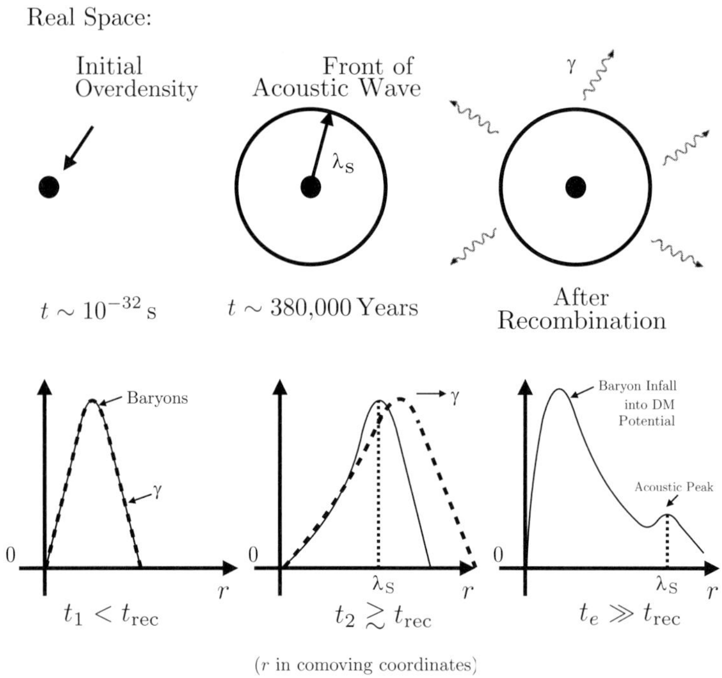

FIGURE 4.18 Time evolution of an acoustic wave. Top: 2-D view. Bottom: Section along the radius of the expanding wave.

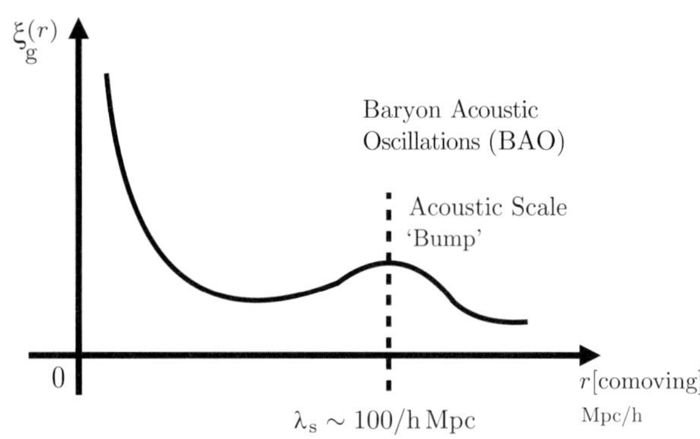

FIGURE 4.19 The acoustic scale appears as a 'bump' in the galaxy correlation function.

of the sound horizon is known from the CMB, and this calibrates the absolute scale of the BAO. Since $\lambda_s^{\rm phys} = a\lambda_s^{\rm com} = \lambda_s^{\rm com}/(1+z)$, it follows that $\lambda_s^{\rm com} = (1+z)\lambda_s^{\rm phys}(z) =$ constant. Therefore, evaluating this at both $z_{\rm BAO} \sim 0.3$ and $z_{\rm rec} = 1{,}100$ we get:

$$\lambda_s^{\rm phys}(z_{\rm BAO}) = \lambda_s^{\rm phys}(z_{\rm rec})\frac{1+z_{\rm rec}}{1+z_{\rm BAO}} \approx 846\lambda_s^{\rm phys}(z_{\rm rec}). \tag{4.109}$$

This says that the physical size of the acoustic scale has grown by approximately a factor of 846 from redshift $z_{\rm rec}$ to redshift $\sim$0.3 where the BAO are typically measured.

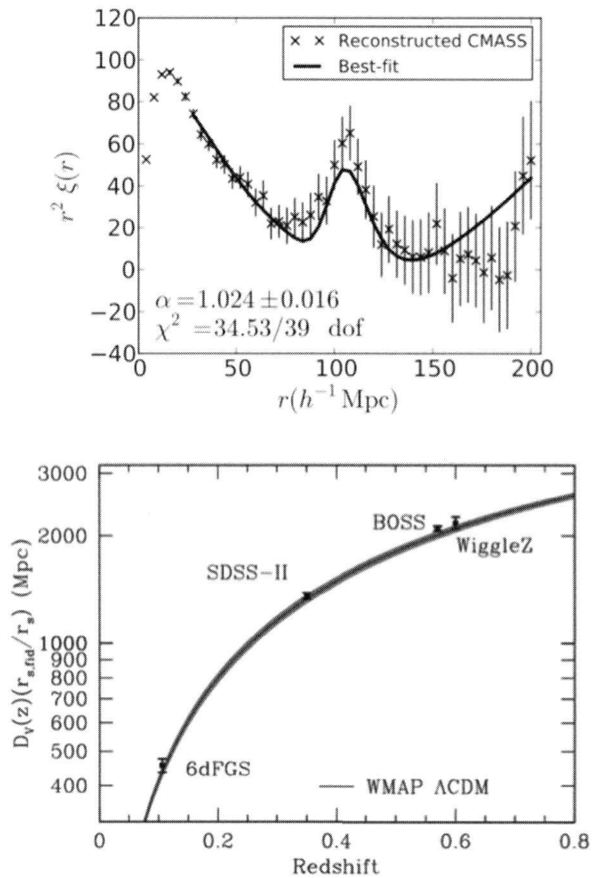

FIGURE 4.20 BAO measurement from the Baryon Oscillation Spectroscopic Survey (BOSS) (on SDSS). The 'ConstantMASS sample consists of 264k massive galaxies covering a volume of 2.2 Gpc^3, measuring the angular diameter distance to redshift z = 0.57 with 1.7% accuracy (Anderson et al., 2012).

Examples of measurements of the BAO scale (achieved for the first time by Eisenstein et al., 2005) are shown in Figure 4.20. The corresponding constraints on the cosmological parameters $(\Omega_m, \Omega_\Lambda)$ are shown as the green ellipses in the bottom plot of Figure 4.10.

4.8 Bayesian Inference

We now turn to discussing some of the statistical underpinnings and algorithmical tools that form the basis of modern precision cosmology. Further details (and references) can be found in e.g. Trotta (2008).

4.8.1 Bayes' Theorem

There are 2 different ways of understanding what probability is. The classical ('frequentist') notion of probability is that probabilities are tied to the frequency of outcomes over a long series of trials. Repeatability of an experiment is the key concept. The

Bayesian outlook[4] is that probability expresses a degree of belief in a proposition, based on the available knowledge of the experimenter. Information is the key concept. Bayesian probability theory is more general than frequentist theory, as the former can deal with unique situations that the latter cannot handle (e.g. 'What is the probability that it will rain tomorrow?').

Let $A, B, C, \ldots$ denote propositions (e.g. that a coin toss gives tails). The 'joint probability' of A and B is the probability of A and B happening together, and is denoted by $P(A, B)$. The 'conditional probability' of A given B is the probability of A happening given that B has happened, and is denoted by $P(A|B)$.

The sum rule is that:

$$P(A) + P(\overline{A}) = 1, \tag{4.110}$$

where $\overline{A}$ denotes the proposition 'not A'.

The product rule is that:

$$P(A, B) = P(A|B)P(B). \tag{4.111}$$

By inverting the order of A and B we obtain that

$$P(B, A) = P(B|A)P(A) \tag{4.112}$$

and because $P(A, B) = P(B, A)$, we obtain Bayes' theorem by equating Equations 4.111 and 4.112:

$$P(A|B) = \frac{P(B|A)P(A)}{P(B)}. \tag{4.113}$$

The marginalisation rule follows from the 2 rules above:

$$P(A) = P(A, B_1) + P(A, B_2) + \cdots = \sum_i P(A, B_i) \tag{4.114}$$

$$= \sum_i P(A|B_i)P(B_i), \tag{4.115}$$

where the sum is over all possible outcomes for proposition B.

Two propositions (or events) are said to be independent if and only if

$$P(A, B) = P(A)P(B). \tag{4.116}$$

4.8.2 Inferential Use of Bayes' Theorem

We replace in Bayes' theorem, Equation 4.113, $A \to \Theta$ (the parameters) and $B \to d$ (the observed data, or samples), obtaining

$$p(\Theta|d) = \frac{p(d|\Theta)p(\Theta)}{p(d)}. \tag{4.117}$$

On the left-hand side, $p(\Theta|d)$ is the posterior probability density function for Θ ('posterior PDF'), and it represents our degree of belief about the value of Θ after we have seen the data, d. The quantity $p(\Theta|d)d\Theta$ is the posterior probability for the value of the parameter Θ to lie in the interval $[\Theta, \Theta + d\Theta]$.

[4] So-called after Rev Thomas Bayes (1701(?)–1761), who was the first to introduce this idea in a paper published posthumously in 1763 (Bayes and Price, 1763).

On the right-hand side, $p(d|\Theta) = \mathcal{L}(\Theta)$ is the likelihood function. It is the probability of the observed data given a certain value of the parameter, seen as a function of the parameter.

The quantity $p(\Theta)$ is the prior probability distribution function ('prior PDF'). It represents our degree of belief in the value of Θ before we see the data (hence the name). This is an essential ingredient of Bayesian statistics.

In the denominator, $p(d)$ is a normalising constant (often called 'the evidence' by physicists, and 'model likelihood' by statisticians), that ensures that the posterior is normalised to unity:

$$p(d) = \int d\Theta p(d|\Theta)p(\Theta). \tag{4.118}$$

The evidence is important for Bayesian model selection. The interpretation is as follows. Bayes' theorem relates the posterior probability for Θ (i.e. what we know about the parameter after seeing the data) to the likelihood and the prior (i.e. what we knew about the parameter before we saw the data). It can be thought of as a general rule to update our knowledge about a quantity (here, Θ) from the prior to the posterior. A result known as *Cox theorem* shows that Bayes' theorem is the unique generalization of Boolean algebra in the presence of uncertainty.

Remember that, in general, $p(\Theta|d) \neq p(d|\Theta)$, i.e. the posterior $p(\Theta|d)$ and the likelihood $p(d|\Theta)$, are 2 different quantities with different meanings. This is because they mean 2 different things: the likelihood is the probability of making the observation if we know what the parameter is; the posterior is the probability of the parameter given that we have made a certain observation. The 2 quantities are related by Bayes' theorem.

An example of Bayesian reasoning used in a court of law might clarify the difference (inspired by MacKay, 2003). A woman is found dead in her house and her husband is arrested on suspicion of murder. The police find that he has a history of beating her. The husband's lawyer argues that only 1 in 1,000 wife beaters is guilty of murdering his partner, hence the probability that his client is guilty is 1 per mille. To reason about this problem in a Bayesian way, let us denote by m the fact that the wife has been murdered, by g the statement that her husband is guilty of murdering her, and by b the fact that the husband used to beat his wife. As before, an overline indicates the negation of the proposition.

The probability quoted by the lawyer is $P(g|b) = 10^{-3}$, i.e. the probability that a randomly selected wife beater murders his wife. However, what we need to evaluate is $P(g|b, m)$, i.e. the probability that the husband is guilty given that his wife has been murdered (m). Bayes' theorem gives:

$$P(g|b, m) = \frac{P(b|g, m)P(g|m)}{P(b|m)}. \tag{4.119}$$

By using the marginalisation rule, Equation 4.114, the denominator can be rewritten as

$$P(b|m) = P(b|g, m)P(g|m) + P(b|\overline{g}, m)P(\overline{g}|m). \tag{4.120}$$

Thus

$$P(g|b, m) = \left(1 + \frac{P(b|\overline{g}, m)P(\overline{g}|m)}{P(b|g, m)P(g|m)}\right)^{-1}. \tag{4.121}$$

We need to estimate the probabilities entering the above equation. $P(b|g, m)$ is the probability that the husband is a wife beater, given that he has murdered his wife. It

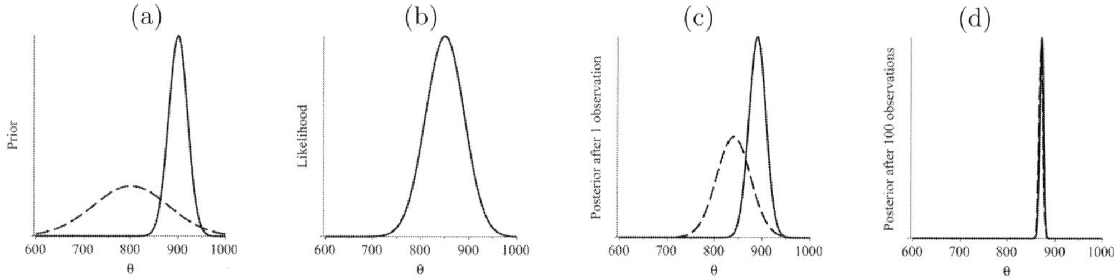

FIGURE 4.21 Converging views in Bayesian inference. Two scientists having different prior beliefs $p(\Theta)$ about the value of a quantity Θ (panel (a), the 2 curves representing 2 different priors) observe 1 datum with likelihood $\mathcal{L}(\Theta)$ (panel (b)), after which their posteriors $p(\Theta|d)$ (panel (c), obtained via Bayes' theorem, Equation 4.113) represent their updated states of knowledge on the parameter. This posterior then becomes the prior for the next observation. After observing 100 data points, the 2 posteriors have become essentially indistinguishable (d).

stands to reason that this should be quite high, perhaps as high as 0.9 (i.e. if somebody murders his wife, it is unlikely that this is the first time he has laid finger on her). $P(b|\overline{g}, m)$ is the probability of the husband being a wife beater without having been the murderer. This is around 1% (sadly) in Western societies. $P(g|m)$ is the probability that the husband is the murderer (independently of whether he is a wife beater). This is about 30%, obtained by looking at historical data on trials involving wives being murdered (sadly, about a third of women who are murdered are killed by their partner/husband). And finally $P(\overline{g}|m = 1) = 1 - P(g|m) = 0.7$. This gives

$$P(g|b, m) = \frac{1}{1 + \frac{0.01 \times 0.7}{0.9 \times 0.3}} = 0.97. \tag{4.122}$$

So we conclude that, given the above information, there is a staggering 97% probability that the husband is guilty, and not 1 per mille as wrongly argued by his lawyer. Every self-respecting Bayesian analysis ought to perform a sensitivity analysis on the choice of the relevant prior probabilities. If we replace $P(b|g, m) = 0.5$ in the above equation (i.e. we now give an even chance to the hypothesis that somebody who is guilty of murdering his wife is a wife beater) we obtain $P(g|b, m) = 0.96$, so still approximately the same very high posterior probability for the husband to be guilty.

Bayesian parameter inference works by updating our state of knowledge about a parameter (or hypothesis) as new data flow in. The posterior from a previous cycle of observations becomes the prior for the next. The price we have to pay is that we have to start somewhere by specifying an initial prior, which is not determined by the theory, but needs to be given by the user. The prior should represent fairly the state of knowledge of the user about the quantity of interest. Eventually, the posterior converges to a unique (objective) result even if different scientists start from different priors (provided their priors are non-0 in regions of parameter space where the likelihood is large) (See Figure 4.21 for an illustration).

There is a vast literature about how to select a prior in an appropriate way. Some aspects are fairly obvious: if your parameter Θ describes a quantity that has e.g. to be strictly positive (such as the number of photons in a detector, or a mass), then the prior will be 0 for values $\Theta < 0$. A standard (but by no means trivial) choice is to take a uniform prior (also called 'flat prior') on Θ, defined as:

$$p(\Theta) = \begin{cases} \frac{1}{(\Theta_{\max} - \Theta_{\min})} & \text{for } \Theta_{\min} \le \Theta \le \Theta_{\max} \\ 0 & \text{otherwise} \end{cases} . \tag{4.123}$$

With this choice of prior in Bayes' theorem, Equation 4.117, the posterior becomes functionally identical to the likelihood up to a proportionality constant:

$$p(\Theta|d) \propto p(d|\Theta) = \mathcal{L}(\Theta). \tag{4.124}$$

In this case, all the usual results about the likelihood carry over, e.g. the maximum likelihood value is also the maximum *a posteriori* value; a 1σ symmetric confidence interval is numerically identically to the symmetric posterior 68.3% credible region (i.e. region containing 68.3% of the posterior probability mass), but the interpretation is very different. In particular, the probability content of an interval around the mean for the posterior should be interpreted as a statement about our degree of belief in the value of Θ (differently from confidence intervals for the likelihood). Under a change of variable, $\Psi = \Psi(\Theta)$, the prior transforms according to:

$$p(\Psi) = p(\Theta)\left|\frac{d\Theta}{d\Psi}\right|. \tag{4.125}$$

In particular, a uniform prior on Θ is no longer uniform in Ψ if the variable transformation is nonlinear. Furthermore, what might sound like a fairly obvious 'non-informative' prior choice, i.e. a uniform prior in each of D variables, $\Theta_i \sim U[0,1]$ (for $i = 1,\ldots,D$), is in reality highly concentrated in the full D-dimensional parameter space. The prior mass concentrates in a shell of radius $r = (D/3)^{1/2}$ with constant variance, resulting in almost all the prior samples being clustered in a thin shell that occupies a tiny fraction of the available prior space. This *concentration of measure* phenomenon renders uniform priors quite dangerous in many dimensions.

4.8.3 Markov Chain Monte Carlo Methods

Until the early 2000s, brute-force, grid-based likelihood scans were the standard approach to constraining cosmological parameters. Perhaps the last example of this method is Wang et al. (2003), where a 9-D parameter space was explored. The problem of evaluating the likelihood on a regularly spaced grid is that the number of likelihood evaluations (and hence CPU time) scales exponentially with the resolution, and it is only practical for a parameter space of limited dimensionality ($D \lesssim 10$).

Although others had applied Markov chain Monte Carlo (MCMC) to cosmological problems before, Lewis and Bridle (2002) became the seminal paper, widely regarded as having established the use of MCMC for cosmological parameter estimation. This was in large part due to a user-friendly, publicly available numerical package which has since become the standard tool for this kind of analysis (`cosmoMC`[5]). `cosmoMC` originally used one of the simplest MCMC algorithms, Metropolis-Hastings, to obtain samples from the posterior distribution. The advantages of MCMC are manifold: it explores the full posterior distribution much more efficiently that grid-based methods (typically, the computational effort scales approximately linearly with D for reasonably well-behaved posteriors); marginalisation to 1 or 2 dimensions is trivial from the MCMC samples; and the computation of posterior distributions for derived quantities (e.g. the age of the universe) is straightforward, requiring only the evaluation of the function of interest over the posterior samples.

The purpose of the MCMC algorithm is to construct a sequence of points in parameter space (called a 'chain'), whose density is proportional to the posterior PDf of Equation 4.117. Developing a full theory of Markov chains is beyond our scope (see e.g. Gilks

[5] Available from `http://cosmologist.info/cosmomc`.

et al., 1996; Robert and Casella, 2004). For our purposes it suffices to say that a Markov chain is defined as a sequence of random variables $\{X^{(0)}, X^{(1)}, \ldots, X^{(M-1)}\}$ such that the probability of the $(t+1)$th element in the chain only depends on the value of the tth element. The crucial property of a Markov chain is that it converges to a stationary state (i.e. which does not change with t), where successive elements of the chain are samples from the target distribution, here the posterior $p(\Theta|d)$. The generation of the elements of the chain is probabilistic in nature, and is described by a *transition probability* $T(\Theta^{(t)}, \Theta^{(t+1)})$, giving the probability of going from point $\Theta^{(t)}$ to point $\Theta^{(t+1)}$ in parameter space. A sufficient condition to obtain a Markov chain is that the transition probability satisfies the *detailed balance condition*

$$p(\Theta^{(t)}|d)T(\Theta^{(t)}, \Theta^{(t+1)}) = p(\Theta^{(t+1)}|d)T(\Theta^{(t+1)}, \Theta^{(t)}). \tag{4.126}$$

This means that the ratio of the probabilities for jumping from $\Theta^{(t)}$ to $\Theta^{(t+1)}$ is inversely proportional to the ratio of the posterior probabilities at the 2 points. Some suitable and effective algorithms to construct such a $T(\Theta^{(t)}, \Theta^{(t+1)})$ are:

- **Metropolis algorithm:** This is arguably the simplest MCMC algorithm.
  (i) Start from a point $\Theta^{(0)}$ drawn from the prior density, and compute its posterior, $p_0 \equiv p(\Theta^{(0)}|d)$.
  (ii) Propose a candidate point $\Theta^{(c)}$ by drawing from the *proposal distribution* $q(\Theta^{(0)}, \Theta^{(c)})$. The proposal distribution might be, for example, a Gaussian of fixed width σ centred around the current point. The distribution q must satisfy the symmetry condition $q(x, y) = q(y, x)$.
  (iii) Evaluate the posterior at the candidate point, $p_c = p(\Theta^{(c)}|d)$. Accept the candidate point with probability

$$\alpha(\Theta^{(0)}, \Theta^{(c)}) = \min\left(\frac{p_c}{p_0}, 1\right). \tag{4.127}$$

  This accept/reject step can be performed by generating a random number u from the uniform distribution $(0, 1)$ and accepting the candidate sample if $u < \alpha$, and rejecting if otherwise.
  (iv) If the candidate point is accepted, add it to the chain and move there. Otherwise stay at the old point (which is thus counted twice in the chain). Loop back to (ii).

  From Equation 4.127, the candidate sample is always accepted when it has a larger posterior than the current one (i.e. $p_c > p_0$). In order to evaluate the acceptance function (Equation 4.127) only the unnormalised posterior is required, as the normalisation constant drops out of the ratio. It is easy to show that the Metropolis algorithm satisfies the detailed balance condition, Equation 4.126, with the transition probability given by $T(\Theta^{(t)}, \Theta^{(t+1)}) = q(\Theta^{(t)}, \Theta^{(t+1)})\alpha(\Theta^{(t)}, \Theta^{(t+1)})$.
- **Metropolis–Hastings algorithm:** This generalised the Metropolis algorithm to non-symmetrical proposal distributions by modifying the acceptance function (Equation 4.127) as follows to allow for cases where $q(x, y) \neq q(y, x)$:

$$\alpha(\Theta^{(t)}, \Theta^{(t+1)}) = \min\left(\frac{p(\Theta^{(c)}|d)q(\Theta^{(c)}, \Theta^{(0)})}{p(\Theta^{(0)}|d)q(\Theta^{(0)}, \Theta^{(c)})}, 1\right). \tag{4.128}$$

- **Gibbs sampling:** Gibbs sampling can be seen as a special case of the Metropolis-Hastings algorithm, where the proposal distribution is taken to be the probability for the jth component of Θ, conditional on the current value of all other components, $k \neq j$, i.e.

$$q(\Theta^{(t)}, \Theta^{(t+1)}) = p(\Theta_j^{(t+1)} | \Theta_{k\neq j}^{(t)}, d). \quad (4.129)$$

Substituting the above proposal distribution in Equation 4.128 and using the relationship $p(x) = p(x_j|x_{k\neq j})p(x_{k\neq j})$ one finds that the acceptance probability α is 1, i.e. every step of the Gibbs sampler is accepted.

- **Hamiltonian Monte Carlo:** The Hamiltonian Monte Carlo algorithm exploits an analogy with dynamical systems in statistical physics to suppress random walk behaviour, thus speeding up convergence of the chain while improving the efficiency of the exploration for high–dimensional parameter spaces (up to thousands of dimensions).

  Alongside the variables of interest Θ, a set of *conjugate momentum variables* y (of the same dimensionality as Θ) is introduced. One then builds the Hamiltonian

$$H(\Theta, y) = \frac{1}{2}\sum_i y_i^2 + \Psi(\Theta), \quad (4.130)$$

where the 'potential' term is minus the log posterior of interest, i.e.

$$\Psi(\Theta) \equiv -\ln p(\Theta|d). \quad (4.131)$$

The idea is to draw samples from the joint distribution

$$p(\Theta, y|d) = \exp(-H) = p(\Theta|d)\prod_i \exp\left(-\frac{y_i^2}{2}\right). \quad (4.132)$$

Samples over the target distribution $p(\Theta|d)$ are then obtained by marginalising over the uninteresting (and unphysical) variables y. In order to generate samples from Equation 4.132, one draws a sample from the Gaussian-distributed kinetic energy term for a fixed value of Θ. This stochastic step can be viewed as a Gibbs sampling step, as the value of y is replaced by drawing from a distribution conditional on the current value of Θ. Then the system is evolved for a 'time' τ along a path of constant energy in the phase space defined by (Θ, y), i.e. one follows Hamilton's equations (deterministic step)

$$\frac{\mathrm{d}\Theta_i}{\mathrm{d}t} = \frac{\partial H}{\partial y_i} = y_i \quad (4.133a)$$

$$\frac{\mathrm{d}y_i}{\mathrm{d}t} = -\frac{\partial H}{\partial \Theta_i} = \frac{\partial \ln p(\Theta|d)}{\partial \Theta_i} \quad (4.133b)$$

until a new point (Θ', y') is reached. Evolution along a path of exactly constant energy would generate a new sample from Equation 4.132, a procedure called 'stochastic dynamics Monte Carlo'. Numerical errors due to the discretisation of Equation 4.133 leads, however, to a change of energy between the start and the end point along the Hamiltonian trajectory, $\delta H = H(\Theta', y') - H(\Theta, y)$. This requires that the new candidate point (Θ', y') be subjected to a Metropolis accept/reject step, i.e. it is accepted with probability

$$\alpha = \min\left(\exp\left(-\delta H\right), 1\right). \quad (4.134)$$

- **Mixture strategies MCMC:** If each one of the transition probabilities $T_i(x, y)$ gives an invariant target distribution, then the transition probability given by the combination of the different transitions

$$T(x, y) = \sum_i \beta_i T_i(x, y), \tag{4.135}$$

with $\beta_i > 0$ and $\sum_i \beta_i = 1$, is also invariant. The same holds true for detailed balance. Such hybrid methods are sometimes called 'hybrid kernel MCMC'. This technique can be useful to improve sampling efficiency for multimodal distributions with several disconnected regions of high posterior density.

- **Importance sampling:** Sometimes we have MCMC samples from a distribution Q which is not quite the posterior $p(\Theta|d)$ we are interested in. Perhaps we have performed an MCMC run with an approximate likelihood function, or new data have become available and we want to update the previous posterior Q to the posterior including the new data p. The expectation value of any function of the parameters can be written as (where $\langle \cdot \rangle$ denotes expectation with respect to the distribution p we are interested in)

$$\langle f(\Theta) \rangle = \frac{\int d\Theta f(\Theta) w(\Theta) Q(\Theta|d)}{\int d\Theta w(\Theta) Q(\Theta|d)}, \tag{4.136}$$

where we have defined the *importance weight* $w(\Theta) \equiv p(\Theta|d)/Q(\Theta|d)$, the ratio of the distribution we are interested in, and the approximate distribution. Equation 4.136 means that we can estimate $\langle f(\Theta) \rangle$ as an expectation value under Q of the quantity $f(\Theta)w(\Theta)$. If we have MCMC samples from Q it follows that

$$\langle f(\Theta) \rangle \approx \frac{1}{M} \sum_{t=0}^{M-1} f(\Theta^{(t)}) u_t, \tag{4.137}$$

where $u_t \equiv w(\Theta^{(t)}) / \sum_{i=0}^{M-1} w(\Theta^{(i)})$ are the normalised importance weights.

There are several important practical issues in working with MCMC methods. Especially for high-dimensional parameter spaces with multimodal posteriors it is important *not* to use MCMC techniques as a black box, since poor exploration of the posterior can lead to serious mistakes in the final inference if it remains undetected. Some of the most relevant aspects are:

(i) Initial samples in the chain must be discarded, since the Markov process is not yet sampling from the equilibrium distribution ('burn-in period'). The burn-in can be assessed by looking at the evolution of the posterior density as a function of the number of steps in the chain. When the chain is started at a random point in parameter space, the posterior probability is typically small and becomes larger at every step as the chain approaches the region where the fit to the data is better. When the chain has moved into the neighbourhood of the posterior peak the curve of the log posterior as a function of the step number flattens and the chain begins sampling from its equilibrium distribution. Samples obtained before reaching this point must be discarded.

(ii) A difficult problem is presented by the assessment of *chain convergence*, which aims at establishing when the MCMC process has gathered enough samples so that the Monte Carlo estimate is sufficiently accurate. Useful diagnostic tools include the Raftery and Lewis statistics (Raftery, 1995) and the Gelman and Rubin criterion (Gelman and Rubin, 1992).

(iii) Bear in mind that MCMC is a *local algorithm*, which can be trapped around local maxima of the posterior density, thus missing regions of even higher posterior altogether. Considerable experimentation is sometimes required to find an implementation of the MCMC algorithm that is well suited to the exploration of the parameter space of interest. Experimenting with different algorithms (each of which has its own strength and weaknesses) is highly recommended.

(iv) Successive samples in a chain are in general correlated. Although this is not prejudicial for a correct statistical inference, it is often interesting to obtain *independent samples* from the posterior. This can be achieved by 'thinning' the chain by an appropriate factor, i.e. by selecting only one sample every K. A discussion of sample independence and how to assess it can be found in Dunkley et al. (2005), along with a convergence test based on the samples' power spectrum.

4.8.4 The Nested Sampling Algorithm

A powerful and efficient alternative to classical MCMC methods has emerged in the last few years in the form of the 'nested sampling' algorithm, invented by Skilling (2004). The original motivation for nested sampling was to compute the evidence integral of Equation 4.118. The recent development of the multimodal nested sampling technique (Feroz and Hobson, 2008; Corsaro and De Ridder, 2014; Handley et al., 2015) has delivered an extremely powerful and versatile algorithm for sampling complex, multimodal distributions. As samples from the posterior are generated as a by-product of the evidence computation, nested sampling can also be used to obtain parameter constraints in the same run as computing the Bayesian evidence.

The gist of nested sampling is that the multidimensional evidence integral is recast into a 1-dimensional integral. This is accomplished by defining the prior volume X as $\mathrm{d}X \equiv p(\Theta|\mathcal{M})\mathrm{d}\Theta$ so that

$$X(\lambda) = \int_{\mathcal{L}(\Theta)>\lambda} p(\Theta|\mathcal{M})\mathrm{d}\Theta, \tag{4.138}$$

where $\mathcal{L}(\Theta) \equiv p(d|\Theta, \mathcal{M})$ is the likelihood function and the integral is over the parameter space enclosed by the iso-likelihood contour $\mathcal{L}(\Theta) = \lambda$. So $X(\lambda)$ gives the volume of parameter space above a certain level λ of the likelihood. Then the Bayesian evidence, Equation 4.118, can be written as

$$p(d|\mathcal{M}) = \int_0^1 \mathcal{L}(X)\mathrm{d}X, \tag{4.139}$$

where $\mathcal{L}(X)$ is the inverse of Equation 4.138. Samples from $\mathcal{L}(X)$ can be obtained by uniformly drawing samples from the likelihood volume within the iso-contour surface defined by λ. Then, the 1-dimensional integral of Equation 4.139 can be obtained by simple quadrature, thus

$$p(d|\mathcal{M}) \approx \sum_i \mathcal{L}(X_i)W_i, \tag{4.140}$$

where the weights are $W_i = \frac{1}{2}(X_{i-1} - X_{i+1})$ (see Skilling, 2004); Mukherjee et al., 2006, for details).

4.8.5 The Gaussian Linear Model

Useful analytical insight can often be garnered by approximating the inference problem with a Gaussian linear model. This is particularly appropriate when the posterior distribution is unimodal and well described by a Gaussian approximation. For the case of CMB data, this applies when using 'normal parameters' (Kosowsky et al., 2002), a combination of the physical quantities which are nearly linear in the CMB power spectrum response. This means that the log-likelihood function is approximately linear in $\delta\Theta$, a departure of the cosmological parameters from a chosen reference value.

Under these circumstances, the posterior can be computed analytically (see Kunz et al., 2006, and especially Kelly, 2007, for a far more complete treatment, including missing data and errors on both the dependent and independent variables). We consider the following *linear model*

$$y = F\Theta + \epsilon, \tag{4.141}$$

where the dependent variable y is a d-dimensional vector of observations (the *data*), $\Theta = \{\Theta_1, \Theta_2, \ldots, \Theta_n\}$ is a vector of dimension n of unknown cosmological parameters that we wish to determine, and F is a $d \times n$ matrix of known constants which specify the relation between the input variables Θ and the dependent variables y ('design matrix'). When observations $y_i(x)$ are fitted with a linear model of the form $f(x) = \sum_{j=1}^{n} \Theta_j X^j(x)$, the matrix F is given by the basis functions X^j evaluated at the locations x_i of the observations $F_{ij} = X^j(x_i)$. Furthermore, ϵ is a d-dimensional vector of random variables with 0 mean (the *noise*). If we assume that ϵ follows a multivariate Gaussian distribution with uncorrelated covariance matrix $C \equiv \mathrm{diag}(\tau_1^2, \tau_2^2, \ldots, \tau_d^2)$, then the likelihood function takes the form

$$p(y|\Theta) = \frac{1}{(2\pi)^{d/2}\prod_j \tau_j} \exp\left[-\frac{1}{2}(b - A\Theta)^t(b - A\Theta)\right], \tag{4.142}$$

where we have defined $A_{ij} = F_{ij}/\tau_i$ and $b_i = y_i/\tau_i$ where A is a $d \times n$ matrix and b is a d-dimensional vector. The likelihood function can be cast in the form

$$p(y|\Theta) = \mathcal{L}_0 \exp\left[-\frac{1}{2}(\Theta - \Theta_0)^t L(\Theta - \Theta_0)\right], \tag{4.143}$$

with the likelihood Fisher matrix L (an $n \times n$ matrix) given by

$$L \equiv A^t A \tag{4.144}$$

and a normalisation constant

$$\mathcal{L}_0 \equiv \frac{1}{(2\pi)^{d/2}\prod_j \tau_j} \exp\left[-\frac{1}{2}(b - A\Theta_0)^t(b - A\Theta_0)\right]. \tag{4.145}$$

Here Θ_0 denotes the parameter value which maximises the likelihood, given by

$$\Theta_0 = L^{-1}A^t b. \tag{4.146}$$

Assuming as a prior PDF a multinormal Gaussian distribution with 0 mean and the $n \times n$ dimensional prior Fisher information matrix P (recall that that the Fisher information matrix is the inverse of the covariance matrix), i.e.

$$p(\Theta) = \frac{|P|^{1/2}}{(2\pi)^{n/2}} \exp\left[-\frac{1}{2}\Theta^t P\Theta\right], \tag{4.147}$$

where $|P|$ denotes the determinant of the matrix P, the posterior distribution for Θ is given by multinormal Gaussian with Fisher information matrix $\mathcal{F}$

$$\mathcal{F} = L + P \tag{4.148}$$

and mean $\bar{\Theta}$ given by

$$\bar{\Theta} = \mathcal{F}^{-1} L \Theta_0. \tag{4.149}$$

Recalling this standard result for Gaussian integrals:

$$\int \exp\left[-\frac{1}{2}(\mathbf{x}-\mathbf{m})^t \mathbf{\Sigma}^{-1}(\mathbf{x}-\mathbf{m})\right] \mathrm{d}\mathbf{x} = \sqrt{\det(2\pi\mathbf{\Sigma})}, \tag{4.150}$$

the evidence, Equation 4.118, is given by

$$\begin{aligned} p(y) &= \mathcal{L}_0 \frac{|\mathcal{F}|^{-1/2}}{|P|^{-1/2}} \exp\left[-\frac{1}{2}\Theta_0^t(L - L\mathcal{F}^{-1}L)\Theta_0\right] \\ &= \mathcal{L}_0 \frac{|\mathcal{F}|^{-1/2}}{|P|^{-1/2}} \exp\left[-\frac{1}{2}(\Theta_0^t L \Theta_0 - \bar{\Theta}^t \mathcal{F} \bar{\Theta}).\right]. \end{aligned} \tag{4.151}$$

4.9 Conclusion

In less than 100 years, cosmology has made giant leaps. Theoretical models have motivated the observational exploration of the cosmos in many different domains: redshift, wavelength, and time. The technological advancements of the last 30 years – in particular, the advent of fast and large charge-coupled device sensors and large field-of-view telescopes – have provided the means of charting the universe to its largest observables scales. In the meantime, CMB measurements have reached the ultimate cosmic variance limit in temperature, and are now pursuing the holy grail of B-mode polarization detection – a smoking gun of inflation. Our SNIa sample is set to increase by orders of magnitude within the next few years, opening up new problems in terms of their characterisation and exploitation in an era when spectral information for all the candidates will not be available.

On the computational and statistical inference side, cosmologists have enthusiastically embraced the challenge, oftentimes leading the development of new algorithms and tools. At its best, modern cosmology thrives at the confluence of well-understood theoretical models and large, accurate datasets. Today's cosmologists happily marry the hands-on approach inherited from the astronomical tradition with the sophisticated outlook on statistical tools that is second nature for a generation of scientists trained in an epoch of data deluge.

There is little doubt that future cosmology will require new approaches of the kind that we are seeing burgeoning today: machine learning, approximate Bayesian computation, fast likelihood evaluators, methods apt to explore very large dimensional parameter spaces, hierarchical Bayesian modelling, Gaussian processes, and other nonparametric methods are only some of the techniques that will allow us to take the field to the next level of accuracy and precision. With upcoming overwhelmingly large data sets from sources such as the Large Synoptic Survey Telescope and the Square Kilometre Array, these methods could not be more timely, nor more urgently needed.

BIBLIOGRAPHY

Anderson, L., Aubourg, E., Bailey, S., et al. 2012. The clustering of galaxies in the SDSS-III Baryon Oscillation Spectroscopic Survey: baryon acoustic oscillations in the Data Release 9 spectroscopic galaxy sample. *Mon. Not. R. Astron. Soc.* 427, 3435.

Bayes, T., and Price, R. 1763. An essay towards solving a problem in the doctrine of chances. *Phil. Trans. Roy. Soc.* 53, 370. Reprinted in *Biometrika* 45, 293–315, 1958.

Betoule, M., Kessler, R., Guy, J., et al. 2014. Improved cosmological constraints from a joint analysis of the SDSS-II and SNLS supernova samples. *Astron. Astrophys.* 568, A22.

Corsaro, E., and De Ridder, J. 2014. DIAMONDS: A new Bayesian nested sampling tool. Application to peak bagging of solar-like oscillations. *Astron. Astrophys.* 571, A71

Davis, T. M., and Lineweaver, C. H. 2004. Expanding confusion: common misconceptions of cosmological horizons and the superluminal expansion of the universe. *Publ. Astron. Soc. Aust.* 21, 97.

Dunkley, J., Bucher, M., Ferreira, P. G., et al. 2005. Fast and reliable Markov chain Monte Carlo technique for cosmological parameter estimation *Mon. Not. Roy. Astron. Soc.* 356, 925.

Eisenstein, D. J., Zehavi, I., Hogg, D. W., et al. 2005. Detection of the baryon acoustic peak in the large-scale correlation function of SDSS luminous red galaxies. *Astrophys. J.* 633, 560.

Feroz, F., and Hobson, M. P. 2008. Multimodal nested sampling: an efficient and robust alternative to Markov Chain Monte Carlo methods for astronomical data analyses. *Mon. Not. R. Astron. Soc.* 384, 449.

Fixsen, D. J., Cheng, E. S., Gales, J. M., et al. 1996. The cosmic microwave background spectrum from the full COBE FIRAS data set. *Astrophys. J.* 473, 576.

Gelman, A., and Rubin, D. 1992. Inference from iterative simulation using multiple sequences. *Stat. Sci.* 7, 457.

Gilks, W., Richardson, S., and Spiegelhalter, D., eds. 1996. *Markov chain Monte Carlo in practice.* Chapman & Hall/CRC, Boca Raton, Florida, USA.

Handley, W. J., Hobson, M. P., and Lasenby, A. N. 2015. POLYCHORD: nested sampling for cosmology. *Mon. Not. R. Astron. Soc.* 450, L61.

Hu, W., and Dodelson, S. 2002. Cosmic microwave background anisotropies. *Ann. Rev. Astron. and Astrophys.* 40, 171.

Hubble, E. 1929. A relation between distance and radial velocity among extra-galactic nebulae. Proc. Natl. Academy of Sciences, 15, 168.

Kelly, B. C. 2007. Some aspects of measurement error in linear regression of astronomical data. *Astrophys. J.* 665, 1489.

Kosowsky, A., Milosavljevic, M., and Jimenez, R. 2002. Efficient cosmological parameter estimation from microwave background anisotropies. *Phys. Rev.* D 66, 063007.

Kunz, M., Trotta, R., and Parkinson, D. 2006. Measuring the effective complexity of cosmological models. *Phys. Rev.* D 74, 023503.

Lewis, A., and Bridle, S. 2002. Cosmological parameters from CMB and other data: a Monte Carlo approach. *Phys. Rev.* D 66, 103511.

MacKay, D. 2003. *Information theory, inference, and learning algorithms.* Cambridge University Press, Cambridge, UK.

March, M. C., Trotta, R., Berkes, P., et al. 2011. Improved constraints on cosmological parameters from Type Ia supernova data. *Mon. Not. R. Astron. Soc.* 418, 2308.

Mukherjee, P., Parkinson, D., and Liddle, A. R. 2006. A nested sampling algorithm for cosmological model selection. *Astrophys. J.* 638, L51.

Penzias, A. A., and Wilson, R. W. 1965. A measurement of excess antenna temperature at 4080 Mc/s. *Astrophys. J.* 142, 419.

Phillips, M. M. 1993. The absolute magnitudes of Type IA supernovae. *Astrophys. J. Lett.* 413, L105.

Planck Collaboration, Ade, P. A. R., Aghanim, N., et al. 2015. Planck 2015 results. XIII. Cosmological parameters. *Astron. Astrophys.* 594, A13, 63.

Raftery, A. 1995. Bayesian model selection in social research. *Sociol. Methodol.* 25, 111.

Rest, A., Scolnic, D., Foley, R. J., et al. 2014. Cosmological constraints from measurements of Type Ia supernovae discovered during the first 1.5 years of the Pan-STARRS1 survey. *Astrophys. J.* 795, 44.

Riess, A. G., Macri, L., Casertano, S., et al. 2011. A 3% solution: determination of the Hubble constant with the Hubble Space Telescope and Wide Field Camera 3. *Astrophys. J.* 730, 119.

Robert, C. P., and Casella, G. 2004. *Monte Carlo statistical methods*, 2nd ed. Springer, New York, USA.

Schneider, P. 2006. *Extragalactic astronomy and cosmology: an introduction.* Springer, Heidelberg, Germany.

Skilling, J. 2004. in *Bayesian inference and maximum entropy methods in science and engineering: Proceedings of the 28th International Workshop, No. 735*, edited by Fischer, R., Preuss, R., and von Toussaint, U., 395–405. American Institute of Physics.

Smoot, G. F., Bennett, C. L., Kogut, A., et al. 1992. Structure in the COBE differential microwave radiometer first-year maps. *Astrophys. J. Lett.* 396, L1.

Trotta, R. 2008. Bayes in the sky: Bayesian inference and model selection in cosmology. *Comtemp. Phys.* 49, 71.

Wang, X., Tegmark, M., Jain, B., et al. 2003. Last stand before WMAP: cosmological parameters from lensing, CMB, and galaxy clustering. *Phys. Rev.* D 68, 123001.

5 An Introduction to Objective Bayesian Statistics

JOSÉ M. BERNARDO

Abstract

The field of statistics includes two major paradigms: frequentist and Bayesian.[1] Bayesian methods provide a *complete* paradigm for both statistical inference and decision-making under uncertainty. Bayesian methods may be derived from an axiomatic system and provide a *coherent* methodology which makes it possible to incorporate relevant initial information, and which solves many of the difficulties which frequentist methods are known to face. If no prior information is to be assumed, a situation often met in scientific reporting and public decision-making, a formal initial prior function must be mathematically derived from the assumed model. This leads to *objective* Bayesian methods, objective in the precise sense that their results, like frequentist results, only depend on the assumed model and the data obtained. The Bayesian paradigm is based on an interpretation of probability as a *rational conditional measure of uncertainty*, which closely matches the sense of the word 'probability' in ordinary language. Statistical inference about a quantity of interest is described as the modification of the uncertainty about its value in the light of evidence, and Bayes' theorem specifies how this modification should precisely be made.

5.1 Introduction

Scientific experimental or observational results generally consist of (possibly many) sets of data of the general form $D = \{\boldsymbol{x}_1, \ldots, \boldsymbol{x}_n\}$, where $\boldsymbol{x}_i$ are somewhat 'homogeneous' (possibly multidimensional) observations. Statistical methods are then typically used to derive conclusions on both the nature of the process which has produced those observations, and on the expected behaviour at future instances of the same process. A central element of *any* statistical analysis is the specification of a *probability model* which is assumed to describe the mechanism which has generated the observed data D as a function of a (possibly multidimensional) parameter (vector) $\boldsymbol{\omega} \in \Omega$, sometimes referred to as the *state of nature*, about whose value only limited information (if any) is available. All derived statistical conclusions are obviously conditional on the assumed probability model.

Unlike most other branches of mathematics, frequentist methods of statistical inference suffer from the lack of an axiomatic basis; as a consequence, their proposed desiderata are often mutually incompatible, and the analysis of the same data may well lead to incompatible results when different, apparently intuitive procedures are tried.[2] In marked contrast, the Bayesian approach to statistical inference is firmly based on axiomatic foundations which provide a unifying logical structure and guarantee the mutual consistency of the methods proposed. Bayesian methods constitute a *complete* paradigm for statistical inference, a scientific revolution in Kuhn's sense.

Bayesian statistics only require the *mathematics* of probability theory and the *interpretation* of probability which most closely corresponds to the standard use of this word

[1] This chapter includes updated sections of the article 'Bayesian statistics', prepared by the author for the Encyclopedia of Life Support Systems, a 2003 online UNESCO publication.

[2] See Lindley, 1972, and Jaynes, 1976, for many instructive examples.

in everyday language: it is no accident that some of the more important seminal books on Bayesian statistics, such as Laplace (1812), Jeffreys (1939), and de Finetti (1970), are actually titled *Probability Theory*. The practical consequences of adopting the Bayesian paradigm are far reaching. Indeed, Bayesian methods (i) reduce statistical inference to problems in probability theory, thereby minimizing the need for completely new concepts, and (ii) serve to discriminate among conventional, typically frequentist statistical techniques, either by providing a logical justification to some (and making explicit the conditions under which they are valid), or proving the logical inconsistency of others.

The main result from these foundations is the mathematical *need* to describe by means of probability distributions all uncertainties present in the problem. In particular, unknown parameters in probability models *must* have a joint probability distribution which describes the available information about their values; this is often regarded as *the* characteristic element of a Bayesian approach. Notice that (in sharp contrast to conventional statistics) *parameters are treated as random variables* within the Bayesian paradigm. This is not a description of their variability (parameters are typically *fixed unknown* quantities) but a description of the *uncertainty* about their true values.

A most important particular case arises when either no relevant prior information is readily available or that information is subjective and an 'objective' analysis is desired, one that is exclusively based on accepted model assumptions and well-documented public prior information. This is addressed by *reference analysis*, which uses information-theoretic concepts to derive formal *reference* prior functions which, when used in Bayes' theorem, lead to posterior distributions encapsulating inferential conclusions on the quantities of interest solely based on the assumed model and the observed data.

In this chapter it is assumed that probability distributions may be described through their probability density functions, and no distinction is made between a random quantity and the particular values that it may take. Bold italic roman fonts are used for *observable* random vectors (typically data) and bold italic greek fonts are used for unobservable random vectors (typically parameters); lowercase is used for variables and uppercase calligraphic is used for their dominion sets. Moreover, the standard mathematical convention of referring to *functions*, say f and g of $\boldsymbol{x} \in \mathcal{X}$, respectively, by $f(\boldsymbol{x})$ and $g(\boldsymbol{x})$, will be used throughout. Thus, $\pi(\boldsymbol{\theta} \mid D, C)$ and $p(\boldsymbol{x} \mid \boldsymbol{\theta}, C)$ respectively represent general *probability densities* of the unknown parameter $\boldsymbol{\theta} \in \Theta$ given data D and conditions C, and of the observable random vector $\boldsymbol{x} \in \mathcal{X}$ conditional on $\boldsymbol{\theta}$ and C. Hence, $\pi(\boldsymbol{\theta} \mid D, C) \geq 0$, $\int_\Theta \pi(\boldsymbol{\theta} \mid D, C)\, d\boldsymbol{\theta} = 1$, and $p(\boldsymbol{x} \mid \boldsymbol{\theta}, C) \geq 0$, $\int_\mathcal{X} p(\boldsymbol{x} \mid \boldsymbol{\theta}, C)\, d\boldsymbol{x} = 1$. This admittedly imprecise notation will greatly simplify the exposition. If the random vectors are discrete, these functions naturally become probability mass functions, and integrals over their values become sums. Density functions of specific distributions are denoted by appropriate names. Thus, if x is a random quantity with a normal distribution of mean μ and standard deviation σ, its probability density function will be denoted $\mathrm{N}(x \mid \mu, \sigma)$.

Bayesian methods make frequent use of the concept of logarithmic divergence, a general measure of the goodness of the approximation of a probability density $p(\boldsymbol{x})$ by another density $\hat{p}(\boldsymbol{x})$. The Kullback–Leibler or *logarithmic divergence* of a probability density $\hat{p}(\boldsymbol{x})$ of the random vector $\boldsymbol{x} \in \mathcal{X}$ from its true probability density $p(\boldsymbol{x})$ is defined as $\kappa\{\hat{p}(\boldsymbol{x}) \mid p(\boldsymbol{x})\} = \int_\mathcal{X} p(\boldsymbol{x}) \log\{p(\boldsymbol{x})/\hat{p}(\boldsymbol{x})\}\, d\boldsymbol{x}$. It may be shown that (i) the logarithmic divergence is nonnegative (and it is zero if, and only if, $\hat{p}(\boldsymbol{x}) = p(\boldsymbol{x})$ almost everywhere), and (ii) that $\kappa\{\hat{p}(\boldsymbol{x}) \mid p(\boldsymbol{x})\}$ is invariant under one-to-one transformations of $\boldsymbol{x}$.

This chapter contains a brief summary of the mathematical foundations of Bayesian statistical methods (Section 2), an overview of the paradigm (Section 3), a discussion of objective Bayesian methods (Section 4), and a description of useful objective inference summaries, including estimation and hypothesis testing (Section 5).[3]

5.2 Foundations

A central element of the Bayesian paradigm is the use of probability distributions to describe *all* relevant unknown quantities, interpreting the probability of an event as a conditional measure of uncertainty, on a $[0, 1]$ scale, about the occurrence of the event in some specific conditions. The limiting extreme values 0 and 1, which are typically inaccessible in applications, respectively describe impossibility and certainty of the occurrence of the event. This interpretation of probability includes and extends all other probability interpretations. Two independent arguments prove the mathematical inevitability of the use of probability distributions to describe uncertainties; these are summarised later in this section.

5.2.1 Probability as a Rational Measure of Conditional Uncertainty

Bayesian statistics uses the word *probability* in precisely the same sense in which this word is used in everyday language, as a *conditional measure of uncertainty* associated with the occurrence of a particular event, given the available information and the accepted assumptions. Thus, $\Pr(E \mid C)$ is a measure of (presumably rational) belief in the occurrence of the *event* E under *conditions* C. It is important to stress that probability is *always* a function of 2 arguments, the event E whose uncertainty is being measured and the conditions C under which the measurement takes place; 'absolute' probabilities do not exist. In typical applications, one is interested in the probability of some event E given the available *data* D, the set of *assumptions* A which one is prepared to make about the mechanism which has generated the data, and the relevant contextual *knowledge* K which might be available. Thus, $\Pr(E \mid D, A, K)$ is to be interpreted as a measure of (presumably rational) belief in the occurrence of the *event* E, given data D, assumptions A and any other available knowledge K, as a measure of how 'likely' is the occurrence of E in these conditions. Sometimes, but certainly not always, the probability of an event under given conditions may be associated with the relative frequency of 'similar' events in 'similar' conditions. The following examples are intended to illustrate the use of probability as a conditional measure of uncertainty.

Example 1: Probabilistic diagnosis. A human population is known to contain 0.2% of people infected by a particular virus. A person, randomly selected from that population, is subject to a test which is from laboratory data known to yield positive results in 98% of infected people and in 1% of uninfected, so that, if V denotes the event that a person carries the virus and $+$ denotes a positive result, $\Pr(+ \mid V) = 0.98$ and $\Pr(+ \mid \overline{V}) = 0.01$. Suppose that the result of the test turns out to be positive. Clearly, one is then interested in $\Pr(V \mid +, A, K)$, the probability that the person carries the virus, given the positive result, the assumptions A about the probability mechanism generating the test results, and the available knowledge K of the prevalence of the infection in the population under study (described here by $\Pr(V \mid K) = 0.002$). An elementary exercise in probability algebra, which involves Bayes' theorem in its simplest form (see Section 3),

[3] Good introductions to objective Bayesian statistics include Lindley (1965), Zellner (1971), and Box and Tiao (1973). For more advanced monographs, see Berger (1985) and Bernardo and Smith (1994).

yields $\Pr(V \mid +, A, K) = 0.164$. Notice that the 4 probabilities involved in the problem have the same interpretation: they are all conditional measures of uncertainty. Besides, $\Pr(V \mid +, A, K)$ is both a measure of the uncertainty associated with the event that the particular person who tested positive is actually infected, and an estimate of the proportion of people in that population (about 16.4%) who would eventually prove to be infected among those which yielded a positive test.

Example 2: Estimation of a proportion. A survey is conducted to estimate the proportion θ of individuals in a population who share a given property. A random sample of n elements is analysed, r of which are found to possess that property. One is then typically interested in using the results from the sample to establish regions of $[0, 1]$ where the unknown value of θ may plausibly be expected to lie; this information is provided by probabilities of the form $\Pr(a < \theta < b \mid r, n, A, K)$, a conditional measure of the uncertainty about the event that θ belongs to (a, b) given the information provided by the data (r, n), the assumptions A made on the behaviour of the mechanism which has generated the data (a random sample of n Bernoulli trials), and any relevant knowledge K on the values of θ which might be available. For example, after a political survey in which 720 citizens out of a random sample of 1500 have declared their support for a particular political measure, one may conclude that $\Pr(\theta < 0.5 \mid 720, 1500, A, K) = 0.933$, indicating a probability of about 93% that a referendum on that issue would be lost. Similarly, after a screening test for an infection where 100 people have been tested, none of which turned out to be infected, one may conclude that $\Pr(\theta < 0.01 \mid 0, 100, A, K) = 0.844$, or a probability of about 84% that the proportion of infected people is smaller than 1%.

Example 3: Measurement of a physical constant. A team of scientists, intending to establish the unknown value of a physical constant μ, obtains data $D = \{x_1, \ldots, x_n\}$ which are considered to be measurements of μ subject to error. The probabilities of interest are then typically of the form $\Pr(a < \mu < b \mid x_1, \ldots, x_n, A, K)$, the probability that the unknown value of μ (fixed in nature, but unknown to the scientists) lies within an interval (a, b) given the information provided by the data D, the assumptions A made on the behaviour of the measurement mechanism, and whatever knowledge K might be available on the value of the constant μ. Again, these probabilities are conditional measures of uncertainty which describe the (necessarily probabilistic) conclusions of the scientists on the true value of μ, given available information and accepted assumptions. For example, after a classroom experiment to measure the gravitational field with a pendulum, a student may report (in m/sec^2) something like $\Pr(9.788 < g < 9.829 \mid D, A, K) = 0.95$, meaning that, under accepted knowledge K and assumptions A, the observed data D indicate that the true value of g lies within 9.788 and 9.829 with probability 0.95, a conditional uncertainty measure on a $[0, 1]$ scale. This is naturally compatible with the fact that the value of the gravitational field at the laboratory may well be known with high precision from available literature or from precise previous experiments, but the student may have been instructed not to use that information as part of the accepted knowledge K. Under some conditions, it is also true that if the same procedure were actually used by many other students with similarly obtained data sets, their reported intervals would actually cover the true value of g in approximately 95% of the cases, thus providing a frequentist calibration of the student's probability statement.

Example 4: Prediction. An experiment is made to count the number r of times that an event E takes place in each of n replications of a well-defined situation. It is observed that E does take place r_i times in replication i, and it is desired to forecast the number of times r that E will take place in a similar future situation. This is a prediction problem on the value of an observable (discrete) quantity r, given the information provided by data D, accepted assumptions A on the probability mechanism which generates the r_i's, and any relevant available knowledge K. Computation of the probabilities $\{\Pr(r \mid r_1, \ldots, r_n, A, K)\}$, for $r = 0, 1, \ldots$, is thus required. For example, the quality-assurance engineer of a firm which produces automobile restraint systems may report, after observing that the entire production of airbags in each of $n = 10$ consecutive months has yielded no complaints from their clients, that $\Pr(r = 0 \mid r_1 = \cdots = r_{10} = 0, A, K) = 0.953$. This should be regarded as a measure, on a $[0, 1]$ scale, of the conditional uncertainty, given observed data, accepted assumptions, and contextual knowledge associated with the event that no airbag complaint will come from next month's production and, if conditions remain constant, this is also an estimate of the proportion of months expected to share this desirable property.

A similar problem may naturally be posed with continuous observables. For instance, after measuring some continuous magnitude in each of n randomly chosen elements within a population, it may be desired to forecast the proportion of items in the whole population whose magnitude satisfies some precise specifications. As an example, after measuring the breaking strengths $\{x_1, \ldots, x_{10}\}$ of 10 randomly chosen safety belt webbings to verify whether they satisfy the requirement of remaining above 26 kN, the quality-assurance engineer may report something like $\Pr(x > 26 \mid x_1, \ldots, x_{10}, A, K) = 0.9987$. This should be regarded as a measure, on a $[0, 1]$ scale, of the conditional uncertainty (given observed data, accepted assumptions, and contextual knowledge) associated with the event that a randomly chosen safety belt webbing will support no less than 26 kN. If production conditions remain constant, it will also be an estimate of the proportion of safety belts which will conform to this particular specification.

Often, additional information of future observations is provided by related covariates. For instance, after observing the outputs $\{\boldsymbol{y}_1, \ldots, \boldsymbol{y}_n\}$ which correspond to a sequence $\{\boldsymbol{x}_1, \ldots, \boldsymbol{x}_n\}$ of different production conditions, it may be desired to forecast the output $\boldsymbol{y}$ which would correspond to a particular set $\boldsymbol{x}$ of production conditions. For instance, the viscosity of commercial condensed milk is required to be within specified values a and b; after measuring the viscosities $\{y_1, \ldots, y_n\}$ which correspond to samples of condensed milk produced under different physical conditions $\{\boldsymbol{x}_1, \ldots, \boldsymbol{x}_n\}$, production engineers require probabilities of the form $\Pr(a < y < b \mid \boldsymbol{x}, (y_1, \boldsymbol{x}_1), \ldots, A, K)$. This is a conditional measure of the uncertainty (always given observed data, accepted assumptions, and contextual knowledge) associated with the event that condensed milk produced under conditions $\boldsymbol{x}$ will actually satisfy the required viscosity specifications.

5.2.2 Statistical Inference and Decision Theory

Decision theory not only provides a precise methodology to deal with decision problems under uncertainty, but its solid axiomatic basis also provides a powerful reinforcement to the logical force of the Bayesian approach. We now summarise the basic argument.

A decision problem exists whenever there are two or more possible courses of action; let $\mathcal{A}$ be the class of possible actions. Moreover, for each $a \in \mathcal{A}$, let Θ_a be the set of *relevant events* which may affect the result of choosing a, and let $c(a, \boldsymbol{\theta}) \in \mathcal{C}_a$, $\boldsymbol{\theta} \in \Theta_a$, be the *consequence* of having chosen action a when event $\boldsymbol{\theta}$ takes place. The class of pairs $\{(\Theta_a, \mathcal{C}_a), a \in \mathcal{A}\}$ describes the *structure* of the decision problem. Without loss of generality, it may be assumed that the possible actions are mutually exclusive, for otherwise one would work with the appropriate Cartesian product.

Different sets of principles have been proposed to capture a minimum collection of logical rules that could sensibly be required for 'rational' decision-making. These all consist of axioms with a strong intuitive appeal; examples include the *transitivity* of preferences (if $a_1 > a_2$ given C, and $a_2 > a_3$ given C, then $a_1 > a_3$ given C), and the *sure-thing principle* (if $a_1 > a_2$ given C and E, and $a_1 > a_2$ given C and not E, then $a_1 > a_2$ given C). Notice that these rules are not intended as a description of actual human decision-making, but as a *normative* set of principles to be followed by someone who aspires to achieve coherent decision-making.

There are naturally different options for the set of acceptable principles (see *e.g.* Ramsey 1926; Savage, 1954; DeGroot, 1970; Bernardo and Smith, 1994, ch. 2, and references therein), but all of them lead basically to the same conclusions, namely:

(i) Preferences among consequences should be measured with a real-valued *utility* function $U(c) = U(a, \boldsymbol{\theta})$ which specifies, on some numerical scale, their desirability.

(ii) The uncertainty of relevant events should be measured with a set of *probability* distributions $\{(\pi(\boldsymbol{\theta} \mid C, a), \boldsymbol{\theta} \in \Theta_a, a \in \mathcal{A}\}$ describing their plausibility given the conditions C under which the decision must be taken.

(iii) The desirability of the available actions is measured by their corresponding *expected utility*

$$\overline{U}(a \mid C) = \int_{\Theta_a} U(a, \boldsymbol{\theta})\, \pi(\boldsymbol{\theta} \mid C, a)\, d\boldsymbol{\theta}, \quad a \in \mathcal{A}.$$

It is often convenient to work in terms of the nonnegative *loss* function defined by

$$L(a, \boldsymbol{\theta}) = \sup_{a \in \mathcal{A}} \{U(a, \boldsymbol{\theta})\} - U(a, \boldsymbol{\theta})$$

which directly measures, as a function of $\boldsymbol{\theta}$, the 'penalty' for choosing a wrong action. The relative undesirability of available actions $a \in \mathcal{A}$ is then measured by their *expected loss*

$$\overline{L}(a \mid C) = \int_{\Theta_a} L(a, \boldsymbol{\theta})\, \pi(\boldsymbol{\theta} \mid C, a)\, d\boldsymbol{\theta}, \quad a \in \mathcal{A}.$$

Notice that, in particular, arguments (i)–(iii) establish the need to quantify the uncertainty about all relevant unknown quantities (the actual values of the $\boldsymbol{\theta}$s), and specify that this quantification *must* have the mathematical structure of probability distributions. These probabilities are conditional on the circumstances C under which the decision is to be taken, which typically, but not necessarily, include the results D of some relevant experimental or observational data.

It has been argued that these developments (which are not questioned when decisions have to be made) do not apply to problems of statistical inference, where no specific decision-making is envisaged. However, there are 2 powerful counterarguments to this. Indeed, a problem of statistical inference is typically considered worth analysing because it *may* eventually help make sensible decisions – a lump of arsenic is poisonous because it *may* kill someone, not because it has actually killed someone (Ramsey, 1926) – and it has been shown (Bernardo, 1979a) that statistical inference on $\boldsymbol{\theta}$ actually *has* the mathematical structure of a decision problem, where the class of alternatives is the functional space

$$\mathcal{A} = \Big\{\pi(\boldsymbol{\theta} \mid D); \quad \pi(\boldsymbol{\theta} \mid D) > 0, \ \int_{\Theta} \pi(\boldsymbol{\theta} \mid D)\, d\boldsymbol{\theta} = 1\Big\}$$

of the conditional probability distributions of $\boldsymbol{\theta}$ given the data, and the utility function is a measure of the amount of information about $\boldsymbol{\theta}$ which the data may be expected to provide.

5.2.3 Exchangeability and Representation Theorem

Available data often take the form of a set $\{\boldsymbol{x}_1, \ldots, \boldsymbol{x}_n\}$ of 'homogeneous' (possibly multidimensional) observations, in the precise sense that only their *values* matter and not the *order* in which they appear. Formally, this is captured by the notion of *exchangeability*. The set of random vectors $\{\boldsymbol{x}_1, \ldots, \boldsymbol{x}_n\}$ is exchangeable if their joint distribution is invariant under permutations. An infinite sequence $\{\boldsymbol{x}_j\}$ of random vectors is exchangeable if all its finite subsequences are exchangeable. Notice that, in particular, any random sample from any model is exchangeable in this sense. The concept of exchangeability, introduced by de Finetti (1937), is central to modern statistical thinking. Indeed, the general *representation theorem* implies that if a set of observations is assumed to be a subset of an exchangeable sequence, then it constitutes *a random sample* from some probability model $\{p(\boldsymbol{x} \mid \boldsymbol{\omega}), \boldsymbol{\omega} \in \boldsymbol{\Omega}\}$, $\boldsymbol{x} \in \mathcal{X}$, described in terms of (labelled by) some *parameter vector* $\boldsymbol{\omega}$; furthermore, this parameter $\boldsymbol{\omega}$ is *defined* as the limit (as $n \to \infty$) of some function of the observations. Available information about the value of $\boldsymbol{\omega}$ in prevailing conditions C is *necessarily* described by *some* probability distribution $\pi(\boldsymbol{\omega} \mid C)$.

For example, in the case of a sequence $\{x_1, x_2, \ldots\}$ of dichotomous exchangeable random quantities $x_j \in \{0, 1\}$, de Finetti's representation theorem establishes that the joint distribution of $(x_1, \ldots, x_n)$ has an *integral representation* of the form

$$p(x_1, \ldots, x_n \mid C) = \int_0^1 \prod_{i=1}^{n} \theta^{x_i} (1-\theta)^{1-x_i} \, \pi(\theta \mid C) \, d\theta, \quad \theta = \lim_{n \to \infty} \frac{r}{n},$$

where $r = \sum x_j$ is the number of positive trials. This is nothing but the joint distribution of a set of (conditionally) independent Bernoulli trials with parameter θ, over which some probability distribution $\pi(\theta \mid C)$ is therefore proven to exist. More generally, for sequences of arbitrary random quantities $\{\boldsymbol{x}_1, \boldsymbol{x}_2, \ldots\}$, exchangeability leads to integral representations of the form

$$p(\boldsymbol{x}_1, \ldots, \boldsymbol{x}_n \mid C) = \int_{\boldsymbol{\Omega}} \prod_{i=1}^{n} p(\boldsymbol{x}_i \mid \boldsymbol{\omega}) \, \pi(\boldsymbol{\omega} \mid C) \, d\boldsymbol{\omega},$$

where $\{p(\boldsymbol{x} \mid \boldsymbol{\omega}), \boldsymbol{\omega} \in \Omega\}$ denotes some probability *model*, $\boldsymbol{\omega}$ is the limit as $n \to \infty$ of some function $f(\boldsymbol{x}_1, \ldots, \boldsymbol{x}_n)$ of the observations, and $\pi(\boldsymbol{\omega} \mid C)$ is some probability distribution over $\boldsymbol{\Omega}$. This formulation includes 'nonparametric' (distribution-free) modelling, where $\boldsymbol{\omega}$ may index, for instance, all continuous probability distributions on $\mathcal{X}$. Notice that $\pi(\boldsymbol{\omega} \mid C)$ does *not* describe a possible variability of $\boldsymbol{\omega}$ (since $\boldsymbol{\omega}$ will typically be a fixed *unknown* vector), but a description on the uncertainty associated with its actual value.

Under appropriate conditioning, exchangeability is a very general assumption, a powerful extension of the traditional concept of a *random sample*. Indeed, many statistical analyses directly assume data (or subsets of data) to be a random sample of conditionally independent observations from some probability model, so that $p(\boldsymbol{x}_1, \ldots, \boldsymbol{x}_n \mid \boldsymbol{\omega}) = \prod_{i=1}^{n} p(\boldsymbol{x}_i \mid \boldsymbol{\omega})$; but *any* random sample is exchangeable, since $\prod_{i=1}^{n} p(\boldsymbol{x}_i \mid \boldsymbol{\omega})$ is obviously invariant under permutations. Notice that the observations in a random sample are only independent *conditional* on the parameter value $\boldsymbol{\omega}$; as nicely put by Lindley, the mantra that the observations $\{\boldsymbol{x}_1, \ldots, \boldsymbol{x}_n\}$ in a random sample are independent is ridiculous when they are used to infer $\boldsymbol{x}_{n+1}$. Notice also that, under exchangeability, the general representation theorem provides an *existence theorem* for a probability distribution $\pi(\boldsymbol{\omega} \mid C)$ on the parameter space Ω, and that this is an argument which only depends on mathematical probability theory.

Another important consequence of exchangeability is that it provides a formal *definition* of the parameter $\boldsymbol{\omega}$ which labels the model as the limit, as $n \to \infty$, of *some* function $f(\boldsymbol{x}_1, \ldots, \boldsymbol{x}_n)$ of the observations; the function f obviously depends both on the assumed model and the chosen parametrization. For instance, in the case of a sequence of Bernoulli trials, the parameter θ is *defined* as the limit, as $n \to \infty$, of the relative frequency r/n. It follows that, under exchangeability, the sentence 'the true value of $\boldsymbol{\omega}$' has a well-defined meaning, if only asymptotically verifiable. Moreover, if 2 different models have parameters which are functionally related by their definition, then the corresponding posterior distributions may be meaningfully compared, for they refer to functionally related quantities. For instance, if a finite subset $\{x_1, \ldots, x_n\}$ of an exchangeable sequence of integer observations is assumed to be a random sample from a Poisson distribution $\text{Po}(x \mid \lambda)$, so that $\text{E}[x \mid \lambda] = \lambda$, then λ is *defined* as $\lim_{n\to\infty}\{\bar{x}_n\}$, where $\bar{x}_n = \sum_j x_j/n$. Similarly, if for some fixed, non-0 integer r, the same data are assumed to be a random sample for a negative binomial $\text{Nb}(x \mid r, \theta)$, so that $\text{E}[x \mid \theta, r] = r(1-\theta)/\theta$, then θ is *defined* as $\lim_{n\to\infty}\{r/(\bar{x}_n + r)\}$. It follows that $\theta \equiv r/(\lambda + r)$ and, hence, θ and $r/(\lambda + r)$ may be treated as the *same* (unknown) quantity whenever this might be needed as, for example, when comparing the relative merits of these alternative probability models.

5.3 The Bayesian Paradigm

The statistical analysis of some observed data D typically begins with some informal *descriptive* evaluation, which is used to suggest a tentative, formal *probability model* $\{p(D \mid \boldsymbol{\omega}),\, \boldsymbol{\omega} \in \Omega\}$ assumed to represent, for some (unknown) value of $\boldsymbol{\omega}$, the probabilistic mechanism which has generated the observed data D. The arguments outlined in Section 2 establish the logical need to assess a *prior* probability distribution $\pi(\boldsymbol{\omega} \mid K)$ over the parameter space Ω, describing the available knowledge K about the value of $\boldsymbol{\omega}$ prior to the data being observed. It then follows from standard probability theory that, if the probability model is correct, all available information about the value of $\boldsymbol{\omega}$ after the data D have been observed is contained in the corresponding *posterior* distribution whose probability density, $\pi(\boldsymbol{\omega} \mid D, A, K)$, is immediately obtained from Bayes' theorem,

$$\pi(\boldsymbol{\omega} \mid D, A, K) = \frac{p(D \mid \boldsymbol{\omega})\, \pi(\boldsymbol{\omega} \mid K)}{\int_{\Omega} p(D \mid \boldsymbol{\omega})\, \pi(\boldsymbol{\omega} \mid K)\, d\boldsymbol{\omega}},$$

where A stands for the assumptions made on the probability model. It is this systematic use of Bayes' theorem to incorporate the information provided by the data that justifies the adjective *Bayesian* by which the paradigm is usually known. It is obvious from Bayes' theorem that any value of $\boldsymbol{\omega}$ with 0 prior density will have 0 posterior density. Thus, it is typically assumed (by appropriate restriction, if necessary, of the *parameter space* Ω) that prior distributions are *strictly positive* (as Savage put it, keep the mind open, or at least ajar). To simplify the presentation, the accepted assumptions A and the available knowledge K are often omitted from the notation, but the fact that *all* statements about $\boldsymbol{\omega}$ given D are *also* conditional to A and K should always be kept in mind.

Example 5: Bayesian inference with a finite parameter space. Let $p(D \mid \theta)$, with $\theta \in \Theta = \{\theta_1, \ldots, \theta_m\}$, be the probability mechanism which is assumed to have generated the observed data D, so that θ may only take a finite number m of different values. Using the finite form of Bayes' theorem, and omitting the prevailing conditions from the notation, the posterior probability of θ_i after data D have been observed is

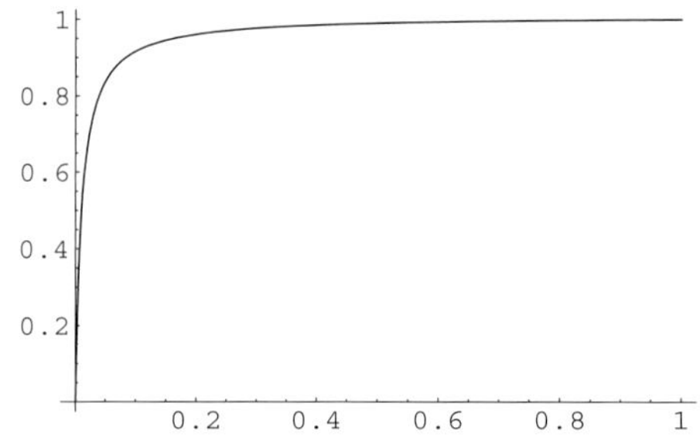

FIGURE 5.1 Posterior probability of infection $\Pr(V \mid +)$ given a positive test, as a function of the prior probability of infection $\Pr(V)$.

$$\Pr(\theta_i \mid D) = \frac{p(D \mid \theta_i)\, \Pr(\theta_i)}{\sum_{j=1}^{m} p(D \mid \theta_j)\, \Pr(\theta_j)}, \quad i = 1, \ldots, m.$$

For any prior distribution $p(\theta) = \{\Pr(\theta_1), \ldots, \Pr(\theta_m)\}$ describing available knowledge about the value of θ, $\Pr(\theta_i \mid D)$ measures how likely θ_i should be judged, given both the initial knowledge described by the prior distribution and the information provided by the data D.

An important, frequent application of this simple technique is provided by probabilistic diagnosis. For example, consider the simple situation where a particular test designed to detect a virus is known from laboratory research to give a positive result in 98% of infected people and in 1% of those uninfected. Then, the posterior probability that a person who tested positive is infected is given by $\Pr(V \mid +) = (0.98\,p)/\{0.98\,p + 0.01\,(1-p)\}$ as a function of $p = \Pr(V)$, the prior probability of a person being infected (the prevalence of the infection in the population under study). Figure 5.1 shows $\Pr(V \mid +)$ as a function of $\Pr(V)$.

As one would expect, the posterior probability is only 0 if the prior probability is 0 (so that it is known that the population is free of infection) and it is only 1 if the prior probability is 1 (so that it is known that the population is universally infected). Notice that if the infection is rare, then the posterior probability of a randomly chosen person being infected will be relatively low even if the test is positive. Indeed, for, say $\Pr(V) = 0.002$, one finds $\Pr(V \mid +) = 0.164$, so that in a population where only 0.2% of individuals are infected, only 16.4% of those testing positive within a random sample will actually prove to be infected: most positives would actually be false positives.

In Section 5.3.1, we describe in some detail the learning process of Bayes' theorem, discuss its implementation in the presence of nuisance parameters, show how it can be used to forecast the value of future observations, and analyse its large sample behaviour.

5.3.1 The Learning Process

In the Bayesian paradigm, the process of learning from the data is systematically implemented by making use of Bayes' theorem to combine the available prior information with the information provided by the data to produce the required posterior distribution. Computation of posterior densities is often facilitated by noting that Bayes' theorem may be simply expressed as

$$\pi(\boldsymbol{\omega} \mid D) \propto p(D \mid \boldsymbol{\omega})\, \pi(\boldsymbol{\omega}),$$

(where $\propto$ stands for 'proportional to' and where, for simplicity, the accepted assumptions A and the available knowledge K have been omitted from the notation), since the missing proportionality constant $[\int_\Omega p(D\,|\,\boldsymbol{\omega})\,\pi(\boldsymbol{\omega})\,d\boldsymbol{\omega}]^{-1}$ may always be deduced from the fact that $\pi(\boldsymbol{\omega}\,|\,D)$, a probability density, must integrate to 1. Hence, to identify the form of a posterior distribution it suffices to identify a *kernel* of the corresponding probability density, that is a function $k(\boldsymbol{\omega})$ such that $\pi(\boldsymbol{\omega}\,|\,D) = c(D)\,k(\boldsymbol{\omega})$ for some $c(D)$ which does not involve $\boldsymbol{\omega}$. In the examples which follow, this technique will often be used.

An *improper prior function* is defined as a positive function $\pi(\boldsymbol{\omega})$ such that $\int_\Omega \pi(\boldsymbol{\omega})\,d\boldsymbol{\omega}$ is not finite. The formal expression of Bayes' theorem remains valid if $\pi(\boldsymbol{\omega})$ is an improper prior function provided that $\int_\Omega p(D\,|\,\boldsymbol{\omega})\,\pi(\boldsymbol{\omega})\,d\boldsymbol{\omega}$ is finite, thus leading to a well-defined *proper* posterior density $\pi(\boldsymbol{\omega}\,|\,D) \propto p(D\,|\,\boldsymbol{\omega})\,\pi(\boldsymbol{\omega})$. In particular, as justified in Section 4, it also remains philosophically valid if $\pi(\boldsymbol{\omega})$ is an appropriately chosen *reference* (typically improper) prior function.

Considered as a function of $\boldsymbol{\omega}$, $l(\boldsymbol{\omega}, D) = p(D\,|\,\boldsymbol{\omega})$ is often referred to as the *likelihood function*. Thus, Bayes' theorem is simply expressed in words by the statement that *the posterior is proportional to the likelihood times the prior*. It follows from Bayes theorem that, provided the *same* prior $\pi(\boldsymbol{\omega})$ is used, 2 different data sets D_1 and D_2, with possibly different probability models $p_1(D_1\,|\,\boldsymbol{\omega})$ and $p_2(D_2\,|\,\boldsymbol{\omega})$ but yielding *proportional* likelihood functions, produce identical posterior distributions for $\boldsymbol{\omega}$. This mathematical fact has been proposed as a principle on its own, the *likelihood principle*, and is seen by many as an obvious requirement for reasonable statistical inference. In particular, for any given prior $\pi(\boldsymbol{\omega})$, the posterior distribution does not depend on the set of possible data values, or the *sample space*. Notice, however, that the likelihood principle only applies to inferences about the parameter vector $\boldsymbol{\omega}$ once the data have been obtained. Consideration of the sample space is essential, for instance, in model criticism, in the design of experiments, in the derivation of predictive distributions, and in the construction of objective Bayesian procedures.

Naturally, the terms prior and posterior are only *relative* to a particular set of data. As one would expect from the coherence induced by probability theory, if data $D = \{\boldsymbol{x}_1, \ldots, \boldsymbol{x}_n\}$ are sequentially presented, the final result will be the same whether data are globally or sequentially processed. Indeed, $\pi(\boldsymbol{\omega}\,|\,\boldsymbol{x}_1, \ldots, \boldsymbol{x}_{i+1}) \propto p(\boldsymbol{x}_{i+1}\,|\,\boldsymbol{\omega})\,\pi(\boldsymbol{\omega}\,|\,\boldsymbol{x}_1, \ldots, \boldsymbol{x}_i)$, for $i = 1, \ldots, n-1$, so that the 'posterior' at a given stage becomes the 'prior' at the next.

In most situations, the posterior distribution is 'sharper' than the prior so that, typically, the density $\pi(\boldsymbol{\omega}\,|\,\boldsymbol{x}_1, \ldots, \boldsymbol{x}_{i+1})$ will be more concentrated around the true value of $\boldsymbol{\omega}$ than $\pi(\boldsymbol{\omega}\,|\,\boldsymbol{x}_1, \ldots, \boldsymbol{x}_i)$. However, this is not always the case: occasionally, a 'surprising' observation increases, rather than decreases, the uncertainty about the value of $\boldsymbol{\omega}$. For instance, in probabilistic diagnosis, a sharp posterior probability distribution (over the possible causes $\{\omega_1, \ldots, \omega_k\}$ of a syndrome) describing a 'clear' diagnosis of disease ω_i (that is, a posterior with a large probability for ω_i) typically update to a less concentrated posterior probability distribution over $\{\omega_1, \ldots, \omega_k\}$ if a new clinical analysis yields data unlikely under ω_i.

For a given probability model, one may find that a particular function of the data $\boldsymbol{t} = \boldsymbol{t}(D)$ is a *sufficient* statistic in the sense that, given the model, $\boldsymbol{t}(D)$ contains all information about $\boldsymbol{\omega}$ which is available in D. Formally, $\boldsymbol{t} = \boldsymbol{t}(D)$ is sufficient if (and only if) there exist nonnegative functions f and g such that the likelihood function may be factorized in the form $p(D\,|\,\boldsymbol{\omega}) = f(\boldsymbol{\omega}, \boldsymbol{t})g(D)$. A sufficient statistic always exists, for $\boldsymbol{t}(D) = D$ is obviously sufficient; however, a much simpler sufficient statistic, with a fixed dimensionality which is independent of the sample size, often exists. In fact this is known to be the case whenever the probability model belongs to the *generalized exponential family*, which includes many of the more frequently used probability models. It is easily established that if $\boldsymbol{t}$ is sufficient, the posterior distribution of $\boldsymbol{\omega}$ only depends on the data D through $\boldsymbol{t}(D)$, and may be directly computed in terms of $p(\boldsymbol{t}\,|\,\boldsymbol{\omega})$, so that, $\pi(\boldsymbol{\omega}\,|\,D) = p(\boldsymbol{\omega}\,|\,\boldsymbol{t}) \propto p(\boldsymbol{t}\,|\,\boldsymbol{\omega})\,\pi(\boldsymbol{\omega})$.

Naturally, for fixed data and model assumptions, different priors lead to different posteriors. Indeed, Bayes' theorem may be described as a data-driven probability transformation machine which maps prior distributions (describing prior knowledge) into posterior distributions (representing combined prior and data knowledge). It is important to analyse whether sensible changes in the prior would induce noticeable changes in the posterior. Posterior distributions based on reference 'noninformative' priors play a central role in this *sensitivity analysis* context. Investigation of the sensitivity of the posterior to changes in the prior is an important ingredient of the comprehensive analysis of the sensitivity of the final results to *all* accepted assumptions which any responsible statistical study should contain.

Example 6: Inference on a binomial parameter. If the data D consist of n Bernoulli observations with parameter θ which contain r positive trials, $p(D\,|\,\theta, n) = \theta^r(1-\theta)^{n-r}$, and $\boldsymbol{t}(D) = \{r, n\}$ is sufficient. Suppose that prior knowledge about θ is described by a beta distribution $\mathrm{Be}(\theta\,|\,\alpha,\beta)$, so that $\pi(\theta\,|\,\alpha,\beta) \propto \theta^{\alpha-1}(1-\theta)^{\beta-1}$. Using Bayes' theorem, the posterior density of θ is $\pi(\theta\,|\,r,n,\alpha,\beta) \propto \theta^r(1-\theta)^{n-r}\,\theta^{\alpha-1}(1-\theta)^{\beta-1}$, *i.e.* proportional to $\theta^{r+\alpha-1}(1-\theta)^{n-r+\beta-1}$, a kernel of the beta distribution $\mathrm{Be}(\theta\,|\,r+\alpha, n-r+\beta)$.

Suppose, for example, that in light of precedent surveys, available information on the proportion θ of citizens who would vote for a particular political measure in a referendum is described by a beta distribution $\mathrm{Be}(\theta\,|\,50, 50)$, so that it is judged to be equally likely that the referendum would be won or lost, and it is judged that the probability that either side wins less than 60% of the vote is 0.95.

A random survey of size 1500 is then conducted, where only 720 citizens declare to be in favour of the proposed measure. Using the results above, the corresponding posterior distribution is then $\mathrm{Be}(\theta\,|\,770, 830)$. These prior and posterior densities are plotted in Figure 5.2; it may be appreciated that, as one would expect, the effect of the data is to drastically reduce the initial uncertainty on the value of θ and, hence, on the referendum outcome. More precisely, $\Pr(\theta < 0.5\,|\,720, 1500, H, K) = 0.933$ (shaded region in Figure 5.2) so that, after the information from the survey has been included, the probability that the referendum will be lost should be judged to be about 93%.

The general situation where the vector of interest is not the whole parameter vector $\boldsymbol{\omega}$, but some function $\boldsymbol{\theta} = \boldsymbol{\theta}(\boldsymbol{\omega})$ of possibly lower dimension than $\boldsymbol{\omega}$, will now be

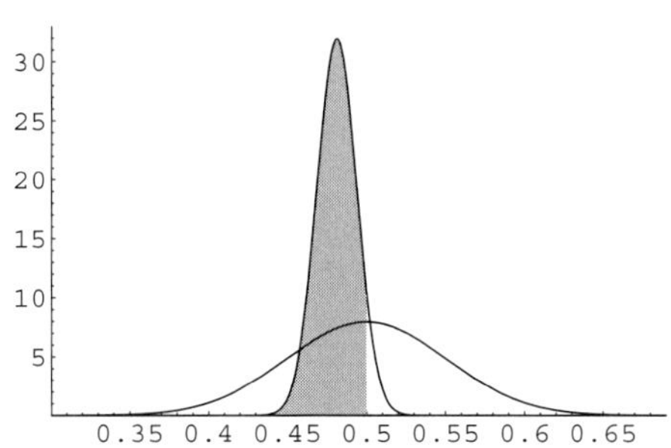

FIGURE 5.2 Prior and posterior densities of the proportion θ of citizens who would vote in favour of the proposal in a referendum.

considered. Let D be some observed data, let $\{p(D \mid \boldsymbol{\omega}), \boldsymbol{\omega} \in \Omega\}$ be a probability model assumed to describe the probability mechanism which has generated D, let $\pi(\boldsymbol{\omega})$ be a probability distribution describing any available information on the value of $\boldsymbol{\omega}$, and let $\boldsymbol{\theta} = \boldsymbol{\theta}(\boldsymbol{\omega}) \in \Theta$ be a function of the original parameters over whose value inferences based on the data D are required. Any valid conclusion on the value of the *vector of interest* $\boldsymbol{\theta}$ will then be contained in its posterior probability distribution $\pi(\boldsymbol{\theta} \mid D)$ which is conditional on the observed data D and will naturally also depend, although not explicitly shown in the notation, on the assumed model $\{p(D \mid \boldsymbol{\omega}), \boldsymbol{\omega} \in \Omega\}$, and on the available prior information encapsulated by $\pi(\boldsymbol{\omega})$. The required posterior distribution $p(\boldsymbol{\theta} \mid D)$ is found by standard use of probability calculus–indeed, by Bayes' theorem, $\pi(\boldsymbol{\omega} \mid D) \propto p(D \mid \boldsymbol{\omega})\, \pi(\boldsymbol{\omega})$. Moreover, consider some other function of the original parameters $\boldsymbol{\lambda} = \boldsymbol{\lambda}(\boldsymbol{\omega}) \in \Lambda$ such that $\boldsymbol{\psi} = \{\boldsymbol{\theta}, \boldsymbol{\lambda}\}$ is a one-to-one transformation of $\boldsymbol{\omega}$, and let $J(\boldsymbol{\omega}) = (\partial \boldsymbol{\psi} / \partial \boldsymbol{\omega})$ be the corresponding Jacobian matrix. Naturally, the introduction of $\boldsymbol{\lambda}$ is not necessary if $\boldsymbol{\theta}(\boldsymbol{\omega})$ is a one-to-one transformation of $\boldsymbol{\omega}$. Using standard change-of-variable probability techniques, the posterior density of $\boldsymbol{\psi}$ is

$$\pi(\boldsymbol{\psi} \mid D) = \pi(\boldsymbol{\theta}, \boldsymbol{\lambda} \mid D) = \left[\frac{\pi(\boldsymbol{\omega} \mid D)}{\mid J(\boldsymbol{\omega}) \mid} \right]_{\boldsymbol{\omega} = \boldsymbol{\omega}(\boldsymbol{\psi})}$$

and the required posterior of $\boldsymbol{\theta}$ is the appropriate *marginal* density, obtained by integration over the *nuisance parameter* $\boldsymbol{\lambda}$,

$$\pi(\boldsymbol{\theta} \mid D) = \int_{\Lambda} \pi(\boldsymbol{\theta}, \boldsymbol{\lambda} \mid D)\, d\boldsymbol{\lambda}.$$

Notice that elimination of unwanted nuisance parameters, a simple integration within the Bayesian paradigm is, however, a difficult (often polemic) problem for frequentist statistics.

Sometimes, the range of possible values of $\boldsymbol{\omega}$ is effectively restricted by contextual considerations. If $\boldsymbol{\omega}$ is known to belong to $\Omega_c \subset \Omega$, the prior distribution is only positive in Ω_c and, using Bayes' theorem, it is immediately found that the restricted posterior is

$$\pi(\boldsymbol{\omega} \mid D, \boldsymbol{\omega} \in \Omega_c) = \frac{\pi(\boldsymbol{\omega} \mid D)}{\int_{\Omega_c} \pi(\boldsymbol{\omega} \mid D)}, \quad \boldsymbol{\omega} \in \Omega_c,$$

and obviously vanishes if $\boldsymbol{\omega} \notin \Omega_c$. Thus, to incorporate a restriction on the possible values of the parameters, it suffices to *renormalise* the unrestricted posterior distribution to the set $\Omega_c \subset \Omega$ of parameter values which satisfy the required condition. Incorporation of known constraints on the parameter values, a simple renormalisation within the Bayesian paradigm, is another difficult problem for conventional statistics.[4]

Example 7: Inference on normal parameters. Let $D = \{x_1, \ldots, x_n\}$ be a random sample from a normal distribution $\mathrm{N}(x \mid \mu, \sigma)$. The corresponding likelihood function is immediately found to be proportional to $\sigma^{-n} \exp[-n\{s^2 + (\bar{x} - \mu)^2\}/(2\sigma^2)]$, where $n\bar{x} = \sum_i x_i$, and $ns^2 = \sum_i (x_i - \bar{x})^2$. It may be shown (see Section 4) that absence of initial information on the value of both μ and σ may formally be described by a joint prior function which is uniform in both μ and $\log(\sigma)$, that is, by the (improper) prior function $\pi(\mu, \sigma) = \sigma^{-1}$. Using Bayes' theorem, the corresponding joint posterior is

$$\pi(\mu, \sigma \mid D) \propto \sigma^{-(n+1)} \exp[-n\{s^2 + (\bar{x} - \mu)^2\}/(2\sigma^2)].$$

[4] For further details on the elimination of nuisance parameters see Liseo (2005).

Thus, using the gamma integral in terms of $\lambda = \sigma^{-2}$ to integrate out σ,

$$\pi(\mu \,|\, D) \propto \int_0^\infty \sigma^{-(n+1)} \exp\left[-\frac{n}{2\sigma^2}[s^2 + (\bar{x} - \mu)^2] \right] d\sigma \propto [s^2 + (\bar{x} - \mu)^2]^{-n/2},$$

which is recognized as a kernel of the Student density $\text{St}(\mu \,|\, \bar{x}, s/\sqrt{n-1}, n-1)$. Similarly, integrating out μ,

$$\pi(\sigma \,|\, D) \propto \int_{-\infty}^\infty \sigma^{-(n+1)} \exp\left[-\frac{n}{2\sigma^2}[s^2 + (\bar{x} - \mu)^2] \right] d\mu \propto \sigma^{-n} \exp\left[-\frac{ns^2}{2\sigma^2} \right].$$

Changing variables to the precision $\lambda = \sigma^{-2}$ results in $\pi(\lambda \,|\, D) \propto \lambda^{(n-3)/2} e^{ns^2\lambda/2}$, a kernel of the gamma density $\text{Ga}(\lambda \,|\, (n-1)/2, ns^2/2)$. In terms of the standard deviation σ this becomes $\pi(\sigma \,|\, D) = p(\lambda \,|\, D)|\partial\lambda/\partial\sigma| = 2\sigma^{-3}\text{Ga}(\sigma^{-2} \,|\, (n-1)/2, ns^2/2)$, a square-root inverted gamma density.

A frequent example of this scenario is provided by laboratory measurements made where central limit conditions apply, so that (assuming no experimental bias) those measurements may be treated as a random sample from a normal distribution centred at the quantity μ which is being measured, and with some (unknown) standard deviation σ. Suppose, for example, that in an elementary physics classroom experiment to measure the gravitational field g with a pendulum, a student obtains $n = 20$ measurements of g yielding (in m/sec^2) a mean $\bar{x} = 9.8087$, and a standard deviation $s = 0.0428$. Using no other information, the corresponding posterior distribution is $\pi(g \,|\, D) = \text{St}(g \,|\, 9.8087, 0.0098, 19)$, represented in the top panel of Figure 5.3. In particular, $\Pr(9.788 < g < 9.829 \,|\, D) = 0.95$, so that, with the information provided by this experiment, the gravitational field at the location of the laboratory may be expected to lie between 9.788 and 9.829 with probability 0.95.

Formally, the posterior distribution of g should be restricted to $g > 0$; however, as immediately obvious from Figure 5.3, this would not have any appreciable effect, due to the fact that the likelihood function is actually concentrated on positive g values.

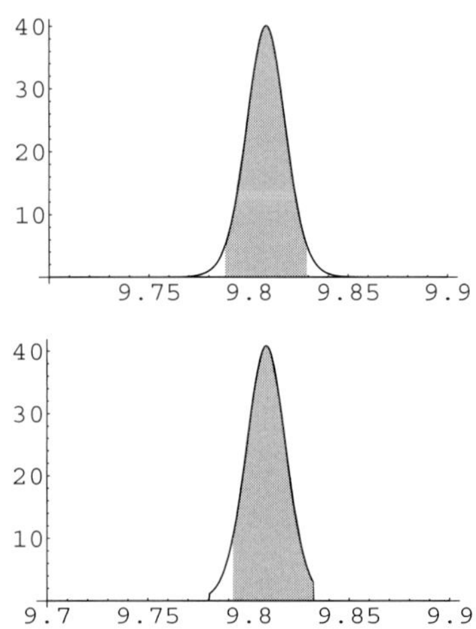

FIGURE 5.3 Posterior probability density $\pi(g \,|\, m, s, n)$ of the value g of the gravitational field, given $n = 20$ normal measurements with mean $m = 9.8087$ and standard deviation $s = 0.0428$, with no additional information (top), and with the value of g restricted to $G_c = \{g;\, 9.7803 < g < 9.8322\}$ (bottom). Shaded areas represent 95% credible regions.

Suppose now that the student is further instructed to incorporate into the analysis the fact that the value of the gravitational field g at the laboratory is known to lie between $9.7803\, m/sec^2$ (average value at the equator) and $9.8322\, m/sec^2$ (average value at the poles). The updated posterior distribution will then be

$$\pi(g \mid D, g \in G_c) = \frac{\mathrm{St}(g \mid m, s/\sqrt{n-1}, n)}{\int_{g \in G_c} \mathrm{St}(g \mid m, s/\sqrt{n-1}, n)}, \quad g \in G_c,$$

represented in the bottom panel of Figure 5.3, where $G_c = \{g;\, 9.7803 < g < 9.8322\}$. One-dimensional numerical integration proves that $\Pr(g > 9.792 \mid D, g \in G_c) = 0.95$. Moreover, if inferences about the standard deviation σ of the measurement procedure are also requested, the corresponding posterior distribution is easily found to be $\pi(\sigma \mid D) = 2\sigma^{-3}\mathrm{Ga}(\sigma^{-2} \mid 9.5, 0.0183)$. This has a mean $\mathrm{E}[\sigma \mid D] = 0.0458$ and yields a 0.95 credible interval $\Pr(0.0334 < \sigma < 0.0642 \mid D) = 0.95$.

5.3.2 *Predictive Distributions*

Let $D = \{\boldsymbol{x}_1, \ldots, \boldsymbol{x}_n\}$, $\boldsymbol{x}_i \in \mathcal{X}$, be a set of exchangeable observations, and consider now a situation where it is desired to predict the value of a future observation $\boldsymbol{x} \in \mathcal{X}$ generated by the same random mechanism that has generated the data D. It follows from the foundations arguments discussed in Section 2 that the solution to this prediction problem is simply encapsulated by the *predictive* distribution $p(\boldsymbol{x} \mid D)$ describing the uncertainty on the value that $\boldsymbol{x}$ will take, given the information provided by D and any other available knowledge. Suppose that contextual information suggests the assumption that data D may be considered a random sample from a distribution in the family $\{p(\boldsymbol{x} \mid \boldsymbol{\omega}), \boldsymbol{\omega} \in \Omega\}$, and let $\pi(\boldsymbol{\omega})$ be a prior distribution describing available information on the value of $\boldsymbol{\omega}$. Since $p(\boldsymbol{x} \mid \boldsymbol{\omega}, D) = p(\boldsymbol{x} \mid \boldsymbol{\omega})$, it then follows from standard probability theory that

$$p(\boldsymbol{x} \mid D) = \int_{\Omega} p(\boldsymbol{x} \mid \boldsymbol{\omega})\, \pi(\boldsymbol{\omega} \mid D)\, d\boldsymbol{\omega},$$

which is an average of the probability distributions of $\boldsymbol{x}$ conditional on the (unknown) value of $\boldsymbol{\omega}$, weighted with the posterior distribution of $\boldsymbol{\omega}$ given D.

If the assumptions on the probability model are correct, the posterior predictive distribution $p(\boldsymbol{x} \mid D)$ will converge, as the sample size increases, to the distribution $p(\boldsymbol{x} \mid \boldsymbol{\omega})$ which has generated the data. Indeed, the best technique to assess the quality of the inferences about $\boldsymbol{\omega}$ encapsulated in $\pi(\boldsymbol{\omega} \mid D)$ is to check against the observed data the predictive distribution $p(\boldsymbol{x} \mid D)$ generated by $\pi(\boldsymbol{\omega} \mid D)$.[5]

Example 8: Prediction in a Poisson process. Let $D = \{r_1, \ldots, r_n\}$ be a random sample from a Poisson distribution $\mathrm{Pn}(r \mid \lambda)$ with parameter λ, so that $p(D \mid \lambda) \propto \lambda^t e^{-\lambda n}$, where $t = \sum r_i$. It may be shown (see Section 4) that absence of initial information on the value of λ may be formally described by the (improper) prior function $\pi(\lambda) = \lambda^{-1/2}$. Using Bayes' theorem, the corresponding posterior is

$$\pi(\lambda \mid D) \propto \lambda^t e^{-\lambda n}\, \lambda^{-1/2} \propto \lambda^{t-1/2} e^{-\lambda n},$$

a kernel of the gamma density $\mathrm{Ga}(\lambda \mid , t+1/2, n)$, with mean $(t+1/2)/n$. The corresponding predictive distribution is the Poisson-Gamma mixture

[5] For a good introduction to Bayesian predictive inference, see Geisser (1993).

$$p(r \mid D) = \int_0^\infty \mathrm{Pn}(r \mid \lambda)\, \mathrm{Ga}(\lambda \mid, t + \frac{1}{2}, n)\, d\lambda = \frac{n^{t+1/2}}{\Gamma(t+1/2)}\, \frac{1}{r!}\, \frac{\Gamma(r+t+1/2)}{(1+n)^{r+t+1/2}}\,.$$

Suppose, for example, that in a firm producing automobile restraint systems the entire production in each of 10 consecutive months has yielded no complaints from clients. With no additional information on the average number λ of complaints per month, the quality-assurance department of the firm may report that the probability that r complaints will be received in the next month of production is given by the expression for $p(r \mid D)$, with $t = 0$ and $n = 10$. In particular, $p(r = 0 \mid D) = 0.953$, $p(r = 1 \mid D) = 0.043$, and $p(r = 2 \mid D) = 0.003$. Many other situations may be described with the same model. For instance, if meteorological conditions remain similar in a given area, $p(r = 0 \mid D) = 0.953$ would describe the chances of no flash floods next year, given 10 years without flash floods in the area.

Example 9: Prediction in a normal process. Consider now prediction of a continuous variable. Let $D = \{x_1, \ldots, x_n\}$ be a random sample from a normal distribution $\mathrm{N}(x \mid \mu, \sigma)$. As mentioned in Example 7, absence of initial information on the values of both μ and σ is formally described by the improper prior function $\pi(\mu, \sigma) = \sigma^{-1}$, and this leads to the joint posterior density described there. The corresponding (posterior) predictive distribution is

$$p(x \mid D) = \int_0^\infty \int_{-\infty}^\infty \mathrm{N}(x \mid \mu, \sigma)\, \pi(\mu, \sigma \mid D)\, d\mu d\sigma = \mathrm{St}(x \mid \bar{x}, s\sqrt{\frac{n+1}{n-1}}, n-1).$$

If μ is known to be positive, the appropriate prior function will be the restricted function $\pi(\mu, \sigma) = \sigma^{-1}$ if $\mu > 0$ and 0 otherwise. However, the expression for $p(x \mid D)$ still holds, provided the likelihood function $p(D \mid \mu, \sigma)$ is concentrated on positive μ values. Suppose, for example, that in the firm producing automobile restraint systems, the observed breaking strengths of $n = 10$ randomly chosen safety belt webbings have mean $\bar{x} = 28.011$ kN and standard deviation $s = 0.443$ kN, and that the relevant engineering specification requires breaking strengths to be larger than 26 kN. If the data may truly be assumed to be a random sample from a normal distribution, the likelihood function is only appreciable for positive μ values, and only the information provided by this small sample is to be used, then it may be claimed that the probability that a safety belt randomly chosen from the same batch as the sample tested would satisfy the required specification is $\Pr(x > 26 \mid D) = 0.9987$. Besides, if production conditions remain constant, 99.87% of the safety belt webbings may be expected to have acceptable breaking strengths.

5.3.3 Asymptotic Behaviour

The behaviour of posterior distributions when the sample size is large is now considered. This is important for at least 2 different reasons: (i) asymptotic results provide useful first-order approximations when actual samples are relatively large, and (ii) objective Bayesian methods typically depend on the asymptotic properties of the assumed model. Let $D = \{\boldsymbol{x}_1, \ldots, \boldsymbol{x}_n\}$, $\boldsymbol{x} \in \mathcal{X}$, be a random sample of size n from $\{p(\boldsymbol{x} \mid \boldsymbol{\omega}), \boldsymbol{\omega} \in \Omega\}$. It may be shown that, as $n \to \infty$, the posterior distribution of a *discrete* parameter $\boldsymbol{\omega}$ typically converges to a degenerate distribution which gives probability 1 to the true value of $\boldsymbol{\omega}$, and that the posterior distribution of a *continuous* parameter $\boldsymbol{\omega}$ typically converges to a normal distribution centered at its *maximum likelihood estimate* $\hat{\boldsymbol{\omega}}$ (MLE), with a variance matrix which decreases with n as $1/n$.

Consider first the situation where $\Omega = \{\boldsymbol{\omega}_1, \boldsymbol{\omega}_2, \ldots\}$ consists of a *countable* (possibly infinite) set of values, such that the probability model which corresponds to the true parameter value $\boldsymbol{\omega}_t$ is *distinguishable* from the others in the sense that the logarithmic divergence $\kappa\{p(\boldsymbol{x}\,|\,\boldsymbol{\omega}_i)\,|\,p(\boldsymbol{x}\,|\,\boldsymbol{\omega}_t)\}$ of each of the $p(\boldsymbol{x}\,|\,\boldsymbol{\omega}_i)$ from $p(\boldsymbol{x}\,|\,\boldsymbol{\omega}_t)$ is strictly positive. Taking logarithms in Bayes' theorem and using the strong law of large numbers on the n conditionally independent and identically distributed random quantities $z_1, \ldots, z_n$, where $z_j = \log[p(\boldsymbol{x}_j\,|\,\boldsymbol{\omega}_i)/p(\boldsymbol{x}_j\,|\,\boldsymbol{\omega}_t)]$, $j = 1, \ldots, n$, it may be shown that

$$\lim_{n\to\infty} \Pr(\boldsymbol{\omega}_t\,|\,\boldsymbol{x}_1, \ldots, \boldsymbol{x}_n) = 1, \qquad \lim_{n\to\infty} \Pr(\boldsymbol{\omega}_i\,|\,\boldsymbol{x}_1, \ldots, \boldsymbol{x}_n) = 0, \quad i \neq t.$$

Thus, under appropriate regularity conditions, the posterior probability of the true parameter value converges to 1 as the sample size grows.

Consider now the situation where $\boldsymbol{\omega}$ is a k-dimensional *continuous* parameter. Using Bayes' theorem as $\pi(\boldsymbol{\omega}\,|\,\boldsymbol{x}_1, \ldots, \boldsymbol{x}_n) \propto \exp\{\log[\pi(\boldsymbol{\omega})] + \sum_{j=1}^n \log[p(\boldsymbol{x}_j\,|\,\boldsymbol{\omega})]\}$, expanding $\sum_j \log[p(\boldsymbol{x}_j\,|\,\boldsymbol{\omega})]$ about its maximum (the MLE $\hat{\boldsymbol{\omega}}$), and assuming regularity conditions (to ensure that terms of order higher than quadratic may be ignored and that the sum of the terms from the likelihood dominate the term from the prior) it is found that the posterior density of $\boldsymbol{\omega}$ is the approximate k-variate normal

$$\pi(\boldsymbol{\omega}\,|\,\boldsymbol{x}_1, \ldots, \boldsymbol{x}_n) \approx \mathrm{N}_k\{\hat{\boldsymbol{\omega}}, \boldsymbol{S}(D, \hat{\boldsymbol{\omega}})\}, \qquad \boldsymbol{S}^{-1}(D, \boldsymbol{\omega}) = \left(-\sum_{l=1}^{n} \frac{\partial^2 \log[p(\boldsymbol{x}_l\,|\,\boldsymbol{\omega})]}{\partial\omega_i \partial\omega_j} \right).$$

A simpler but somewhat poorer approximation may be obtained using the strong law of large numbers on the sums above to establish that $\boldsymbol{S}^{-1}(D, \hat{\boldsymbol{\omega}}) \approx n\,\boldsymbol{F}(\hat{\boldsymbol{\omega}})$, where $\boldsymbol{F}(\boldsymbol{\omega})$ is *Fisher's information matrix*, with general element

$$\boldsymbol{F}_{ij}(\boldsymbol{\omega}) = -\int_{\mathcal{X}} p(\boldsymbol{x}\,|\,\boldsymbol{\omega})\; \frac{\partial^2 \log[p(\boldsymbol{x}\,|\,\boldsymbol{\omega})]}{\partial\omega_i \partial\omega_j}\, d\boldsymbol{x},$$

so that

$$\pi(\boldsymbol{\omega}\,|\,\boldsymbol{x}_1, \ldots, \boldsymbol{x}_n) \approx N_k(\boldsymbol{\omega}\,|\,\hat{\boldsymbol{\omega}}, n^{-1}\,\boldsymbol{F}^{-1}(\hat{\boldsymbol{\omega}})).$$

Thus, under appropriate regularity conditions, the posterior density of the parameter vector $\boldsymbol{\omega}$ approaches, as the sample size grows, a multivariate normal density centred at the MLE $\hat{\boldsymbol{\omega}}$, with a variance matrix which decreases with n as n^{-1}.

Example 10: Asymptotic approximation with binomial data. Let $D = (x_1, \ldots, x_n)$ consist of n independent Bernoulli trials with parameter θ, so that $p(D\,|\,\theta, n) = \theta^r(1-\theta)^{n-r}$. This likelihood function is maximised at $\hat{\theta} = r/n$, and Fisher's information function is $F(\theta) = \theta^{-1}(1-\theta)^{-1}$. Thus, using the results above, the posterior distribution of θ will be the approximate normal,

$$\pi(\theta\,|\,r, n) \approx \mathrm{N}(\theta\,|\,\hat{\theta}, s(\hat{\theta})/\sqrt{n}), \quad s(\theta) = \{\theta(1-\theta)\}^{1/2}$$

with mean $\hat{\theta} = r/n$ and variance $\hat{\theta}(1-\hat{\theta})/n$. This provides a reasonable approximation to the exact posterior if (i) the prior $\pi(\theta)$ is relatively 'flat' in the region where the likelihood function matters, and (ii) both r and n are moderately large. If, say, $n = 1{,}500$ and $r = 720$, this leads to $\pi(\theta\,|\,D) \approx \mathrm{N}(\theta\,|\,0.480, 0.013)$, and to $\Pr(\theta > 0.5\,|\,D) \approx 0.940$, which may be compared with the exact value $\Pr(\theta > 0.5\,|\,D) = 0.933$ obtained from the posterior distribution which corresponds to the prior $\mathrm{Be}(\theta\,|\,50, 50)$.

It follows from the *joint* posterior asymptotic behaviour of $\boldsymbol{\omega}$ and from the properties of the multivariate normal distribution that if the parameter vector is decomposed into $\boldsymbol{\omega} = (\boldsymbol{\theta}, \boldsymbol{\lambda})$ and Fisher's information matrix is correspondingly partitioned, so that

$$\boldsymbol{F}(\boldsymbol{\omega}) = \boldsymbol{F}(\boldsymbol{\theta}, \boldsymbol{\lambda}) = \begin{pmatrix} \boldsymbol{F}_{\theta\theta}(\boldsymbol{\theta}, \boldsymbol{\lambda}) & \boldsymbol{F}_{\theta\lambda}(\boldsymbol{\theta}, \boldsymbol{\lambda}) \\ \boldsymbol{F}_{\lambda\theta}(\boldsymbol{\theta}, \boldsymbol{\lambda}) & \boldsymbol{F}_{\lambda\lambda}(\boldsymbol{\theta}, \boldsymbol{\lambda}) \end{pmatrix}$$

and

$$\boldsymbol{S}(\boldsymbol{\theta}, \boldsymbol{\lambda}) = \boldsymbol{F}^{-1}(\boldsymbol{\theta}, \boldsymbol{\lambda}) = \begin{pmatrix} \boldsymbol{S}_{\theta\theta}(\boldsymbol{\theta}, \boldsymbol{\lambda}) & \boldsymbol{S}_{\theta\lambda}(\boldsymbol{\theta}, \boldsymbol{\lambda}) \\ \boldsymbol{S}_{\lambda\theta}(\boldsymbol{\theta}, \boldsymbol{\lambda}) & \boldsymbol{S}_{\lambda\lambda}(\boldsymbol{\theta}, \boldsymbol{\lambda}) \end{pmatrix},$$

then the *marginal* posterior distribution of $\boldsymbol{\theta}$ will be

$$\pi(\boldsymbol{\theta} \mid D) \approx \mathrm{N}\{\boldsymbol{\theta} \mid \hat{\boldsymbol{\theta}},\, n^{-1}\, \boldsymbol{S}_{\theta\theta}(\hat{\boldsymbol{\theta}}, \hat{\boldsymbol{\lambda}})\},$$

while the *conditional* posterior distribution of $\boldsymbol{\lambda}$ given $\boldsymbol{\theta}$ will be

$$\pi(\boldsymbol{\lambda} \mid \boldsymbol{\theta}, D) \approx \mathrm{N}\{\boldsymbol{\lambda} \mid \hat{\boldsymbol{\lambda}} - \boldsymbol{F}_{\lambda\lambda}^{-1}(\boldsymbol{\theta}, \hat{\boldsymbol{\lambda}})\boldsymbol{F}_{\lambda\theta}(\boldsymbol{\theta}, \hat{\boldsymbol{\lambda}})(\hat{\boldsymbol{\theta}} - \boldsymbol{\theta}),\, n^{-1}\, \boldsymbol{F}_{\lambda\lambda}^{-1}(\boldsymbol{\theta}, \hat{\boldsymbol{\lambda}})\}.$$

Notice that $\boldsymbol{F}_{\lambda\lambda}^{-1} = \boldsymbol{S}_{\lambda\lambda}$ if (and only if) $\boldsymbol{F}$ is block diagonal, i.e. if (and only if) $\boldsymbol{\theta}$ and $\boldsymbol{\lambda}$ are asymptotically independent.

Example 11: Asymptotic approximation with normal data. Let $D = (x_1, \ldots, x_n)$ be a random sample from a normal distribution $\mathrm{N}(x \mid \mu, \sigma)$. The corresponding likelihood function $p(D \mid \mu, \sigma)$ is maximised at $(\hat{\mu}, \hat{\sigma}) = (\bar{x}, s)$, and Fisher's information matrix is diagonal, with $F_{\mu\mu} = \sigma^{-2}$. Hence, the posterior distribution of μ is approximately $\mathrm{N}(\mu \mid \bar{x}, s/\sqrt{n})$; this may be compared with the exact result $\pi(\mu \mid D) = \mathrm{St}(\mu \mid \bar{x}, s/\sqrt{n-1}, n-1)$ obtained previously under the assumption of no prior knowledge.

5.4 Reference Analysis

Under the Bayesian paradigm, the outcome of any inference problem (the posterior distribution of the quantity of interest) combines the information provided by the data with relevant available prior information. In many situations, however, either the available prior information on the quantity of interest is too vague to warrant the effort required to have it formalised in the form of a probability distribution, or it is too subjective to be useful in scientific communication or public decision-making. It is therefore important to be able to identify the mathematical form of a 'noninformative' prior, a prior that would have a minimal effect, relative to the data, on the posterior inference. More formally, suppose that the probability mechanism which has generated the available data D is assumed to be $p(D \mid \boldsymbol{\omega})$, for some $\boldsymbol{\omega} \in \Omega$, and that the quantity of interest is some real-valued function $\theta = \theta(\boldsymbol{\omega})$ of the model parameter $\boldsymbol{\omega}$. Without loss of generality, it may be assumed that the probability model is of the form

$$\mathcal{M} = \{p(D \mid \theta, \boldsymbol{\lambda}), D \in \mathcal{D}, \theta \in \Theta, \boldsymbol{\lambda} \in \Lambda\},$$

where $\boldsymbol{\lambda}$ is some appropriately chosen nuisance parameter vector. As described in Section 3, to obtain the required posterior distribution of the quantity of interest $\pi(\theta \mid D)$ it is necessary to specify a *joint* prior $\pi(\theta, \boldsymbol{\lambda})$. It is now required to identify the form of that joint prior $\pi_\theta(\theta, \boldsymbol{\lambda} \mid \mathcal{M}, \mathcal{P})$, the θ-*reference prior*, which would have a *minimal effect* on the corresponding posterior distribution of θ,

$$\pi(\theta \mid D) \propto \int_{\Lambda} p(D \mid \theta, \boldsymbol{\lambda})\, \pi_\theta(\theta, \boldsymbol{\lambda} \mid \mathcal{M}, \mathcal{P})\, d\boldsymbol{\lambda},$$

within the class $\mathcal{P}$ of all the prior distributions compatible with whatever information about $(\theta, \boldsymbol{\lambda})$ one is prepared to assume, which may just be the class $\mathcal{P}_0$ of *all* strictly positive priors. To simplify the notation, when there is no danger of confusion the reference prior $\pi_\theta(\theta, \boldsymbol{\lambda} \,|\, \mathcal{M}, \mathcal{P})$ is often simply denoted by $\pi(\theta, \boldsymbol{\lambda})$, but keep in mind its dependence on the quantity of interest θ, the assumed model $\mathcal{M}$, and the class $\mathcal{P}$ of priors compatible with assumed knowledge.

To use a conventional expression, the reference prior 'would let the data speak for themselves' about the likely value of θ. Properly defined, reference *posterior* distributions have an important role to play in scientific communication, for they provide the answer to a central question in the sciences: conditional on the assumed model $p(D \,|\, \theta, \boldsymbol{\lambda})$, and on any further assumptions of the value of θ on which there might be universal agreement, the reference posterior $\pi(\theta \,|\, D)$ should specify what *could* be said about θ *if* the only available information about θ were some well-documented data D and whatever information (if any) one is prepared to assume by restricting the prior to belong to an appropriate class $\mathcal{P}$.

Much work has been done to formulate 'reference' priors which would make this idea mathematically precise.[6] Section 5.4.1 concentrates on an approach that is based on information theory to derive reference distributions which may be argued to provide the most advanced general procedure available; this was initiated by Bernardo (1979b, 1981) and further developed by Berger and Bernardo (1989, 1992a,b,c), Bernardo (1997, 2005b, 2011, 2015), Bernardo and Ramón (1998), Berger, Bernardo, and Sun (2009a,b, 2012, 2015), and references therein. In the formulation described below, far from ignoring prior knowledge, the reference posterior exploits certain well-defined features of a *possible* prior, namely those describing a situation where relevant knowledge about the quantity of interest (beyond that universally accepted, as specified by the choice of $\mathcal{P}$) may be held to be negligible compared to the information about that quantity which repeated experimentation (from a specific data-generating mechanism $\mathcal{M}$) might possibly provide. Reference analysis is appropriate in contexts where the set of inferences which could be drawn in this *possible* situation is considered to be pertinent.

Any statistical analysis contains a fair number of subjective elements; these include (among others) the data selected, the model assumptions, and the choice of the quantities of interest. Reference analysis may be argued to provide an 'objective' Bayesian solution to statistical-inference problems in just the same sense that conventional statistical methods claim to be 'objective': in that the solutions only depend on model assumptions and observed data.

5.4.1 Reference Distributions

One parameter. Consider the experiment which consists of the observation of data D, generated by a random mechanism $p(D \,|\, \theta)$ which only depends on a real-valued parameter $\theta \in \Theta$, and let $\boldsymbol{t} = \boldsymbol{t}(D) \in T$ be *any* sufficient statistic (which may well be the complete data set D). In Shannon's general information theory, the *amount of information* $I^\theta\{T, \pi(\theta)\}$ which may be expected to be provided by D, or (equivalently) by $\boldsymbol{t}(D)$, about the value of θ is defined by

$$I^\theta\{T, \pi(\theta)\} = \kappa\,\{p(\boldsymbol{t})\pi(\theta) \,|\, p(\boldsymbol{t} \,|\, \theta)\pi(\theta)\} = \mathrm{E}_{\boldsymbol{t}}\left[\int_\Theta \pi(\theta \,|\, \boldsymbol{t}) \log \frac{\pi(\theta \,|\, \boldsymbol{t})}{\pi(\theta)}\, d\theta\right],$$

the expected logarithmic divergence of the prior from the posterior. This is naturally a *functional* of the prior $\pi(\theta)$: the larger the prior information, the smaller the information

[6] For historical details, see Bernardo and Smith (1994, section 5.6.2), Kass and Wasserman (1996), Bernardo (2005b) and references therein.

which the data may be expected to provide. The functional $I^\theta\{T, \pi(\theta)\}$ is concave, nonnegative, and invariant under one-to-one transformations of θ. Consider now the amount of information $I^\theta\{T^k, \pi(\theta)\}$ about θ which may be expected from the experiment which consists of k conditionally independent replications $\{\boldsymbol{t}_1, \ldots, \boldsymbol{t}_k\}$ of the original experiment. As $k \to \infty$, such an experiment would provide any *missing information* about θ which could possibly be obtained within this framework; thus, as $k \to \infty$, the functional $I^\theta\{T^k, \pi(\theta)\}$ will approach the missing information about θ associated with the prior $p(\theta)$. Intuitively, a θ-'noninformative' prior is one which *maximises the missing information* about θ. Formally, if $\pi_k(\theta)$ denotes the prior density which maximises $I^\theta\{T^k, \pi(\theta)\}$ in the class $\mathcal{P}$ of prior distributions which are compatible with accepted assumptions on the value of θ (which may well be the class $\mathcal{P}_0$ of *all* strictly positive proper priors) then the θ-reference prior $\pi(\theta \mid \mathcal{M}, \mathcal{P})$ is the limit as $k \to \infty$ (in a sense to be made precise) of the sequence of priors $\{\pi_k(\theta), k = 1, 2, \ldots\}$.

Notice that this limiting procedure is *not* some kind of asymptotic approximation, but an essential element of the *definition* of a reference prior. In particular, this definition implies that reference distributions only depend on the *asymptotic* behaviour of the assumed probability model, a feature which simplifies their actual derivation.

Example 12: Maximum entropy. If θ may only take a finite number of values, so that the parameter space is $\Theta = \{\theta_1, \ldots, \theta_m\}$ and $\pi(\theta) = \{p_1, \ldots, p_m\}$, with $p_i = \Pr(\theta = \theta_i)$, and there is no topology associated with the parameter space Θ, so that the θ_i's are just labels with no quantitative meaning, then the missing information associated with $\{p_1, \ldots, p_m\}$ reduces to

$$\lim_{k\to\infty} I^\theta\{T^k, \pi(\theta)\} = H(p_1, \ldots, p_m) = -\sum_{i=1}^m p_i \log(p_i),$$

that is, the entropy of the prior distribution $\{p_1, \ldots, p_m\}$.

Thus, in the non-quantitative finite case, the reference prior $\pi(\theta \mid \mathcal{M}, \mathcal{P})$ is that with maximum entropy in the class $\mathcal{P}$ of priors compatible with accepted assumptions. Consequently, the reference prior algorithm contains 'maximum entropy' priors as the particular case which obtains when the parameter space is a finite set of labels, the only case where the original concept of entropy as a measure of uncertainty is unambiguous and well behaved. In particular, if $\mathcal{P}$ is the class $\mathcal{P}_0$ of all priors over $\{\theta_1, \ldots, \theta_m\}$, then the reference prior is the uniform prior over the set of possible θ values, $\pi(\theta \mid \mathcal{M}, \mathcal{P}_0) = \{1/m, \ldots, 1/m\}$.

If the parameter values are not simple labels, but have a quantitative meaning, the situation is far more complex.[7]

Formally, the *reference prior function* $\pi(\theta \mid \mathcal{M}, \mathcal{P})$ of a univariate parameter θ is defined to be the limit of the sequence of the proper priors $\pi_k(\theta)$ which maximise $I^\theta\{T^k, \pi(\theta)\}$ in the precise sense that, for any value of the sufficient statistic $\boldsymbol{t} = \boldsymbol{t}(D)$, the *reference posterior*, the limit $\pi(\theta \mid \boldsymbol{t})$ of the corresponding sequence of posteriors $\{\pi_k(\theta \mid \boldsymbol{t})\}$, may be obtained from $\pi(\theta \mid \mathcal{M}, \mathcal{P})$ by formal use of Bayes' theorem, so that $\pi(\theta \mid \boldsymbol{t}) \propto p(\boldsymbol{t} \mid \theta)\, \pi(\theta \mid \mathcal{M}, \mathcal{P})$.

Reference prior *functions* are often simply called reference priors, even though they are usually *not* probability distributions. They should *not* be considered as expressions of belief, but technical devices to obtain (proper) posterior distributions which are a limiting form of the posteriors which could have been obtained from possible prior beliefs which

[7] See Berger, Bernardo, and Sun (2012) for details.

were relatively uninformative with respect to the quantity of interest when compared with the information which data could provide.

If (i) the sufficient statistic $\boldsymbol{t} = \boldsymbol{t}(D)$ is a consistent estimator $\tilde{\theta}$ of a continuous parameter θ, and (ii) the class $\mathcal{P}$ contains *all* strictly positive priors, then the reference prior may be shown to have a simple form in terms of any *asymptotic* approximation to the posterior distribution of θ. Notice that, by construction, an *asymptotic* approximation to the posterior does *not* depend on the prior. Specifically, if the posterior density $\pi(\theta \mid D)$ has an asymptotic approximation of the form $\pi(\theta \mid \tilde{\theta}, n)$, the (unrestricted) reference prior is simply

$$\pi(\theta \mid \mathcal{M}, \mathcal{P}_0) \propto \pi(\theta \mid \tilde{\theta}, n)\big|_{\tilde{\theta}=\theta} .$$

One-parameter reference priors are *invariant* under re-parameterization in the sense that if $\psi = \psi(\theta)$ is a one-to-one function of θ, then the ψ-reference prior is simply the appropriate probability transformation of the θ-reference prior.

Example 13: The Jeffreys prior. If θ is univariate and continuous, and the posterior distribution of θ given $\{x_1 \ldots, x_n\}$ is asymptotically normal with standard deviation $s(\tilde{\theta})/\sqrt{n}$, then using the expression above the reference prior function is $\pi(\theta) \propto s(\theta)^{-1}$. Under regularity conditions (often satisfied in practice, see Section 3.3), the posterior distribution of θ is asymptotically normal with variance $n^{-1}\, F^{-1}(\hat{\theta})$, where $F(\theta)$ is Fisher's information function and $\hat{\theta}$ is the MLE of θ. Hence, the reference prior function in these conditions is $\pi(\theta \mid \mathcal{M}, \mathcal{P}_0) \propto F(\theta)^{1/2}$, which is known as the Jeffreys prior. It follows that the reference prior algorithm contains Jeffreys priors as the particular case which obtains when the probability model only depends on a single, continuous univariate parameter, regularity conditions guarantee asymptotic normality, and there is no additional information, so that the class of possible priors is the set $\mathcal{P}_0$ of all strictly positive priors over Θ. These are precisely the conditions under which there is general agreement on the use of the Jeffreys prior as a 'noninformative' prior.

Example 14: Reference prior for a binomial parameter. Let data $D = \{x_1, \ldots, x_n\}$ consist of a sequence of n independent Bernoulli trials, so that $p(x \mid \theta) = \theta^x(1-\theta)^{1-x}$, $x \in \{0, 1\}$; this is a regular, one-parameter continuous model, whose Fisher's information function is $F(\theta) = \theta^{-1}(1-\theta)^{-1}$. Thus, the reference prior is $\pi(\theta) \propto \theta^{-1/2}(1-\theta)^{-1/2}$, i.e. the (proper) beta distribution $\text{Be}(\theta \mid 1/2, 1/2)$. Since the reference algorithm is invariant under re-parametrization, the reference prior of $\phi(\theta) = 2 \arcsin \sqrt{(\theta)}$ is $\pi(\phi) = \pi(\theta)/|\delta\phi/\delta\theta| = 1$; thus, the reference prior is uniform on the variance-stabilizing transformation $\phi(\theta) = 2\text{arc}\sin\sqrt{\theta}$, a feature generally true under regularity conditions. In terms of θ, the reference posterior is $\pi(\theta \mid D) = \pi(\theta \mid r, n) = \text{Be}(\theta \mid r + 1/2, n - r + 1/2)$, where $r = \sum x_j$ is the number of positive trials.

Suppose, for example, that $n = 100$ randomly selected people have been tested for an infection and that all tested negative, so that $r = 0$. The reference posterior distribution of the proportion θ of people infected is then the beta distribution $\text{Be}(\theta \mid 0.5, 100.5)$, represented in Figure 5.4. It may well be known that the infection was rare, leading to the assumption that $\theta < \theta_0$, for some upper bound θ_0; the (restricted) reference prior would then be of the form $\pi(\theta) \propto \theta^{-1/2}(1-\theta)^{-1/2}$ if $\theta < \theta_0$, and 0 otherwise. However, provided the likelihood is concentrated in the region $\theta < \theta_0$, the corresponding posterior would virtually be identical to $\text{Be}(\theta \mid 0.5, 100.5)$. Thus, just on the basis of the observed experimental results, one may claim that the proportion of infected people is surely smaller than 5% (for the reference posterior probability of the event $\theta > 0.05$ is 0.001), that θ is smaller than 0.01 with probability 0.844 (area of the shaded region in

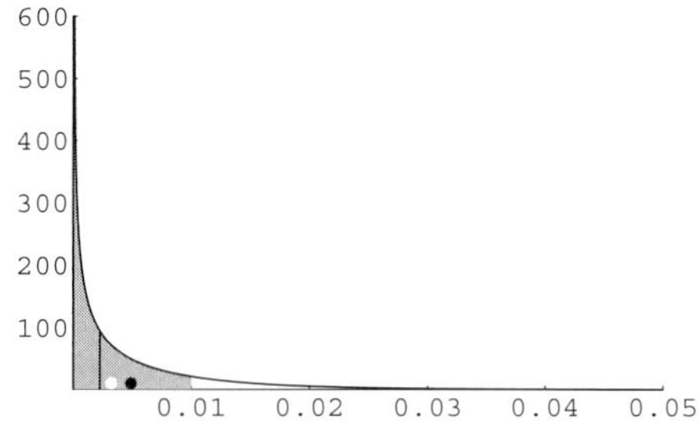

FIGURE 5.4 Posterior distribution of the proportion of infected people in the population, given the results of $n = 100$ tests, none of which were positive.

Figure 5.4), that it is equally likely to be over or below 0.23% (for the median, represented by a vertical line, is 0.0023), and that the probability that a person randomly chosen from the population is infected is 0.005 (the posterior mean, represented in the figure by a black circle), since $\Pr(x = 1 \mid r, n) = \mathrm{E}[\theta \mid r, n] = 0.005$. If a particular point estimate of θ is required (say a number to be quoted in the summary headline) the intrinsic estimator suggests itself (see Section 5); this is found to be $\theta^* = 0.0032$ (represented in the figure with a white circle). Notice that the traditional solution to this problem, based on the asymptotic behaviour of the MLE, here $\hat{\theta} = r/n = 0$ for any n, makes absolutely no sense in this scenario.

One nuisance parameter. The extension of the reference prior algorithm to the case of 2 parameters follows the usual mathematical procedure of reducing the problem to a sequential application of the established procedure for the single-parameter case. Thus, if the probability model is $p(\boldsymbol{t} \mid \theta, \lambda)$, $\theta \in \Theta$, $\lambda \in \Lambda$ and a θ-reference prior $\pi_\theta(\theta, \lambda \mid \mathcal{M}, \mathcal{P})$ is required, the reference algorithm proceeds in 2 steps:

(i) Conditional on θ, $p(\boldsymbol{t} \mid \theta, \lambda)$ only depends on the nuisance parameter λ and, hence, the 1-parameter algorithm may be used to obtain the *conditional* reference prior $\pi(\lambda \mid \theta, \mathcal{M}, \mathcal{P})$.

(ii) If $\pi(\lambda \mid \theta, \mathcal{M}, \mathcal{P})$ is proper, this may be used to integrate out the nuisance parameter, thus obtaining $p(\boldsymbol{t} \mid \theta) = \int_\Lambda p(\boldsymbol{t} \mid \theta, \lambda)\, \pi(\lambda \mid \theta, \mathcal{M}, \mathcal{P})\, d\lambda$, a 1-parameter model to which the 1-parameter algorithm may be applied to obtain $\pi(\theta \mid \mathcal{M}, \mathcal{P})$. The θ-reference prior is then $\pi_\theta(\theta, \lambda \mid \mathcal{M}, \mathcal{P}) = \pi(\lambda \mid \theta, \mathcal{M}, \mathcal{P})\, \pi(\theta \mid \mathcal{M}, \mathcal{P})$, and the required reference posterior is $\pi(\theta \mid \boldsymbol{t}) \propto p(\boldsymbol{t} \mid \theta)\, \pi(\theta \mid \mathcal{M}, \mathcal{P})$.

If the conditional reference prior is *not* proper, then the procedure is performed within an increasing sequence $\{\Lambda_i\}$ of subsets converging to Λ over which $\pi(\lambda \mid \theta)$ is integrable. This makes it possible to obtain a corresponding sequence of θ-reference posteriors $\{\pi_i(\theta \mid \boldsymbol{t}\}$ for the quantity of interest θ, and the required reference posterior is the corresponding intrinsic limit $\pi(\theta \mid \boldsymbol{t}) = \lim_i \pi_i(\theta \mid \boldsymbol{t})$.

A θ-reference prior is then defined as a positive function $\pi_\theta(\theta, \lambda)$ which may be formally used in Bayes' theorem as a prior to obtain the reference posterior, i.e. such that, for any sufficient $\boldsymbol{t} \in T$ (which may well be the whole data set D) $\pi(\theta \mid \boldsymbol{t}) \propto \int_\Lambda p(\boldsymbol{t} \mid \theta, \lambda)\, \pi_\theta(\theta, \lambda)\, d\lambda$. The approximating sequences should be *consistently* chosen within a given model. Thus, given a probability model $\{p(\boldsymbol{x} \mid \boldsymbol{\omega}), \boldsymbol{\omega} \in \Omega\}$ an appropriate approximating sequence $\{\Omega_i\}$ should be chosen for the whole parameter space Ω. Thus, if the analysis is done in terms of, say, $\boldsymbol{\psi} = \{\psi_1, \psi_2\} \in \Psi(\Omega)$, the approximating sequence should be chosen such that $\Psi_i = \psi(\Omega_i)$. A natural approximating sequence in location-scale problems is $\{\mu, \log \sigma\} \in [-i, i]^2$.

The θ-reference prior does *not* depend on the choice of the nuisance parameter λ; thus, for any $\psi = \psi(\theta, \lambda)$ such that (θ, ψ) is a 1-to-1 function of (θ, λ), the θ-reference prior in terms of (θ, ψ) is simply $\pi_\theta(\theta, \psi) = \pi_\theta(\theta, \lambda)/|\partial(\theta, \psi)/\partial(\theta, \lambda)|$, the appropriate probability transformation of the θ-reference prior in terms of (θ, λ). Notice, however, that the reference prior *may* depend on the parameter of interest; thus, the θ-reference prior may differ from the ϕ-reference prior unless either ϕ is a piecewise 1-to-1 transformation of θ, or ϕ is asymptotically independent of θ. This is an expected consequence of the fact that the conditions under which the missing information about θ is maximised are not generally the same as the conditions which maximise the missing information about an arbitrary function $\phi = \phi(\theta, \lambda)$.

The *non-existence* of a unique 'noninformative' prior which would be appropriate for any inference problem within a given model was established by Dawid, Stone, and Zidek (1973), when they showed that this is incompatible with *consistent marginalisation.* Indeed, if given the model $p(D \,|\, \theta, \lambda)$, the reference posterior of the quantity of interest θ, $\pi(\theta \,|\, D) = \pi(\theta \,|\, \boldsymbol{t})$, only depends on the data through a statistic $\boldsymbol{t}$ whose sampling distribution, $p(\boldsymbol{t} \,|\, \theta, \lambda) = p(\boldsymbol{t} \,|\, \theta)$, only depends on θ, one would expect the reference posterior to be of the form $\pi(\theta \,|\, \boldsymbol{t}) \propto \pi(\theta)\, p(\boldsymbol{t} \,|\, \theta)$ for some prior $\pi(\theta)$. However, examples were found where this cannot be the case if a *unique* joint 'noninformative' prior were used for all possible quantities of interest.

Example 15: Regular, 2-dimensional continuous reference prior functions. If the joint posterior distribution of (θ, λ) is asymptotically normal, then the θ-reference prior may be derived in terms of the corresponding Fisher's information matrix, $\boldsymbol{F}(\theta, \lambda)$. Thus, if

$$\boldsymbol{F}(\theta, \lambda) = \begin{pmatrix} F_{\theta\theta}(\theta, \lambda) & F_{\theta\lambda}(\theta, \lambda) \\ F_{\theta\lambda}(\theta, \lambda) & F_{\lambda\lambda}(\theta, \lambda) \end{pmatrix}, \quad \text{and} \quad \boldsymbol{S}(\theta, \lambda) = \boldsymbol{F}^{-1}(\theta, \lambda),$$

then the unrestricted θ-reference prior is $\pi_\theta(\theta, \lambda \,|\, \mathcal{M}, \mathcal{P}_0) = \pi(\lambda \,|\, \theta)\, \pi(\theta)$, where

$$\pi(\lambda \,|\, \theta) \propto F_{\lambda\lambda}^{1/2}(\theta, \lambda), \quad \lambda \in \Lambda.$$

If $\pi(\lambda \,|\, \theta)$ is proper,

$$\pi(\theta) \propto \exp\left\{ \int_\Lambda \pi(\lambda \,|\, \theta)\, \log[S_{\theta\theta}^{-1/2}(\theta, \lambda)]\, d\lambda \right\}, \quad \theta \in \Theta.$$

If $\pi(\lambda \,|\, \theta)$ is not proper, integrations are performed on an approximating sequence $\{\Lambda_i\}$ to obtain a sequence $\{\pi_i(\lambda \,|\, \theta)\, \pi_i(\theta)\}$, (where $\pi_i(\lambda \,|\, \theta)$ is the proper renormalisation of $\pi(\lambda \,|\, \theta)$ to Λ_i) and the θ-reference prior $\pi_\theta(\theta, \lambda)$ is defined as its appropriate limit. Moreover, if (i) both $F_{\lambda\lambda}^{1/2}(\theta, \lambda)$ and $S_{\theta\theta}^{-1/2}(\theta, \lambda)$ factorise, so that

$$S_{\theta\theta}^{-1/2}(\theta, \lambda) \propto f_\theta(\theta)\, g_\theta(\lambda), \quad F_{\lambda\lambda}^{1/2}(\theta, \lambda) \propto f_\lambda(\theta)\, g_\lambda(\lambda),$$

and (ii) the parameters θ and λ are variation independent, so that Λ does not depend on θ, then the θ-reference prior is simply $\pi_\theta(\theta, \lambda) = f_\theta(\theta)\, g_\lambda(\lambda)$, even if the conditional reference prior $\pi(\lambda \,|\, \theta) = \pi(\lambda) \propto g_\lambda(\lambda)$ (which will not depend on θ) is actually improper.

Example 16: Reference priors for the normal model. The information matrix which corresponds to a normal model $\mathrm{N}(x \,|\, \mu, \sigma)$ is

$$\boldsymbol{F}(\mu, \sigma) = \begin{pmatrix} \sigma^{-2} & 0 \\ 0 & 2\sigma^{-2} \end{pmatrix}, \quad \boldsymbol{S}(\mu, \sigma) = \boldsymbol{F}^{-1}(\mu, \sigma) = \begin{pmatrix} \sigma^2 & 0 \\ 0 & \frac{1}{2}\sigma^2 \end{pmatrix};$$

hence $F_{\sigma\sigma}^{1/2}(\mu,\sigma) = \sqrt{2}\,\sigma^{-1} = f_\sigma(\mu)\,g_\sigma(\sigma)$, with $g_\sigma(\sigma) = \sigma^{-1}$, and thus $\pi(\sigma\,|\,\mu) = \sigma^{-1}$. Similarly, $S_{\mu\mu}^{-1/2}(\mu,\sigma) = \sigma^{-1} = f_\mu(\mu)\,g_\mu(\sigma)$, with $f_\mu(\mu) = 1$, and thus $\pi(\mu) = 1$. Therefore, the μ-reference prior is $\pi_\mu(\mu,\sigma\,|\,\mathcal{M},\mathcal{P}_0) = \pi(\sigma\,|\,\mu)\,\pi(\mu) = \sigma^{-1}$, as already anticipated. Moreover, as one would expect from the fact that $\boldsymbol{F}(\mu,\sigma)$ is diagonal and also anticipated, it is similarly found that the σ-reference prior is $\pi_\sigma(\mu,\sigma\,|\,\mathcal{M},\mathcal{P}_0) = \sigma^{-1}$, the same as before.

Suppose, however, that the quantity of interest is not the mean μ or the standard deviation σ, but the standardised mean $\phi = \mu/\sigma$. Fisher's information matrix in terms of the parameters ϕ and σ is $\boldsymbol{F}(\phi,\sigma) = J^t\,\boldsymbol{F}(\mu,\sigma)\,J$, where $J = (\partial(\mu,\sigma)/\partial(\phi,\sigma))$ is the Jacobian of the inverse transformation; this yields

$$\boldsymbol{F}(\phi,\sigma) = \begin{pmatrix} 1 & \phi\,\sigma^{-1} \\ \phi\,\sigma^{-1} & \sigma^{-2}(2+\phi^2) \end{pmatrix}, \quad \boldsymbol{S}(\phi,\sigma) = \begin{pmatrix} 1+\frac{1}{2}\phi^2 & -\frac{1}{2}\phi\sigma \\ -\frac{1}{2}\phi\sigma & \frac{1}{2}\sigma^2 \end{pmatrix}.$$

Thus, $S_{\phi\phi}^{-1/2}(\phi,\sigma) \propto (1+\frac{1}{2}\phi^2)^{-1/2}$ and $F_{\sigma\sigma}^{1/2}(\phi,\sigma) \propto \sigma^{-1}(2+\phi^2)^{1/2}$. Hence, using again the results in Example 15, $\pi_\phi(\phi,\sigma\,|\,\mathcal{M},\mathcal{P}_0) = (1+\frac{1}{2}\phi^2)^{-1/2}\sigma^{-1}$. In the original parametrization, this is $\pi_\phi(\mu,\sigma\,|\,\mathcal{M},\mathcal{P}_0) = (1+\frac{1}{2}(\mu/\sigma)^2)^{-1/2}\sigma^{-2}$, which is very different from $\pi_\mu(\mu,\sigma\,|\,\mathcal{M},\mathcal{P}_0) = \pi_\sigma(\mu,\sigma\,|\,\mathcal{M},\mathcal{P}_0) = \sigma^{-1}$. The corresponding reference posterior of ϕ is $\pi(\phi\,|\,x_1,\ldots,x_n) \propto (1+\frac{1}{2}\phi^2)^{-1/2}\,p(t\,|\,\phi)$ where $t = (\sum x_j)/(\sum x_j^2)^{1/2}$, a 1-dimensional (marginally sufficient) statistic whose sampling distribution, $p(t\,|\,\mu,\sigma) = p(t\,|\,\phi)$, only depends on ϕ. Thus, the reference prior algorithm is seen to be consistent under marginalisation.

Many parameters. The reference algorithm is easily generalised to an arbitrary number of parameters. If the model is $p(\boldsymbol{t}\,|\,\omega_1,\ldots,\omega_m)$, a joint reference prior

$$\pi(\theta_m\,|\,\theta_{m-1},\ldots,\theta_1)\times\cdots\times\pi(\theta_2\,|\,\theta_1)\times\pi(\theta_1)$$

may sequentially be obtained for each *ordered* parametrization $\{\theta_1(\boldsymbol{\omega}),\ldots,\theta_m(\boldsymbol{\omega})\}$ of interest, and these are invariant under re-parametrization of any of the $\theta_i(\boldsymbol{\omega})$'s. The choice of the ordered parametrization $\{\theta_1,\ldots,\theta_m\}$ precisely describes the particular prior required, namely that which *sequentially* maximises the missing information about each of the θ_i's, conditional on $\{\theta_1,\ldots,\theta_{i-1}\}$, for $i = m, m-1,\ldots,1$.

Example 17: Stein's paradox. Let D be a random sample from an m-variate normal distribution with mean $\boldsymbol{\mu} = \{\mu_1,\ldots,\mu_m\}$ and unitary variance matrix. The reference prior which corresponds to any permutation of the μ_i's is uniform, and this prior leads indeed to appropriate reference posterior distributions for any of the μ_i's, namely $\pi(\mu_i\,|\,D) = \mathrm{N}(\mu_i\,|\,\bar{x}_i, 1/\sqrt{n})$. Suppose, however, that the quantity of interest is $\theta = \sum_i \mu_i^2$, the distance of $\boldsymbol{\mu}$ to the origin. As shown by Stein (1959), the posterior distribution of θ based on that uniform prior (or in any 'flat' proper approximation) has very undesirable properties; this is due to the fact that a uniform (or nearly uniform) prior, although 'noninformative' with respect to each of the individual μ_i's, is actually highly informative on the sum of their squares, introducing a severe positive bias (Stein's paradox). However, the reference prior which corresponds to a parametrization of the form $\{\theta,\lambda_1,\ldots,\lambda_{m-1}\}$ produces, for any choice of the nuisance parameters $\lambda_i = \lambda_i(\boldsymbol{\mu})$, the reference posterior $\pi(\theta\,|\,D) = \pi(\theta\,|\,t) \propto \theta^{-1/2}\chi^2(nt\,|\,m, n\theta)$, where $t = \sum_i \bar{x}_i^2$, and this posterior is shown to have the appropriate consistency properties.

Far from being specific to Stein's example, the inappropriate behaviour in problems with many parameters of specific marginal posterior distributions derived from multivariate 'flat' priors (proper or improper) is indeed very frequent. Hence, sloppy, uncontrolled use of 'flat' priors (rather than the relevant reference priors), is *very* strongly discouraged.

Limited Information

Although often used in contexts where no universally agreed prior knowledge about the quantity of interest is available, the reference algorithm may be used to specify a prior which incorporates any acceptable prior knowledge; it suffices to maximise the missing information within the class $\mathcal{P}$ of priors which is compatible with such accepted knowledge. Indeed, by progressive incorporation of further restrictions into $\mathcal{P}$, the reference prior algorithm becomes a method of (prior) *probability assessment.* As described below, the problem has a fairly simple analytical solution when those restrictions take the form of known expected values. The incorporation of other type of restrictions usually involves numerical computations.

Example 18: Univariate restricted reference priors. If the probability mechanism which is assumed to have generated the available data only depends on a univariate continuous parameter $\theta \in \Theta \subset \Re$, and the class $\mathcal{P}$ of acceptable priors is a class of proper priors which satisfies some expected value restrictions, so that

$$\mathcal{P} = \left\{\pi(\theta);\quad \pi(\theta) > 0,\ \int_\Theta \pi(\theta)\, d\theta = 1,\quad \int_\Theta g_i(\theta)\, \pi(\theta)\, d\theta = \beta_i,\ i = 1, \ldots, m\right\}$$

then the (restricted) reference prior is

$$\pi(\theta \,|\, \mathcal{M}, \mathcal{P}) \propto \pi(\theta \,|\, \mathcal{M}, \mathcal{P}_0)\, \exp\left[\sum_{j=1}^{m} \gamma_i\, g_i(\theta)\right],$$

where $\pi(\theta \,|\, \mathcal{M}, \mathcal{P}_0)$ is the unrestricted reference prior and the γ_i's are constants (the corresponding Lagrange multipliers), to be determined by the restrictions which define $\mathcal{P}$. Suppose, for instance, that data are considered to be a random sample from a location model centered at θ, and that it is further assumed that $\mathrm{E}[\theta] = \mu_0$ and that $\mathrm{Var}[\theta] = \sigma_0^2$. The unrestricted reference prior for any regular location problem may be shown to be uniform, so that here $\pi(\theta \,|\, \mathcal{M}, \mathcal{P}_0) = 1$. Thus, the restricted reference prior must be of the form $\pi(\theta \,|\, \mathcal{M}, \mathcal{P}) \propto \exp\{\gamma_1\theta + \gamma_2(\theta - \mu_0)^2\}$, with $\int_\Theta \theta\, \pi(\theta \,|\, \mathcal{M}, \mathcal{P})\, d\theta = \mu_0$ and $\int_\Theta (\theta - \mu_0)^2\, \pi(\theta \,|\, \mathcal{M}, \mathcal{P})\, d\theta = \sigma_0^2$. Hence, $\pi(\theta \,|\, \mathcal{M}, \mathcal{P})$ is the normal distribution with the specified mean and variance, $\mathrm{N}(\theta \,|\, \mu_0, \sigma_0)$.

5.4.2 Frequentist Properties

Bayesian methods provide a *direct* solution to the problems typically posed in statistical inference; indeed, posterior distributions precisely state what can be said about unknown quantities of interest *given* available data and prior knowledge. In particular, unrestricted reference posterior distributions state what could be said if no prior knowledge about the quantities of interest were available.

A frequentist analysis of the behaviour of Bayesian procedures under repeated sampling may, however, be illuminating, for this provides some interesting connections between frequentist and Bayesian inference. It is found that the frequentist properties of Bayesian reference procedures are typically excellent, and may be used to provide a form of calibration for reference posterior probabilities.

Point Estimation

It is generally accepted that, as the sample size increases, a 'good' estimator $\tilde{\boldsymbol{\theta}}$ of $\boldsymbol{\theta}$ ought to get the correct value of $\boldsymbol{\theta}$ eventually, that is to be *consistent*. Under appropriate regularity conditions, any Bayes estimator $\boldsymbol{\phi}^*$ of any function $\boldsymbol{\phi}(\boldsymbol{\theta})$ converges in probability to $\boldsymbol{\phi}(\boldsymbol{\theta})$, so that sequences of Bayes estimators are typically *consistent*. Indeed, it is known that if there is a consistent sequence of estimators, then Bayes estimators are consistent. The rate of convergence is often best for reference Bayes estimators.

It is also generally accepted that a 'good' estimator should be *admissible*, that is, *not dominated* by any other estimator in the sense that its expected loss under sampling (conditional to $\boldsymbol{\theta}$) cannot be larger for all $\boldsymbol{\theta}$ values than that corresponding to another estimator. Any *proper* Bayes estimator is admissible; moreover, as established by Wald (1950), a procedure *must* be Bayesian (proper or improper) to be admissible. Most published admissibility results refer to quadratic loss functions, but they often extend to more general loss functions. Reference Bayes estimators are typically admissible with respect to appropriate loss functions.

Notice, however, that many other apparently intuitive frequentist ideas on estimation have been proved to be potentially misleading. For example, given a sequence of n Bernoulli observations with parameter θ resulting in r positive trials, the *best unbiased* estimate of θ^2 is found to be $r(r-1)/\{n(n-1)\}$, which yields $\tilde{\theta}^2 = 0$ when $r = 1$; but to estimate the probability of two positive trials as zero, when one positive trial has been observed, is certainly less than sensible. In marked contrast, any Bayes reference estimator provides a reasonable answer. For example, the intrinsic estimator of θ^2 is simply $(\theta^*)^2$, where θ^* is the intrinsic estimator of θ described in Section 5.1. In particular, if $r = 1$ and $n = 2$ the intrinsic estimator of θ^2 is (as one would naturally expect) $(\theta^*)^2 = 1/4$.

Interval Estimation

As the sample size increases, the frequentist coverage probability of a posterior q-credible region typically converges to q so that, for *large samples*, Bayesian credible intervals may (under regularity conditions) be interpreted as *approximate* frequentist confidence regions: under repeated sampling, a Bayesian q-credible region of $\boldsymbol{\theta}$ based on a large sample covers the true value of $\boldsymbol{\theta}$ approximately $100q\%$ of the time. Detailed results are readily available for univariate problems. For instance, consider the probability model $\{p(D \mid \boldsymbol{\omega}), \boldsymbol{\omega} \in \Omega\}$: let $\theta = \theta(\boldsymbol{\omega})$ be any univariate quantity of interest and let $\boldsymbol{t} = \boldsymbol{t}(D) \in T$ be any sufficient statistic. If $\theta_q(\boldsymbol{t})$ denotes the $100q\%$ quantile of the posterior distribution of θ which corresponds to some unspecified prior, so that

$$\Pr[\theta \leq \theta_q(\boldsymbol{t}) \mid \boldsymbol{t}] = \int_{\theta \leq \theta_q(\boldsymbol{t})} \pi(\theta \mid \boldsymbol{t})\, d\theta = q,$$

then the coverage probability of the q-credible interval $\{\theta; \theta \leq \theta_q(\boldsymbol{t})\}$,

$$\Pr[\theta_q(\boldsymbol{t}) \geq \theta \mid \boldsymbol{\omega}] = \int_{\theta_q(\boldsymbol{t}) \geq \theta} p(\boldsymbol{t} \mid \boldsymbol{\omega})\, d\boldsymbol{t},$$

is such that

$$\Pr[\theta_q(\boldsymbol{t}) \geq \theta \mid \boldsymbol{\omega}] = \Pr[\theta \leq \theta_q(\boldsymbol{t}) \mid \boldsymbol{t}] + O(n^{-1/2}).$$

This *asymptotic* approximation is true for *all* (sufficiently regular) positive priors. However, the approximation is better, actually $O(n^{-1})$, for a particular class of priors known as (first-order) *probability matching* priors.[8] Reference priors are typically found to be probability matching priors, so that they provide this improved asymptotic agreement. As a matter of fact, the agreement (in regular problems) is typically quite good even for relatively small samples.

Example 19: Product of normal means. Consider the case where independent random samples $\{x_1, \ldots, x_n\}$ and $\{y_1, \ldots, y_m\}$ have respectively been taken from the normal densities $\mathrm{N}(x \,|\, \omega_1, 1)$ and $\mathrm{N}(y \,|\, \omega_2, 1)$, and suppose that the quantity of interest is the product of their means, $\phi = \omega_1 \omega_2$ (for instance, one may be interested in inferences about the area ϕ of a rectangular piece of land, given measurements $\{x_i\}$ and $\{y_j\}$ of its sides). Notice that this is a simplified version of a problem that is often encountered in the sciences, where one is interested in the product of several magnitudes, all of which have been measured with error. Using the procedure described in Example 15, with the natural approximating sequence induced by $(\omega_1, \omega_2) \in [-i, i]^2$, the ϕ-reference prior is found to be

$$\pi_\phi(\omega_1, \omega_2 \,|\, \mathcal{M}, \mathcal{P}_0) \propto (n\,\omega_1^2 + m\,\omega_2^2)^{-1/2},$$

different from the uniform prior $\pi_{\omega_1}(\omega_1, \omega_2 \,|\, \mathcal{M}, \mathcal{P}_0) = \pi_{\omega_2}(\omega_1, \omega_2 \,|\, \mathcal{M}, \mathcal{P}_0) = 1$, which should be used to make objective inferences about either ω_1 or ω_2. The prior $\pi_\phi(\omega_1, \omega_2)$ may be shown to provide approximate agreement between Bayesian credible regions and frequentist confidence intervals for ϕ; indeed, this prior (with $m = n$) was originally suggested by Stein in the 1980s to obtain such approximate agreement. The same example was later used by Efron (1986) to stress the fact that, even within a fixed probability model $\{p(D \,|\, \boldsymbol{\omega}), \boldsymbol{\omega} \in \Omega\}$, the prior required to make objective inferences about some function of the parameters $\phi = \phi(\boldsymbol{\omega})$ must generally depend on the function ϕ.[9]

The numerical agreement between reference Bayesian credible regions and frequentist confidence intervals is actually perfect in special circumstances. Indeed, as Lindley (1958) pointed out, this is the case in those problems of inference which may be transformed to location-scale problems.

Example 20: Inference on normal parameters. Let $D = \{x_1, \ldots, x_n\}$ be a random sample from a normal distribution $\mathrm{N}(x \,|\, \mu, \sigma)$. As mentioned before, the reference posterior of the quantity of interest μ is the Student distribution $\mathrm{St}(\mu \,|\, \bar{x}, s/\sqrt{n-1}, n-1)$. Thus, normalising μ, the posterior distribution of $t(\mu) = \sqrt{n-1}(\bar{x} - \mu)/s$, as a function of μ given D, is the standard Student $\mathrm{St}(t \,|\, 0, 1, n-1)$ with $n-1$ degrees of freedom. On the other hand, this function t is recognized to be precisely the conventional t statistic, whose sampling distribution is well known to also be standard Student with $n-1$ degrees of freedom. It follows that, for all sample sizes, posterior reference credible intervals for μ given the data will be numerically identical to frequentist confidence intervals based on the sampling distribution of t.

A similar result is obtained in inferences about the variance. Thus, the reference posterior distribution of $\lambda = \sigma^{-2}$ is the gamma distribution $\mathrm{Ga}(\lambda \,|\, (n-1)/2, ns^2/2)$ and, hence, the posterior distribution of $r = ns^2/\sigma^2$, as a function of σ^2 given D, is a (central) χ^2 with $n-1$ degrees of freedom. But the function r is recognised to be a conventional statistic for this problem, whose sampling distribution is well known to also be χ^2 with $n-1$ degrees of freedom. It follows that, for all sample sizes, posterior reference credible

[8] For details on probability matching priors see Datta and Sweeting (2005) and references therein.

[9] For further details on the reference analysis of this problem, see Berger and Bernardo (1989).

intervals for σ^2 (or any one-to-one function of σ^2) given the data will be numerically identical to frequentist confidence intervals based on the sampling distribution of r.

5.5 Inference Summaries

From a Bayesian viewpoint, the final outcome of a problem of inference about *any* unknown quantity is nothing but the corresponding posterior distribution. Thus, given some data D and conditions C, *all* that can be said about any function $\boldsymbol{\omega}$ of the parameters which govern the model is contained in the posterior distribution $\pi(\boldsymbol{\omega} \mid D, C)$, and *all* that can be said about some function $\boldsymbol{y}$ of future observations from the same model is contained in its posterior predictive distribution $p(\boldsymbol{y} \mid D, C)$. Indeed, Bayesian inference may technically be described as a decision problem where the space of available actions is the class of those posterior probability distributions of the quantity of interest which are compatible with accepted assumptions.

However, to make it easier for the user to assimilate the appropriate conclusions, it is often convenient to *summarise* the information contained in the posterior distribution by (i) providing values of the quantity of interest which, in the light of the data, are likely to be 'close' to its true value and by (ii) measuring the compatibility of the results with hypothetical values of the quantity of interest which might have been suggested in the context of the investigation. In this section, those Bayesian counterparts of traditional *estimation* and *hypothesis testing* problems are briefly considered.

5.5.1 Estimation

In one or two dimensions, a graph of the posterior probability density of the quantity of interest (or the probability mass function in the discrete case) immediately conveys an intuitive, 'impressionist' summary of the main conclusions which may possibly be drawn on its value. Indeed, this is greatly appreciated by users, and may be quoted as an important asset of Bayesian methods. From a plot of its posterior density, the region where (given the data) a univariate quantity of interest is likely to lie is easily distinguished. For instance, all important conclusions about the value of the gravitational field in Example 7 are qualitatively available from Figure 5.3. However, this does not easily extend to more than 2 dimensions and, besides, *quantitative* conclusions (in a simpler form than provided by the mathematical expression of the posterior distribution) are often required.

Point Estimation

Let D be the available data, which are assumed to have been generated by a probability model $\{p(D \mid \boldsymbol{\omega}), \boldsymbol{\omega} \in \Omega\}$, and let $\boldsymbol{\theta} = \boldsymbol{\theta}(\boldsymbol{\omega}) \in \Theta$ be the quantity of interest. A *point estimator* of $\boldsymbol{\theta}$ is some function of the data $\tilde{\boldsymbol{\theta}} = \tilde{\boldsymbol{\theta}}(D)$ which could be regarded as an appropriate proxy for the actual, unknown value of $\boldsymbol{\theta}$. Formally, to choose a point estimate for $\boldsymbol{\theta}$ is a *decision problem*, where the action space is the class Θ of possible $\boldsymbol{\theta}$ values. From a decision-theoretic perspective, to choose a point estimate $\tilde{\boldsymbol{\theta}}$ of some quantity $\boldsymbol{\theta}$ is a *decision* to act as though $\tilde{\boldsymbol{\theta}}$ were $\boldsymbol{\theta}$, not to assert something about the value of $\boldsymbol{\theta}$ (although desire to assert something simple may well be the reason to obtain an estimate). As prescribed by the foundations of decision theory (Section 2), to solve this decision problem it is necessary to specify a *loss function* $L(\tilde{\boldsymbol{\theta}}, \boldsymbol{\theta})$ measuring the consequences of acting *as if* the true value of the quantity of interest were $\tilde{\boldsymbol{\theta}}$, when it is actually $\boldsymbol{\theta}$. The expected posterior loss if $\tilde{\boldsymbol{\theta}}$ were used is

$$\overline{L}[\tilde{\boldsymbol{\theta}} \,|\, D] = \int_{\Theta} L(\tilde{\boldsymbol{\theta}}, \boldsymbol{\theta})\, \pi(\boldsymbol{\theta} \,|\, D)\, d\boldsymbol{\theta},$$

and the corresponding *Bayes estimator* $\boldsymbol{\theta}^*$ is that function of the data, $\boldsymbol{\theta}^* = \boldsymbol{\theta}^*(D)$, which minimises this expectation.

Example 21: Conventional Bayes estimators. For any given model and data, the Bayes estimator obviously depends on the chosen loss function. The loss function is context specific, and should be chosen in terms of the anticipated uses of the estimate; however, a number of conventional loss functions have been suggested for those situations where no particular uses are envisaged. These loss functions produce estimates which may be regarded as simple descriptions of the location of the posterior distribution. For example, if the loss function is quadratic, so that $L(\tilde{\boldsymbol{\theta}}, \boldsymbol{\theta}) = (\tilde{\boldsymbol{\theta}} - \boldsymbol{\theta})^t(\tilde{\boldsymbol{\theta}} - \boldsymbol{\theta})$, then the Bayes estimator is the posterior mean $\boldsymbol{\theta}^* = E[\boldsymbol{\theta} \,|\, D]$, assuming that the mean exists. Similarly, if the loss function is a 0–1 function, so that $L(\tilde{\boldsymbol{\theta}}, \boldsymbol{\theta}) = 0$ if $\tilde{\boldsymbol{\theta}}$ belongs to a ball or radius ϵ centered in $\boldsymbol{\theta}$ and $L(\tilde{\boldsymbol{\theta}}, \boldsymbol{\theta}) = 1$ otherwise, then the Bayes estimator $\boldsymbol{\theta}^*$ tends to the posterior mode as the ball radius ϵ tends to 0, assuming that a unique mode exists. If θ is univariate and the loss function is linear, so that $L(\tilde{\theta}, \theta) = c_1(\tilde{\theta} - \theta)$ if $\tilde{\theta} \geq \theta$, and $L(\tilde{\theta}, \theta) = c_2(\theta - \tilde{\theta})$ otherwise, then the Bayes estimator is the posterior quantile of order $c_2/(c_1 + c_2)$, so that $\Pr[\theta < \theta^*] = c_2/(c_1 + c_2)$. In particular, if $c_1 = c_2$, the Bayes estimator is the posterior median. The results derived for linear loss functions clearly illustrate the fact that any possible parameter value may turn out be the Bayes estimator: it all depends on the loss function describing the consequences of the anticipated uses of the estimate.

Example 22: Intrinsic estimation. Conventional loss functions are typically non-invariant under re-parametrization. It follows that the Bayes estimator $\boldsymbol{\phi}^*$ of a 1-to-1 transformation $\boldsymbol{\phi} = \boldsymbol{\phi}(\boldsymbol{\theta})$ of the original parameter $\boldsymbol{\theta}$ is not necessarily $\boldsymbol{\phi}(\boldsymbol{\theta}^*)$ (the univariate posterior median, which is invariant, is an interesting exception). Moreover, conventional loss functions focus on the 'distance' between the estimate $\tilde{\boldsymbol{\theta}}$ and the true value $\boldsymbol{\theta}$, rather than on the 'distance' between the probability models they label. Inference-oriented loss functions directly focus on how different the probability model $p(D \,|\, \boldsymbol{\theta}, \boldsymbol{\lambda})$ is from its closest approximation within the family $\{p(D \,|\, \tilde{\boldsymbol{\theta}}, \boldsymbol{\lambda}_i), \boldsymbol{\lambda}_i \in \Lambda\}$, and typically produce invariant solutions. An attractive example is the intrinsic discrepancy, $\delta(\tilde{\boldsymbol{\theta}}, \boldsymbol{\theta})$ defined as the minimum logarithmic divergence between a probability model labelled by $\boldsymbol{\theta}$ and a probability model labelled by $\tilde{\boldsymbol{\theta}}$. When there are no nuisance parameters, this is

$$\delta(\tilde{\boldsymbol{\theta}}, \boldsymbol{\theta}) = \min\{\kappa(\tilde{\boldsymbol{\theta}} \,|\, \boldsymbol{\theta}), \kappa(\boldsymbol{\theta} \,|\, \tilde{\boldsymbol{\theta}})\}, \quad \kappa(\boldsymbol{\theta}_i \,|\, \boldsymbol{\theta}) = \int_{T} p(\boldsymbol{t} \,|\, \boldsymbol{\theta}) \log \frac{p(\boldsymbol{t} \,|\, \boldsymbol{\theta})}{p(\boldsymbol{t} \,|\, \boldsymbol{\theta}_i)}\, d\boldsymbol{t},$$

where $\boldsymbol{t} = \boldsymbol{t}(D) \in T$ is any sufficient statistic (which may well be the whole data set D). The definition is easily extended to problems with nuisance parameters; in this case,

$$\delta(\tilde{\boldsymbol{\theta}}, \boldsymbol{\theta}, \boldsymbol{\lambda}) = \min_{\boldsymbol{\lambda}_i \in \Lambda} \delta(\tilde{\boldsymbol{\theta}}, \boldsymbol{\lambda}_i, \boldsymbol{\theta}, \boldsymbol{\lambda})$$

measures the logarithmic divergence from $p(\boldsymbol{t} \,|\, \boldsymbol{\theta}, \boldsymbol{\lambda})$ of its closest approximation with $\boldsymbol{\theta} = \tilde{\boldsymbol{\theta}}$, and the loss function now depends on the complete parameter vector $(\boldsymbol{\theta}, \boldsymbol{\lambda})$. Although not explicitly shown in the notation, the intrinsic discrepancy function typically depends on the sample size n; indeed, when the data consist of a random sample $D = \{\boldsymbol{x}_1, \ldots, \boldsymbol{x}_n\}$ from some model $p(\boldsymbol{x} \,|\, \boldsymbol{\theta})$ then $\kappa(\boldsymbol{\theta}_i \,|\, \boldsymbol{\theta}) = n \int_X p(\boldsymbol{x} \,|\, \boldsymbol{\theta}) \log[p(\boldsymbol{x} \,|\, \boldsymbol{\theta})/p(\boldsymbol{x} \,|\, \boldsymbol{\theta}_i)] d\boldsymbol{x}$ so that the discrepancy associated with the full model is simply n times the discrepancy which corresponds to a single observation. The intrinsic discrepancy is a symmetric, nonnegative loss function with a direct interpretation in information-theoretic terms as the minimum amount of information which is expected to be necessary to distinguish between the model $p(D\,|\,\boldsymbol{\theta},\boldsymbol{\lambda})$ and its closest approximation within the class $\{p(D\,|\,\tilde{\boldsymbol{\theta}},\boldsymbol{\lambda}_i),\boldsymbol{\lambda}_i\in\Lambda\}$. Moreover, it is invariant under one-to-one re-parametrization of the parameter of interest $\boldsymbol{\theta}$, and does not depend on the choice of the nuisance parameter $\boldsymbol{\lambda}$. The intrinsic estimator is naturally obtained by minimising the reference posterior expected intrinsic discrepancy

$$d(\tilde{\boldsymbol{\theta}}\,|\,D) = \int_{\Lambda}\int_{\Theta} \delta(\tilde{\boldsymbol{\theta}},\boldsymbol{\theta},\boldsymbol{\lambda})\,\pi(\boldsymbol{\theta},\boldsymbol{\lambda}\,|\,D)\,d\boldsymbol{\theta}d\boldsymbol{\lambda}.$$

Since the intrinsic discrepancy is invariant under re-parametrization, minimising its posterior expectation produces invariant estimators.[10]

Example 23: Intrinsic estimation of a binomial parameter. In the estimation of a binomial proportion θ, given data $D=(n,r)$, the Bayes reference estimator associated with the quadratic loss (the corresponding posterior mean) is $E[\theta\,|\,D] = (r+\frac{1}{2})/(n+1)$, while the quadratic loss-based estimator of, say, the log-odds $\phi(\theta) = \log[\theta/(1-\theta)]$, is found to be $E[\phi\,|\,D] = \psi(r+\frac{1}{2}) - \psi(n-r+\frac{1}{2})$ (where $\psi(x) = d\log[\Gamma(x)]/dx$ is the digamma function), which is not equal to $\phi(E[\theta\,|\,D])$. The intrinsic loss function in this problem is

$$\delta(\tilde{\theta},\theta) = n\,\min\{\kappa(\tilde{\theta}\,|\,\theta),\kappa(\theta\,|\,\tilde{\theta})\},\quad \kappa(\theta_i\,|\,\theta) = \theta\log\frac{\theta}{\theta_i} + (1-\theta)\log\frac{1-\theta}{1-\theta_i},$$

and the corresponding intrinsic estimator θ^* is obtained by minimising the expected posterior loss $d(\tilde{\theta}\,|\,D) = \int \delta(\tilde{\theta},\theta)\,\pi(\theta\,|\,D)\,d\theta$. The exact value of θ^* may be obtained by numerical minimisation, but a very good approximation is given by $\theta^* \approx (r+\frac{1}{3})/(n+\frac{2}{3})$.

Since intrinsic estimation is an invariant procedure, the intrinsic estimator of the log-odds will simply be the log-odds of the intrinsic estimator of θ. As one would expect, when r and $n-r$ are both large, all Bayes estimators of any well-behaved function $\phi(\theta)$ will cluster around $\phi(E[\theta\,|\,D])$.

Interval Estimation

To describe the inferential content of the posterior distribution of the quantity of interest $\pi(\boldsymbol{\theta}\,|\,D)$ it is often convenient to quote regions $R\subset\Theta$ of given probability under $\pi(\boldsymbol{\theta}\,|\,D)$. For example, the identification of regions containing 50%, 90%, 95%, or 99% of the probability under the posterior may be sufficient to convey the general quantitative messages implicit in $\pi(\boldsymbol{\theta}\,|\,D)$; indeed, this is the intuitive basis of graphical representations of univariate distributions like those provided by boxplots.

Any region $R\subset\Theta$ such that $\int_R \pi(\boldsymbol{\theta}\,|\,D)d\boldsymbol{\theta} = q$ (so that, given data D, the true value of $\boldsymbol{\theta}$ belongs to R with probability q), is said to be a posterior *q-credible region* of $\boldsymbol{\theta}$. Notice that this provides immediately a direct, intuitive statement about the unknown quantity of interest $\boldsymbol{\theta}$ in probability terms, in marked contrast to the circumlocutory statements provided by frequentist confidence intervals. Clearly, for any given q there are generally infinitely many credible regions. A credible region is invariant under re-parametrization; thus, for any q-credible region R of $\boldsymbol{\theta}$, $\boldsymbol{\phi}(R)$ is a q-credible region of $\boldsymbol{\phi}=\boldsymbol{\phi}(\boldsymbol{\theta})$. Sometimes, credible regions are selected to have minimum size (length, area, volume), resulting in *highest probability density* (HPD) regions, where all points in the region have larger probability density than all points outside. However, HPD regions are

[10] For further details on intrinsic point estimation see Bernardo and Juárez (2003) and Bernardo (2006).

not invariant under re-parametrization: the image $\boldsymbol{\phi}(R)$ of an HPD region R will be a credible region for $\boldsymbol{\phi}$, but will not generally be HPD; indeed, there is no compelling reason to restrict attention to HPD credible regions. In one-dimensional problems, posterior quantiles are often used to derive credible regions. Thus, if $\theta_q = \theta_q(D)$ is the $100q\%$ posterior quantile of θ, then $R = \{\theta;\, \theta \leq \theta_q\}$ is a one-sided, typically unique q-credible region, and it is invariant under re-parametrization. Indeed, *probability-centred* q-credible regions of the form $R = \{\theta;\, \theta_{(1-q)/2} \leq \theta \leq \theta_{(1+q)/2}\}$ are easier to compute, and are often quoted in preference to HPD regions.

Example 24: Inference on normal parameters, continued. In the numerical example about the value of the gravitational field described in the top panel of Figure 5.3, the interval $[9.788, 9.829]$ in the unrestricted posterior density of g is an HPD, 95% credible region for g. Similarly, the interval $[9.7803, 9.8322]$ in the lower panel of Figure 5.3 is also a 95% credible region for g, but it is not HPD.

Decision theory may also be used to select credible regions. Thus, *lowest posterior loss* regions are defined as those where all points in the region have smaller posterior expected loss than all points outside. Using the intrinsic discrepancy as a loss function yields *intrinsic credible regions* which, as one would expect from an invariant loss function, are coherent under one-to-one transformations.[11]

The concept of a credible region for a function $\boldsymbol{\theta} = \boldsymbol{\theta}(\boldsymbol{\omega})$ of the parameter vector is trivially extended to prediction problems. Thus, a posterior q-credible region for $\boldsymbol{x} \in \mathcal{X}$ is a subset R of the sample space $\mathcal{X}$ with posterior predictive probability q, so that $\int_R p(\boldsymbol{x} \mid D)d\boldsymbol{x} = q$.

5.5.2 Hypothesis Testing

The reference posterior distribution $\pi(\boldsymbol{\theta} \mid D)$ of the quantity of interest $\boldsymbol{\theta}$ conveys immediate intuitive information on those values of $\boldsymbol{\theta}$ which, given the assumed model, may be taken to be *compatible* with the observed data D, namely, those with a relatively high probability density. Sometimes, a *restriction* $\boldsymbol{\theta} \in \Theta_0 \subset \Theta$ of the possible values of the quantity of interest (where Θ_0 possibly consists of a single value $\boldsymbol{\theta}_0$) is suggested in the course of the investigation as deserving special consideration, either because restricting $\boldsymbol{\theta}$ to Θ_0 would greatly simplify the model, or because additional, context-specific arguments suggest that $\boldsymbol{\theta} \in \Theta_0$. Intuitively, the *hypothesis* $H_0 \equiv \{\boldsymbol{\theta} \in \Theta_0\}$ should be judged to be *compatible* with the observed data D if there are elements in Θ_0 with a relatively high posterior density. However, a more precise conclusion is often required and, once again, this is made possible by adopting a decision-oriented approach. Formally, testing the hypothesis $H_0 \equiv \{\boldsymbol{\theta} \in \Theta_0\}$ is a *decision problem* where the action space has only 2 elements, namely to accept (a_0) or reject (a_1) the proposed restriction. To solve this decision problem, it is necessary to specify an appropriate loss function, $L(a_i, \boldsymbol{\theta})$, measuring the consequences of accepting or rejecting H_0 as a function of the actual value $\boldsymbol{\theta}$ of the vector of interest. Notice that this requires the statement of an *alternative* a_1 to accepting H_0; this is only to be expected, for an action is taken not because it is good, but because it is better than anything else that has been imagined.

Given data D, the optimal action will be to reject H_0 if (and only if) the expected posterior loss of accepting $\int_\Theta L(a_0, \boldsymbol{\theta})\, \pi(\boldsymbol{\theta} \mid D)\, d\boldsymbol{\theta}$ is larger than the expected posterior loss of rejecting $\int_\Theta L(a_1, \boldsymbol{\theta})\, \pi(\boldsymbol{\theta} \mid D)\, d\boldsymbol{\theta}$, that is, if (and only if)

[11] For details, see Bernardo (2005a, 2007, 2011).

$$\int_{\Theta} [L(a_0, \boldsymbol{\theta}) - L(a_1, \boldsymbol{\theta})]\, \pi(\boldsymbol{\theta} \,|\, D)\, d\boldsymbol{\theta} = \int_{\Theta} \Delta L(\boldsymbol{\theta})\, \pi(\boldsymbol{\theta} \,|\, D)\, d\boldsymbol{\theta} > 0.$$

Therefore, only the loss difference $\Delta L(\boldsymbol{\theta}) = L(a_0, \boldsymbol{\theta}) - L(a_1, \boldsymbol{\theta})$, which measures the *advantage* of rejecting H_0 as a function of $\boldsymbol{\theta}$, has to be specified. Thus, as common sense dictates, the hypothesis H_0 should be rejected whenever the expected advantage of rejecting H_0 is positive.

A crucial element in the specification of the loss function is a description of what is actually meant by rejecting H_0. By assumption a_0 means to act *as if* H_0 were true, i.e. as if $\boldsymbol{\theta} \in \Theta_0$, but there are at least 2 options for the alternative action a_1. This may either mean (i) the *negation* of H_0, that is to act as if $\boldsymbol{\theta} \notin \Theta_0$, or, alternatively, it may rather mean (ii) to reject the simplification implied by H_0 and keep the unrestricted model, $\boldsymbol{\theta} \in \Theta$, which is true by assumption. Both options have been analysed in the literature, although it may be argued that the problems of scientific data analysis where hypothesis testing procedures are typically used are better described by the second alternative. Indeed, an established model, identified by $H_0 \equiv \{\boldsymbol{\theta} \in \Theta_0\}$, is often embedded into a more general model, $\{\boldsymbol{\theta} \in \Theta, \Theta_0 \subset \Theta\}$, constructed to include possibly promising departures from H_0, and it is required to verify whether presently available data D are still compatible with $\boldsymbol{\theta} \in \Theta_0$, or whether the extension to $\boldsymbol{\theta} \in \Theta$ is really required.

Example 25: Conventional hypothesis testing. Let $\pi(\boldsymbol{\theta} \,|\, D)$, $\boldsymbol{\theta} \in \Theta$, be the posterior distribution of the quantity of interest, let a_0 be the decision to work under the restriction $\boldsymbol{\theta} \in \Theta_0$, and let a_1 be the decision to work under the complementary restriction $\boldsymbol{\theta} \notin \Theta_0$. Suppose, moreover, that the loss structure has the simple, 0–1 form given by $\{L(a_0, \boldsymbol{\theta}) = 0, L(a_1, \boldsymbol{\theta}) = 1\}$ if $\boldsymbol{\theta} \in \Theta_0$ and, similarly, $\{L(a_0, \boldsymbol{\theta}) = 1, L(a_1, \boldsymbol{\theta}) = 0\}$ if $\boldsymbol{\theta} \notin \Theta_0$, so that the advantage $\Delta L(\boldsymbol{\theta})$ of rejecting H_0 is 1 if $\boldsymbol{\theta} \notin \Theta_0$ and it is -1 otherwise. With this loss function it is immediately found that the optimal action is to reject H_0 if (and only if) $\Pr(\boldsymbol{\theta} \notin \Theta_0 \,|\, D) > \Pr(\boldsymbol{\theta} \in \Theta_0 \,|\, D)$. Notice that this formulation requires that $\Pr(\boldsymbol{\theta} \in \Theta_0) > 0$, that is, that the hypothesis H_0 has a strictly positive prior probability. If $\boldsymbol{\theta}$ is a continuous parameter and Θ_0 has 0 measure (for instance if H_0 consists of a single point $\boldsymbol{\theta}_0$), this requires the use of a non-regular, 'sharp' prior concentrating a positive probability mass on Θ_0.[12]

Example 26: Intrinsic hypothesis testing. Let $\pi(\boldsymbol{\theta} \,|\, D)$, $\boldsymbol{\theta} \in \Theta$, be the posterior distribution of the quantity of interest, and let a_0 be the decision to work under the restriction $\boldsymbol{\theta} \in \Theta_0$, but let a_1 now be the decision to keep the general, unrestricted model $\boldsymbol{\theta} \in \Theta$. In this case, the advantage $\Delta L(\boldsymbol{\theta})$ of rejecting H_0 as a function of $\boldsymbol{\theta}$ may safely be assumed to have the form $\Delta L(\boldsymbol{\theta}) = \delta(\Theta_0, \boldsymbol{\theta}) - \delta^*$, for some $\delta^* > 0$, where (i) $\delta(\Theta_0, \boldsymbol{\theta})$ is some measure of the discrepancy between the assumed model $p(D \,|\, \boldsymbol{\theta})$ and its closest approximation within the class $\{p(D \,|\, \boldsymbol{\theta}_0), \boldsymbol{\theta}_0 \in \Theta_0\}$, such that $\delta(\Theta_0, \boldsymbol{\theta}) = 0$ whenever $\boldsymbol{\theta} \in \Theta_0$, and (ii) δ^* is a context-dependent utility constant which measures the (necessarily positive) advantage of being able to work with the simpler model when it is true. Choices for both $\delta(\Theta_0, \boldsymbol{\theta})$ and δ^* which may be appropriate for general use will now be described.

For reasons similar to those supporting its use in point estimation, an attractive choice for the function $d(\Theta_0, \boldsymbol{\theta})$ is an appropriate extension of the intrinsic discrepancy; when there are no nuisance parameters, this is

$$\delta(\Theta_0, \boldsymbol{\theta}) = \inf_{\boldsymbol{\theta}_0 \in \boldsymbol{\Theta}_0} \min\{\kappa(\boldsymbol{\theta}_0 \,|\, \boldsymbol{\theta}),\, \kappa(\boldsymbol{\theta} \,|\, \boldsymbol{\theta}_0)\},$$

[12] For details see Kass and Raftery (1995) and references therein.

where $\kappa(\boldsymbol{\theta}_0 \,|\, \boldsymbol{\theta}) = \int_T p(\boldsymbol{t} \,|\, \boldsymbol{\theta}) \log\{p(\boldsymbol{t} \,|\, \boldsymbol{\theta})/p(\boldsymbol{t} \,|\, \boldsymbol{\theta}_0)\} d\boldsymbol{t}$, and $\boldsymbol{t} = \boldsymbol{t}(D) \in T$ is any sufficient statistic, which may well be the whole dataset D. As before, if the data $D = \{\boldsymbol{x}_1, \ldots, \boldsymbol{x}_n\}$ consist of a random sample from $p(\boldsymbol{x} \,|\, \boldsymbol{\theta})$, then

$$\kappa(\boldsymbol{\theta}_0 \,|\, \boldsymbol{\theta}) = n \int_X p(\boldsymbol{x} \,|\, \boldsymbol{\theta}) \log \frac{p(\boldsymbol{x} \,|\, \boldsymbol{\theta})}{p(\boldsymbol{x} \,|\, \boldsymbol{\theta}_0)} \, d\boldsymbol{x}.$$

Naturally, the loss function $\delta(\Theta_0, \boldsymbol{\theta})$ reduces to the intrinsic discrepancy $\delta(\boldsymbol{\theta}_0, \boldsymbol{\theta})$ of Example 13 when Θ_0 contains a single element $\boldsymbol{\theta}_0$. Besides, as in the case of estimation, the definition is easily extended to problems with nuisance parameters, with

$$\delta(\boldsymbol{\Theta}_0, \boldsymbol{\theta}, \boldsymbol{\lambda}) = \inf_{\boldsymbol{\theta}_0 \in \boldsymbol{\Theta}_0, \boldsymbol{\lambda}_0 \in \Lambda} \delta(\boldsymbol{\theta}_0, \boldsymbol{\lambda}_0, \boldsymbol{\theta}, \boldsymbol{\lambda}).$$

The hypothesis H_0 should be rejected if the posterior expected advantage of rejecting is

$$d(\Theta_0 \,|\, D) = \int_\Lambda \int_\Theta \delta(\boldsymbol{\Theta}_0, \boldsymbol{\theta}, \boldsymbol{\lambda}) \, \pi(\boldsymbol{\theta}, \boldsymbol{\lambda} \,|\, D) \, d\boldsymbol{\theta} d\boldsymbol{\lambda} > \delta^*,$$

for some $\delta^* > 0$. As an expectation of a nonnegative quantity, $d(\Theta_0, D)$ is obviously nonnegative. Moreover, if $\boldsymbol{\phi} = \boldsymbol{\phi}(\boldsymbol{\theta})$ is a 1-to-1 transformation of $\boldsymbol{\theta}$, then $d(\boldsymbol{\phi}(\Theta_0), D) = d(\Theta_0, D)$ so that, as one should clearly require, the expected intrinsic loss of rejecting H_0 is invariant under re-parametrization.

It may be shown that, as the sample size increases, the expected value of $d(\Theta_0, D)$ under sampling tends to 1 when H_0 is true, and tends to infinity otherwise; thus $d(\Theta_0, D)$ may be regarded as a continuous, positive measure of how inappropriate (in loss of information units) it would be to simplify the model by accepting H_0. In traditional language, $d(\Theta_0, D)$ is a test statistic for H_0 and the hypothesis should be rejected if the value of $d(\Theta_0, D)$ exceeds some critical value δ^*. In sharp contrast to conventional hypothesis testing, this critical value δ^* is found to be a context-specific, positive utility constant δ^*, which may precisely be described as the number of information units which the decision maker is prepared to lose in order to be able to work with the simpler model H_0, and does not depend on the sampling properties of the probability model. The procedure may be used with standard, continuous regular priors even in sharp hypothesis testing, when Θ_0 is a 0-measure set (as would be the case if $\boldsymbol{\theta}$ is continuous and Θ_0 contains a single point $\boldsymbol{\theta}_0$).

Naturally, to implement the test, the utility constant δ^* which defines the rejection region must be chosen. Values of $d(\Theta_0, D)$ of about 1 should be regarded as an indication of no evidence against H_0, since this is precisely the expected value of the test statistic $d(\Theta_0, D)$ under repeated sampling from the null. If follows from its definition that $d(\Theta_0, D)$ is the reference posterior expectation of the log-likelihood ratio against the null. Hence, values of $d(\Theta_0, D)$ of about $\log[12] \approx 2.5$, and $\log[150] \approx 5$ should be respectively regarded as an indication of mild evidence against H_0, and significant evidence against H_0. In the canonical problem of testing a value $\mu = \mu_0$ for the mean of a normal distribution with known variance (see below), these values correspond to the observed sample mean $\bar{x}$ respectively lying 2 or 3 posterior standard deviations from the null value μ_0. Notice that, in sharp contrast to frequentist hypothesis testing, where it is hazily recommended to adjust the significance level for dimensionality and sample size, this provides an absolute scale (in information units) which remains valid for any sample size and any dimensionality.[13]

[13] For further details on intrinsic hypothesis testing see Bernardo and Juárez (2003), Bernardo and Juárez (2007) and Bernardo (2011, 2015).

Example 27: Testing the value of a normal mean. Let the data $D = \{x_1, \ldots, x_n\}$ be a random sample from a normal distribution $\mathrm{N}(x \mid \mu, \sigma)$, where σ is assumed to be known, and consider the problem of testing whether these data are or are not compatible with some specific sharp hypothesis $H_0 \equiv \{\mu = \mu_0\}$ on the value of the mean.

The conventional approach to this problem requires a non-regular prior which places a probability mass, say p_0, on the value μ_0 to be tested, with the remaining $1-p_0$ probability continuously distributed over $\Re$. If this prior is chosen to be $\pi(\mu \mid \mu \neq \mu_0) = \mathrm{N}(\mu \mid \mu_0, \sigma_0)$, Bayes' theorem may be used to obtain the corresponding posterior probability,

$$\Pr[\mu_0 \mid D, \lambda] = \frac{B_{01}(D, \lambda)\, p_0}{(1-p_0) + p_0\, B_{01}(D, \lambda)},$$

$$B_{01}(D, \lambda) = \left(1 + \frac{n}{\lambda}\right)^{1/2} \exp\left[-\frac{1}{2}\frac{n}{n+\lambda}\, z^2\right],$$

where $z = (\bar{x} - \mu_0)/(\sigma/\sqrt{n})$ measures, in standard deviations, the distance between $\bar{x}$ and μ_0, and $\lambda = \sigma^2/\sigma_0^2$ is the ratio of model-to-prior variance. The function $B_{01}(D, \lambda)$, a ratio of (integrated) likelihood functions, is called the Bayes factor in favour of H_0. With a conventional 0–1 loss function, H_0 should be rejected if $\Pr[\mu_0 \mid D, \lambda] < 1/2$. The choices $p_0 = 1/2$ and $\lambda = 1$ or $\lambda = 1/2$, describing particular forms of sharp prior knowledge, have been suggested in the literature for routine use. The conventional approach to sharp hypothesis testing deals with situations of concentrated prior probability; it assumes important prior knowledge about the value of μ and, hence, should not be used unless this is an appropriate assumption. Moreover, the resulting posterior probability is extremely sensitive to the specific prior specification (Bartlett, 1957). In most applications, H_0 is really a hazily defined small region rather than a point. For moderate sample sizes, the posterior probability $\Pr[\mu_0 \mid D, \lambda]$ is an approximation to the posterior probability $\Pr[\mu_0 - \epsilon < \mu < \mu_0 - \epsilon \mid D, \lambda]$ for some small interval around μ_0 which would have been obtained from a regular, continuous prior heavily concentrated around μ_0; however, this approximation always breaks down for sufficiently large sample sizes. One consequence (which is immediately apparent from the last 2 equations) is that, for any fixed value of the pertinent statistic z, the posterior probability of the null, $\Pr[\mu_0 \mid D, \lambda]$, tends to 1 as $n \to \infty$. Far from being specific to this example, this unappealing behaviour of posterior probabilities based on sharp, non-regular priors generally known as Lindley's paradox (Lindley, 1957) is always present in the conventional Bayesian approach to sharp hypothesis testing.

The intrinsic approach may be used without assuming any sharp prior knowledge. The intrinsic discrepancy is $\delta(\mu_0, \mu) = n(\mu - \mu_0)^2/(2\sigma^2)$, a simple transformation of the standardised distance between μ and μ_0. The reference prior is uniform and the corresponding (proper) posterior distribution is $\pi(\mu \mid D) = N(\mu \mid \bar{x}, \sigma/\sqrt{n})$. The expected value of $\delta(\mu_0, \mu)$ with respect to this posterior is $d(\mu_0, D) = (1+z^2)/2$, where $z = (\bar{x}-\mu_0)/(\sigma/\sqrt{n})$ is the standardised distance between $\bar{x}$ and μ_0. As foretold by the general theory, the expected value of $d(\mu_0, D)$ under repeated sampling is 1 if $\mu = \mu_0$, and increases linearly with n if $\mu = \mu_0$. Moreover, in this canonical example, to reject H_0 whenever $|z| > 2$ or $|z| > 3$, that is whenever μ_0 is 2 or 3 posterior standard deviations away from $\bar{x}$, respectively corresponds to rejecting H_0 whenever $d(\mu_0, D)$ is larger than 2.5, or larger than 5.

If σ is unknown, the reference prior is $\pi(\mu, \sigma) = \sigma^{-1}$, and the intrinsic discrepancy becomes

$$\delta(\mu_0, \mu, \sigma) = \frac{n}{2} \log\left[1 + \left(\frac{\mu - \mu_0}{\sigma}\right)^2\right]. \tag{1}$$

The intrinsic test statistic $d(\mu_0, D)$ is found as the expected value of $\delta(\mu_0, \mu, \sigma)$ under the corresponding joint reference posterior distribution; this may be exactly expressed in terms of hypergeometric functions, and is well approximated by

$$d(\mu_0, D) \approx \frac{1}{2} + \frac{n}{2} \log\left(1 + \frac{t^2}{n}\right), \tag{2}$$

where t is the traditional statistic $t = \sqrt{n-1}(\bar{x}-\mu_0)/s$, $ns^2 = \sum_j (x_j - \bar{x})^2$. For instance, for samples sizes 5, 30, and 1,000, and using the utility constant $\delta^* = 5$, the hypothesis H_0 would be rejected whenever $|t|$ is, respectively, larger than 5.025, 3.240, and 3.007.

5.6 Discussion

This chapter focuses on the basic concepts of the Bayesian paradigm, with special emphasis on the derivation of 'objective' methods, where the results only depend on the data obtained and the model assumed. Many technical aspects have been spared; the interested reader is referred to the bibliography for further information. This final section briefly reviews the main arguments for an objective Bayesian approach.

5.6.1 Coherence

By using probability distributions to characterise *all* uncertainties in the problem, the Bayesian paradigm reduces statistical inference to applied probability, thereby ensuring the coherence of the proposed solutions. There is no need to investigate, on a case-by-case basis, whether the solution to a particular problem is logically correct: a Bayesian result is only a *mathematical consequence of explicitly stated assumptions* and hence, unless a logical mistake has been committed in its derivation, it cannot be formally wrong. In marked contrast, conventional statistical methods are plagued with counterexamples. These include, among many others, negative estimators of positive quantities, q-confidence regions ($q < 1$) which consist of the whole parameter space, empty sets of 'appropriate' solutions, and incompatible answers from alternative methodologies simultaneously supported by the theory.

The Bayesian approach does require, however, the specification of a (prior) probability distribution over the parameter space. The sentence, 'A prior distribution does not exist for this problem', is often stated to justify the use of non-Bayesian methods. However, the general representation theorem *proves the existence* of such a distribution whenever the observations are assumed to be exchangeable (and, if they are assumed to be a random sample then, *a fortiori*, they are assumed to be exchangeable). To ignore this fact, and to proceed as if a prior distribution did not exist, just because it is not easy to specify, is mathematically untenable.

5.6.2 Objectivity

It is generally accepted that any statistical analysis is subjective, in the sense that it is always conditional on accepted assumptions (on the structure of the data, on the probability model, and on the outcome space) and those assumptions, although possibly well founded, are definitely *subjective* choices. It is, therefore, mandatory to make all assumptions very explicit.

Users of conventional statistical methods rarely dispute the mathematical foundations of the Bayesian approach, but claim to be able to produce 'objective' answers in contrast to the possibly subjective elements involved in the choice of the prior distribution.

Bayesian methods do indeed require the choice of a prior distribution, and critics of the Bayesian approach systematically point out that, in many important situations, including scientific reporting and public decision-making, the results must exclusively depend on documented data which might be subject to independent scrutiny. This is of course true, but those critics choose to ignore the fact that this particular case is covered within the Bayesian approach by the use of *reference* prior distributions which (i) are mathematically derived from the accepted probability model (and, hence, they are 'objective' insofar as the choice of that model might be objective), and (ii) by construction, they produce posterior probability distributions which, given the accepted probability model, *only* contain the information about their values which data may provide and, *optionally*, any further contextual information over which there might be universal agreement.

5.6.3 Operational Meaning

An issue related to objectivity is that of the operational meaning of reference posterior probabilities; it is found that the analysis of their behaviour under repeated sampling provides a suggestive form of calibration. Indeed, $\Pr[\boldsymbol{\theta} \in R \,|\, D] = \int_R \pi(\boldsymbol{\theta} \,|\, D)\, d\boldsymbol{\theta}$, the reference posterior probability that $\boldsymbol{\theta} \in R$, is *both* a measure of the conditional uncertainty (given the assumed model and the observed data D) about the event that the unknown value of $\boldsymbol{\theta}$ belongs to $R \subset \Theta$, and the limiting proportion of the regions which would cover $\boldsymbol{\theta}$ under repeated sampling, conditional on data 'sufficiently similar' to D. Under broad conditions (to guarantee regular asymptotic behaviour), all large data sets from the same model are 'sufficiently similar' among themselves in this sense and hence, given those conditions, reference posterior credible regions are *approximate* frequentist confidence regions.

The conditions for this approximate equivalence to hold exclude, however, important special cases, like those involving 'extreme' or 'relevant' observations. In special situations, when probability models may be transformed to location-scale models, there is an exact equivalence; in those cases, reference posterior credible intervals are, for any sample size, exact frequentist confidence intervals.

5.6.4 Generality

In sharp contrast to most conventional statistical methods, which may only be exactly applied to a handful of relatively simple, stylized situations, Bayesian methods are defined to be totally general. Indeed, for a given probability model and prior distribution over its parameters, the derivation of posterior distributions is a well-defined mathematical exercise. In particular, Bayesian methods do not require any particular regularity conditions on the probability model, do not depend on the existence of sufficient statistics of finite dimension, do not rely on asymptotic relations, and do not require the derivation of any sampling distribution, nor (*a fortiori*) the existence of a 'pivotal' statistic whose sampling distribution is independent of the parameters.

However, when used in complex models with many parameters, Bayesian methods often require the computation of multidimensional definite integrals and, for many years, this requirement effectively placed practical limits on the complexity of the problems which could be handled. This has dramatically changed in recent years with the general availability of large computing power, and the parallel development of simulation-based numerical integration techniques like *importance sampling* or *Markov chain Monte Carlo* (MCMC). These methods provide a structure within which many complex models may be analysed using generic software. MCMC is numerical integration using Markov chains. Monte Carlo integration proceeds by drawing samples from the required distributions, and computing sample averages to approximate expectations. MCMC methods draw the required samples by running appropriately defined Markov chains for a long time; specific methods to construct those chains include the Gibbs sampler and the Metropolis algorithm, originated in the 1950s in the literature of statistical physics. The development of improved algorithms and appropriate diagnostic tools to establish their convergence remain an active research area.[14]

BIBLIOGRAPHY

Bartlett, M. 1957. A comment on D. V. Lindley's statistical paradox. *Biometrika* **44**, 533–534.

Berger, J. O. 1985. *Statistical decision theory and Bayesian analysis.* Springer, Berlin, Germany.

Berger, J. O., and Bernardo, J. M. 1989. Estimating a product of means: Bayesian analysis with reference priors. *J. Am. Statist. Assoc.* **84**, 200–207.

Berger, J. O., and Bernardo, J. M. 1992a. On the development of reference priors. In *Bayesian statistics 4*, edited by Bernardo, J. M., Berger, J. O., Dawid, A. P., et al., 35–60. Oxford University Press, Oxford.

Berger, J. O., and Bernardo, J. M. 1992b. Ordered group reference priors with applications to a multinomial problem. *Biometrika* **79**, 25–37.

Berger, J. O., and Bernardo, J. M. 1992c. Reference priors in a variance components problem. In *Bayesian analysis in statistics and econometrics*, edited by Bernardo, J. M., Berger, J. O., Dawid, A. P., et al., 35–60. Oxford University Press, Oxford, 323–340.

Berger, J. O., Bernardo, J. M., and Sun, D. 2009a. The formal definition of reference priors. *Ann. Statist.* **37**, 905–938.

Berger, J. O., Bernardo, J. M., and Sun, D. 2009b. Natural induction: an objective Bayesian approach. *Rev. Acad. Sci. Madrid* **A 103**, 125–159 (invited paper with discussion).

Berger, J. O., Bernardo, J. M., and Sun, D. 2012. Objective priors for discrete parameter spaces. *J. Am. Statist. Assoc.* **107**, 636–648.

Berger, J. O., Bernardo, J. M., and Sun, D. (2015). Overall objective priors. *Bayesian Anal.* **10**, 189–246 (with discussion).

Bernardo, J. M. 1979a. Expected information as expected utility. *Ann. Statist.* **7**, 686–690.

Bernardo, J. M. 1979b. Reference posterior distributions for Bayesian inference. *J. R. Statist. Soc. B* **41**, 113-147 (with discussion).

Bernardo, J. M. 1981. Reference decisions. *Sym. Math.* **25**, 85–94.

Bernardo, J. M. 1997. Noninformative priors do not exist. *J. Stat. Plan. Infer.* **65**, 159–189 (with discussion).

Bernardo, J. M. 2005a. Intrinsic credible regions: an objective Bayesian approach to interval estimation. *Test* **14**, 317–384 (with discussion).

Bernardo, J. M. 2005b. Reference analysis. In *Handbook of statistics 25*, edited by Dey, D. K., and Rao, C. R., 17–90. Elsevier, Amsterdam, Netherlands.

Bernardo, J. M. 2006. Intrinsic point estimation of the normal variance. In *Bayesian statistics and its applications*, edited by Upadhyay, S. K., Singh, U., and Dey, D. K., 110–121. Anamaya Publishers, New Delhi, India.

Bernardo, J. M. 2007. Objective Bayesian point and region estimation in location-scale models. *Sort* **14**, 3–44 (with discussion).

Bernardo, J. M. 2011. Integrated objective Bayesian estimation and hypothesis testing. In *Bayesian statistics 9*, edited by Bernardo, J. M., Bayarri, M. J., Berger, J. O., et al., 1–68. Oxford University Press, Oxford, UK (with discussion).

Bernardo, J. M. 2015. Comparing proportions: a modern solution to a classical problem. *Current trends in Bayesian methodology with applications* Dey, D. K., Singh, U, and Loganathan, A., 59–78. CRC Press, Boca Raton, Florida, USA.

Bernardo and Juárez, M. A. 2003. Intrinsic estimation. *Bayesian statistics 7*, edited by Bernardo, J. M., Bayarri, M. J., Berger, J. O., et al., 465–476. Oxford University Press, Oxford, UK.

[14] For an introduction to MCMC methods in Bayesian inference, see Gilks et al. (1996), Mira (2005), and references therein.

Bernardo, and Juárez, M. A. 2007. Comparing normal means: new methods for an old problem. *Bayesian Anal.* **2**, 45–58.

Bernardo, J. M., and Ramón, J. M. 1998. An introduction to Bayesian reference analysis: inference on the ratio of multinomial parameters. *Statistician* **47**, 1–35.

Bernardo, J. M., and Rueda, R. 2002. Bayesian hypothesis testing: a reference approach. *Int. Stat. Rev.* **70** , 351–372.

Bernardo, J. M., and Smith, A. F. M. 1994. *Bayesian theory.* Wiley, Chichester, UK.

Box, G. E. P., and Tiao, G. C. 1973. *Bayesian inference in statistical analysis.* Addison-Wesley, Reading, MA, USA.

Datta, G. S., and Sweeting, T. J. 2005. Probability matching priors. In *Handbook of statistics 25*, edited by Dey and Rao, 91–114.

Dawid, A. P., Stone, M., and Zidek, J. V. 1973. Marginalization paradoxes in Bayesian and structural inference. *J. R. Statist. Soc. B* **35**, 189–233 (with discussion).

de Finetti, B. 1937. La prévision: ses lois logiques, ses sources subjectives. *Ann. Inst. H. Poincaré* **7**, 1–68.

de Finetti, B. 1970. *Teoria delle probabilità.* Einaudi, Turin, Italy.

DeGroot, M. H. 1970. *Optimal statistical decisions.* McGraw-Hill, New York, USA.

Efron, B. 1986. Why isn't everyone a Bayesian? *Amer. Statist.* **40**, 1–11 (with discussion).

Geisser, S. 1993. *Predictive inference: an introduction.* Chapman and Hall, London, UK.

Gilks, W. R., Richardson, S. Y., and Spiegelhalter, D. J., eds. 1996. *Markov chain Monte Carlo in practice.* Chapman and Hall, London, UK.

Jaynes, E. T. 1976. Confidence intervals vs. Bayesian intervals. *Foundations of probability theory, statistical inference and statistical theories of science 2*, edited by Harper, W. L., and Hooker, C. A., 175–257. Reidel, Dordrecht, Netherlands.

Jeffreys, H. 1939. *Theory of probability.* Oxford University Press, Oxford, UK.

Kass, R. E., and Raftery, A. E. 1995. Bayes factors. *J. Am. Statist. Assoc.* **90**, 773–795.

Kass, R. E., and Wasserman, L. 1996. The selection of prior distributions by formal rules. *J. Am. Statist. Assoc.* **91**, 1343–1370.

Laplace, P. S. 1812. *Théorie analytique des probabilités.* Courcier, Paris, France.

Lindley, D. V. 1957. A statistical paradox. *Biometrika* **44**, 187–192.

Lindley, D. V. 1958. Fiducial distribution and Bayes' theorem. *R. Stat. Soc. B* **20**, 102–107.

Lindley, D. V. 1965. Introduction to probability and statistics from a bayesian viewpoint. Cambridge University Press, Cambridge, UK.

Lindley, D. V. 1972. *Bayesian statistics, a review.* SIAM, Philadelphia, Penn., USA.

Liseo, B. (2005). The elimination of nuisance parameters. In *Handbook of statistics 25*, edited by Dey and Rao, 193–219.

Mira, A. 2005. MCMC methods to estimate Bayesian parametric models. In *Handbook of statistics 25*, edited by Dey and Rao, 415–436.

Ramsey, F. P. 1926. Truth and probability. *The foundations of Mathematics and other logical essays* (R. B. Braithwaite, ed.). London: Kegan Paul, 156–198.

Savage, L. J. 1954. *The foundations of statistics.* Wiley, New York, USA.

Stein, C. 1959. An example of wide discrepancy between fiducial and confidence intervals. *Ann. Math. Statist.* **30**, 877–880.

Zellner, A. 1971. *An introduction to Bayesian inference in econometrics.* Wiley, New York, USA.